Lecture Notes in Computer Science 16047

The series Lecture Notes in Computer Science (LNCS), including its subseries Lecture Notes in Artificial Intelligence (LNAI) and Lecture Notes in Bioinformatics (LNBI), has established itself as a medium for the publication of new developments in computer science and information technology research, teaching, and education.

LNCS enjoys close cooperation with the computer science R & D community, the series counts many renowned academics among its volume editors and paper authors, and collaborates with prestigious societies. Its mission is to serve this international community by providing an invaluable service, mainly focused on the publication of conference and workshop proceedings and postproceedings. LNCS commenced publication in 1973.

Robert Wrembel · Gabriele Kotsis · A Min Tjoa ·
Ismail Khalil
Editors

Database and Expert Systems Applications

36th International Conference, DEXA 2025
Bangkok, Thailand, August 25–27, 2025
Proceedings, Part II

 Springer

Editors
Robert Wrembel 🄳
Poznan University of Technology
Poznań, Poland

A Min Tjoa 🄳
Vienna University of Technology
Vienna, Austria

Gabriele Kotsis 🄳
Johannes Kepler University Linz
Linz, Austria

Ismail Khalil 🄳
Johannes Kepler University Linz
Linz, Austria

ISSN 0302-9743   ISSN 1611-3349 (electronic)
Lecture Notes in Computer Science
ISBN 978-3-032-02087-1   ISBN 978-3-032-02088-8 (eBook)
https://doi.org/10.1007/978-3-032-02088-8

Preface

We present the proceedings of the 36th International Conference on Database and Expert Systems Applications (DEXA 2025). DEXA was established in 1990 under the name International Conference on Database and Expert Systems Applications, which has persisted until today. The conference has been running annually without a break for over three decades, serving as a premier international forum for researchers, practitioners, and industry experts in the fundamental fields of data modeling, databases and data storage systems, data engineering, data analytics, data science, and recently - machine learning and artificial intelligence, for standard and big data.

This year, DEXA was held on 25–27 August, 2025 in Bangkok, Thailand. The conference received 123 submissions. From this set, 35 were accepted as regular papers (giving an acceptance rate of 28%).

DEXA 2025 is proud to have accepted also 22 short papers. They offer a vital platform for presenting innovative projects and preliminary results, novel ideas, ongoing research, or concise technical contributions that may not yet be mature enough for a full paper but have significant potential for future development and promise vibrant discussions. The inclusion of short papers encourages broader participation and facilitates the timely dissemination of emerging research.

The selection of all papers was based on evaluations by the Program Committee members. Each paper was single-blindly evaluated by three members. Here we express our gratitude to the PC members of DEXA 2025, for their timely and thorough evaluations.

As in the past, the DEXA 2025 proceedings consist of two volumes with regular and short papers. The accepted papers cover a variety of research topics on both theoretical and practical aspects. The papers cover among others the following topics: (1) large language models, (2) data quality, (3) applications of machine learning and artificial intelligence, (4) classification techniques, (5) image processing, analytics, and vision systems, (6) recommender techniques, (7) data integration techniques, (8) optimization methods, (9) graph applications, (10) data analytics methods, (11) security and privacy, and (12) benchmarks and surveys.

This year, DEXA introduced so-called short invited talks, with the goal to present trends in data and knowledge engineering in a less formal setting. Four such talks were accepted for the conference.

Also, this year, for the first time, the best papers will be published in a special issue of the Data & Knowledge Engineering (DKE, Elsevier) journal, entitled *Integrating Machine Learning and Data Engineering for Advanced Data Science*. This special issue will also include the best papers from the DAWAK 2025 conference. Taking the opportunity, the PC-chair would like to thank the DKE Editor-in-Chief, Carson Woo, for his approval of the special issue. Special gratitude goes to Ismail Khalil - a Steering Committee member of DEXA/DAWAK and the main organiser of these events. His invaluable help in all tasks related to organizing DEXA 2025 materialized in these two volumes of the proceedings and the event itself.

August 2025,

Robert Wrembel
Gabriele Kotsis
A Min Tjoa
Ismail Khalil

Organisation

Program Committee Chair

Robert Wrembel   Poznań University of Technology, Poland

Publicity Chairs

Aziz Nanthaamornphong   Prince of Songkla University, Thailand
Putu Wuri Handayani   University of Indonesia, Indonesia

Steering Committee

Gabriele Kotsis   Johannes Kepler University Linz, Austria
A Min Tjoa   Vienna University of Technology, Austria
Lukas Fischer   Software Competence Center Hagenberg, Austria
Bernhard Moser   Software Competence Center Hagenberg, Austria
Christine Strauss   University of Vienna, Austria
Ismail Khalil   Johannes Kepler University Linz, Austria

Program Committee Members

A´Min Tjoa   Vienna University of Technology, Austria
Abdelkader Hameurlain   IRIT, Toulouse University, France
Abdessamad Imine   Loria, France
Adam Przybylek   Gdańsk University of Technology, Poland
Adriana Marotta   Universidad de la República, Uruguay
Aida Omerovic   SINTEF and NTNU, Norway
Allel Hadjali   LIAS/ENSMA, France
Andreas Ekelhart   Secure Business Austria, Austria
Anne Kayem   University of Exeter, UK
Bala Srinivasan   Monash University, Australia
Bartosz Bebel   Poznań University of Technology, Poland
Bettina Fazzinga   University of Calabria, Italy
Brahim Ouhbi   ENSAM, Morocco
Cedric Du Mouza   CNAM, France

Christian Thomsen	Aalborg University, Denmark
Dawid Wiśniewski	Poznań University of Technology, Poland
Deborah Dahl	Conversational Technologies, USA
Ela Pustulka	FHNW University of Applied Sciences and Arts, Switzerland
Elio Masciari	University of Naples Federico II, Italy
Erich Neuhold	University of Vienna, Austria
Eunika Mercier-Laurent	University of Reims Champagne Ardenne and IFIP, France
Flavio Ferrarotti	Software Competence Centre Hagenberg, Austria
Flavius Frasincar	Erasmus University Rotterdam, The Netherlands
Florence Sedes	IRIT - University of Toulouse, France
Franck Morvan	RIT - University of Toulouse, France
Giovanna Guerrini	University of Genova, Italy
Gheorghe Cosmin	Babeş-Bolyai University, Romania
Silaghi Hamidah Ibrahim	Universiti Putra Malaysia, Malaysia
Hendrik Decker	Ludwig Maximilian University of Munich, Germany
Hiroyuki Toda	Yokohama City University, Japan
Idir Amine Amarouche	USTHB, Algeria
Ionut Iacob	Georgia Southern University, USA
Isao Echizen	University of Tokyo, Japan
Ismael Navas-Delgado	University of Málaga, Spain
Ivan Izonin	Lviv Polytechnic National University, Ukraine
Ivanna Dronyuk	Jan Dlugosz University, Poland
Javier Nieves	Azterlan, Spain
Jean-Paul Kasprzyk	University of Liège, Belgium
Jérôme Darmont	Université Lyon 2, France
Jianwei Zhang	Iwate University, Japan
Johann Gamper	Free University of Bozen-Bolzano, Italy
Jorge Lloret	University of Zaragoza, Spain
Josef Küng	Johannes Kepler University Linz, Austria
Jun Miyazaki	Tokyo Institute of Technology. Japan
Kamonluk Suksen	Chulalongkorn University, Thailand
Karim Benouaret	Université Claude Bernard Lyon 1, France
Lars Moench	University of Hagen, Germany
Laura Erhan	Free University of Bozen-Bolzano, Italy
Laurent d'Orazio	IRISA, France
Lenka Lhotska	CVUT, Czech Republic
Luca Caviglione	IMATI-CNR, Italy
Manfred Hauswirth	TU Berlin, Germany
Manolis Gergatsoulis	Ionian University, Greece

Marcin Paprzycki	IBS PAN and WSM, Poland
Marinette Savonnet	University of Burgundy, France
Markus Endres	Munich University of Applied Sciences, Germany
Massimo Guarascio	ICAR-CNR, Italy
Maude Manouvrier	Université Paris Dauphine - PSL, France
Michal Kratky	VSB-Technical University of Ostrava, Czech Republic
Michael Sheng	Macquarie University, Australia
Mizuho Iwaihara	Waseda University, Japan
Mustafa Atay	Winston-Salem State University, USA
Nazha Selmaoui	University of New Caledonia, New Caledonia
Noura Faci	Université Lyon 1, France
Olivier Teste	IRIT, France
Pavlo Radiuk	Khmelnytskyi National University, Ukraine
Paweł Misiorek	Poznań University of Technology, Poland
Peiquan Jin	University of Science and Technology of China, China
Petra Asprion	FHNW University of Applied Sciences and Arts, Switzerland
Rachid Anane	Coventry University, UK
Riad Mokadem	University of Toulouse, France
Riccardo Albertoni	CNR-IMATI, Italy
Samira Maghool	Università degli Studi di Milano, Italy
Sergio Ilarri	University of Zaragoza, Spain
Soon Chun	City University of New York, USA
Srinath Srinivasa	International Institute of Information Technology Bangalore, India
Stéphane Jean	University of Poitiers, ISAE-ENSMA, LIAS, France
Sven Groppe	University of Lübeck, Germany
Talel Abdessalem	Télécom Paris, France
Toshiyuki Amagasa	University of Tsukuba, Japan
Traian Marius Truta	Northern Kentucky University, USA
Vincenzo Deufemia	University of Salerno, Italy
Vitaliy Yakovyna	University of Warmia and Mazury in Olsztyn, Poland
Wojciech Macyna	Wrocław University of Technology, Poland
Yan Zhu	Southwest Jiaotong University, China
Yang-Sae Moon	Kangwon National University, South Korea

External Reviewers

A. K. M. Tauhidul Islam	Informatica, USA
Amna Rizvi	University of Sydney, Australia
Anand Kumar	Amazon, USA
Andre Kashliev	Eastern Michigan University, USA
Davide Costa	Altilia.ai, Italy
Eleftherios Kalogeros	Ionian University, Greece
Fayçal Saidani	Université Mouloud Mammeri de Tizi Ouzou, Algeria
Feng Yu	Youngstown State University, USA
Francesco Granata	Altilia.ai, Italy
Gautier Filardo	Centre de Recherche de la Gendarmerie Nationale, France
Luca De Grandis	Altilia, Italy
Maryam Mozaffari	Free University of Bozen-Bolzano, Italy
Matthew Damigos	Ionian University, Greece
Muhammad Umair	University of Sydney, Australia
Thilina Lokuruge	University of Sydney, Australia
Vinu Venugopal	International Institute of Information Technology, Bangalore, India

Organisers

From Data Silos to Data Mesh: A Case Study in Financial Data Architecture (Industrial Talk)

Mariusz Sienkiewicz

Director of Supervisory Data Analysis Center, Polish Financial Supervision Authority, Poland

Abstract. Successful data analytics implementation requires seamless access to both data and related metadata. In many organizations, analytics challenges arise from Data Silos, which impede cross-functional access to data and knowledge sharing across the organization. This talk presents practical insights from a data architecture transformation project conducted at a large institution with over 1,400 employees and overseeing over 2,000 market entities. The organization faced significant analytical and operational challenges due to the presence of Data Silos—isolated repositories associated with specific business areas. To address these limitations, the institution initiated a transition to a Data Mesh architecture to improve data availability and enhance analytical capabilities. This talk explains the rationale behind the persistence of silos, evaluates alternative architectural models, and justifies the choice of Data Mesh based on organizational context. Key elements of the transformation include developing a data management framework, implementing a data catalog, creating a data lake to provide data input flexibility, and establishing a common analytics platform based on Data Domains. While the project is still ongoing, the talk describes the methods being implemented and shares early results, key learnings, and practical recommendations for institutions undertaking similar architectural transitions.

Invited Talks

Blending Contextual Data with Heterogeneous Time Dimensions for Improved Time Series Analysis

Anton Dignös

Free University of Bozen-Bolzano, Italy

Abstract. In modern industrial settings, sensors continuously generate vast amounts of time series data critical for automation and process optimization. However, analyzing this data in isolation limits its effectiveness, as it often lacks integration with contextual factors that influence outcomes but are not directly observable. While traditional data fusion techniques aim at combining multimodal data such as images or videos, contextual factors in industrial environments frequently differ not in modality but in temporal structure. We identify four distinct time dimensions - constant, time series, events, and intervals - that commonly characterize contextual data in these settings. By transforming diverse time structures into a unified format, we enable the application of conventional machine learning techniques, enhancing the depth and accuracy of industrial data analysis. This talk presents a case study and initial work on a foundational approach for systematically integrating such temporally heterogeneous contextual factors into time series analysis.

A Hybrid Data Model to Support Transportation Analytics of Emergency Service Vehicles

Carson K. Leung

University of Manitoba, Winnipeg, Canada

Abstract. Using a single type of database solution to support real-world applications is becoming more and more challenging because of the volume and variety of data. For instance, the data collected for the transportation industry comprise both structured and unstructured data. Using solely a single type of database solution—relational database system-only or graph database-only—to store and manage data can be challenging. As real-world applications ask even more complex questions related to data, the database solution should be able to facilitate answering these questions in a reasonable time. Hence, in this talk, I present a hybrid model, which integrates data to support transportation analytics. The model consists of relational databases and non-relational databases (namely, graph databases), pooling their strengths to support the demands of the modern application. I also demonstrate this hybrid data model as a practical solution with a case study on improving emergency services—such as emergency medical services (EMS)—response times by having the support of the presented platform.

Contents – Part II

Optimisation Methods

Graph Applications

Analytics

Security/Privacy

Benchmarks and Surveys

Contents – Part I

Data Quality

Machine Learning/Artificial Intelligence Applications

Classification Techniques

Image Processing, Analytics, and Vision Systems

Relationship Analysis of Image-Text Pair in SNS Posts

Takuto Nabeoka[1]([✉]) [iD], Yijun Duan[2] [iD], and Qiang Ma[2] [iD]

[1] Graduate School of Informatics, Kyoto University, Kyoto, Japan
nabeoka.takuto.x79@kyoto-u.jp
[2] Graduate School of Science and Technology, Kyoto Institute of Technology,
Kyoto, Japan
{yijun,qiang}@kit.ac.jp

Abstract. Social networking services (SNS) contain vast amounts of image-text posts, necessitating effective analysis of their relationships for improved information retrieval. This study addresses the classification of image-text pairs in SNS, overcoming prior limitations in distinguishing relationships beyond similarity. We propose a graph-based method to classify image-text pairs into similar and complementary relationships. Our approach first embeds images and text using CLIP, followed by clustering. Next, we construct an Image-Text Relationship Clustering Line Graph (ITRC-Line Graph), where clusters serve as nodes. Finally, edges and nodes are swapped in a pseudo-graph representation. A Graph Convolutional Network (GCN) then learns node and edge representations, which are fused with the original embeddings for final classification. Experimental results on a publicly available dataset demonstrate the effectiveness of our method.

Keywords: SNS · Multimodal Learning · Clustering · GNN · Relationship Analysis

1 Introduction

The widespread use of social networking services (SNS), such as X (formerly Twitter), has made it easier for anyone to send and receive information. A significant proportion of SNS posts contain images, which are visually engaging and tend to attract greater attention. Studies indicate that approximately 42% of posts on X include images, receiving 22.8% more engagement than text-only posts [8]. This highlights the importance of both images and text in SNS and underscores the necessity of analyzing their relationships to better understand information dissemination and optimize SNS utilization.

Analyzing image-text relationships enables the efficient extraction of relevant information from large datasets, facilitating various applications. For instance, it can support content recommendation in post-assistance and document creation systems, contribute to the development of multimodal datasets, and enhance downstream tasks such as remote sensing [20].

R. Wrembel et al. (Eds.): DEXA 2025, LNCS 16047, pp. 3–18, 2026.
https://doi.org/10.1007/978-3-032-02088-8_1

Unrelated

Over 4 days w/o power since #Irma.
Going old school now!!

Similar

The calm before the storm
last night #irma

Complementary

The backyard #Irma

Fig. 1. Example of X's posts and the image-text pair class defined in DisRel [15]

Sosea et al. [15] developed the DisRel dataset from 4,991 posts collected from X during U.S. disasters in 2017. The posts are manually classified into three categories: "Unrelated", "Similar", and "Complementary" (Fig. 1). In the "Unrelated" example, an image of laundry and a text about going to school are unrelated. The "Similar" example shows an image representing "the calm before the storm," sharing a clear relationship with the text. The "Complementary" example pairs a text about a backyard with an image of a backyard, providing additional context.

Among the different types of relationships, the "Complementary" relationship plays a crucial role in accurately understanding information, as it enhances comprehension through the integration of images and text. However, "Complementary" relationships in image-text pairs are rarely explicitly stated within each modality, making them difficult to detect automatically. In fact, the previous study [15] reports that while the best-performing model achieved an F1 score of 0.82 for identifying "Similar" relationships, it only reached 0.62 for "Complementary" relationships. Therefore, improving the classification accuracy of "Complementary" relationships remains a challenge. Since "Similar" relationships share common information, they can be effectively classified using image recognition or multimodal models. In contrast, "Complementary" relationships require both image and text to capture their full meaning.

This study aims to improve the accuracy of capturing "Complementary" relationships in image-text pair classification, in addition to "Similar" relationships. We propose a graph-based model that incorporates both intra-pair and inter-pair relationships. First, we construct a heterogeneous graph, where each image-text pair is represented as an edge. This edge representation captures internal relationships within pairs. However, this graph tends to be sparse, making it difficult to fully leverage inter-pair information. To overcome this issue, we introduce the ITRC-Graph (Image-Text Relationship Clustering Graph), where clusters of images and texts are treated as nodes. This graph structure enables the consideration of relationships between pairs through its edge representation. Furthermore, to effectively update edge representations, we employ the Graph Convolutional Network (GCN) [7]. Since GCN is designed for node representation learning, we transform the ITRC-Graph into the ITRC-Line Graph (Image-

Text Relationship Clustering Line Graph) by swapping nodes and edges. This allows the GCN model to learn meaningful edge representations. Finally, we fuse the updated edge representations with the original embeddings of each image-text pair and train a classifier to predict relationships efficiently using multiple sources of information. Our proposed method captures not only internal information within pairs but also inter-pair relationships, improving the classification of both similar and complementary relationships in image-text pairs.

The contributions of this study are as follows:

1. We propose a novel task of classifying image-text pairs into similar and complementary relationships using graph-based representations, facilitating applications of graph-based methods.
2. We propose a method to construct the ITRC-Line Graph and learn representations using a Graph Convolutional Network (GCN). We also develop a fusion method that combines image, text, and ITRC-Line Graph information to improve classification accuracy (Sect. 3).
3. We validate our model through experiments on a public dataset (Sect. 4) and analyze the impact of the graph-based approach (Sect. 5). Experimental results demonstrate that the proposed model effectively classifies image-text relationships, especially in the"Complementary" category.

2   Related Works

Previous studies on image-text relationship classification in SNS have explored various approaches. Vempala et al. [17] modeled the interactions between images and text and proposed a dataset that categorizes relationships into four types, along with a classification task. Their approach combines two binary classification tasks: one determining whether the text is explicitly represented in the image and another assessing whether the image adds meaning to the tweet. Sun et al. [16] later refined this dataset by correcting mislabeling issues and improving classification performance using unsupervised clustering techniques. Other classification approaches have also been explored. Otto et al. [13] categorized relationships into eight types based on cross-modal semantic relationships, while Xu et al. [19] developed a dataset with five categories grounded in human perception, focusing on entities and scenes. Additionally, Sosea et al. [15] focused on disaster-related data, classifying relationships into three categories, as shown in Fig. 1. They employed multimodal models, such as ViLBERT [10], a BERT-based model designed to learn the interactions between images and text. In this study, unlike previous research, we propose a classification method that maps image and text data into a shared latent feature space and constructs a graph using clustering. Specifically, we aim to improve the classification accuracy of complementary relationships by utilizing the dataset [15].

For edge classification, Aggarwal et al. [1] formulated edge classification as a problem solvable via heuristic methods using the Jaccard coefficient. Wang et al. [18] introduced Edge2Vec, which maps edge information into a low-dimensional space. Cheng et al. [4] proposed a method that addresses the issue of topological

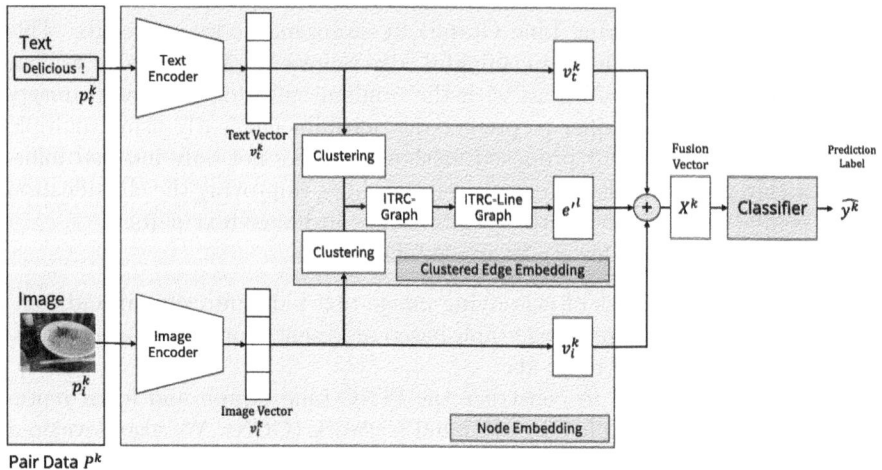

Fig. 2. Overview of Proposed Method

imbalance. In this study, we update edge representations using GCN by adopting an approach that reconstructs edges as nodes, enabling effective learning of edge representations.

3   Proposed Method

3.1   Problem Definition

For a set of N paired posts P, the n-th data sample consists of an image i and a text t, represented as a pair $P_n = \{p_n^t, p_n^i\}$. The output of the image-text relationship analysis is the predicted label $\hat{y}_n$, which represents the estimated relationship between p_n^t and p_n^i. The task is to predict one of C relationship types for each P_n.

3.2   Overview of the Proposed Method

Figure 2 illustrates the overall workflow of the proposed method.

First, in the Node Embedding stage, the image-text pair data is embedded to obtain the respective representation vectors $\mathbf{v}_n^t$ and $\mathbf{v}_n^i$ (Sect. 3.3).

Next, in the Clustered Edge Embedding stage, two major processes are performed. First, the text and image embeddings are clustered to construct the ITRC-Graph G', where each cluster serves as a node (Sect. 3.4). Then, G' is transformed into the ITRC-Line Graph G^*, where nodes and edges are interchanged. Through representation learning using GCN, the node embeddings of

G^* are updated, yielding edge embeddings for G' (Sect. 3.4). To prevent information loss, the intermediate layer embeddings of the G^* learning model are utilized as edge vectors $\mathbf{v}_n^e$ (Sect. 3.4).

Finally, in the fusion stage, the text embedding, the image embedding, and the edge embedding ($\mathbf{v}_n^t, \mathbf{v}_n^i, \mathbf{v}_n^e$) are fused (Sects. 3.5). Then, in the classification stage, the fusion vector is classified using an MLP model (Sect. 3.6).

3.3   Image and Text Encoding

For the entire set of SNS posts P, the image-text pair data $P_n = \{p_n^t, p_n^i\}$ is embedded using an encoder, transforming the text and image into 512-dimensional vectors $\mathbf{v}_n^t$ and $\mathbf{v}_n^i$, respectively. Then, for each pair, a vector set $V_n = \{\mathbf{v}_n^t, \mathbf{v}_n^i\}$ is constructed, and the entire dataset is represented as the set of all pairs' vectors $V = \{V_1, V_2, \ldots, V_N\}$. Here, the encoder employed is CLIP [14].

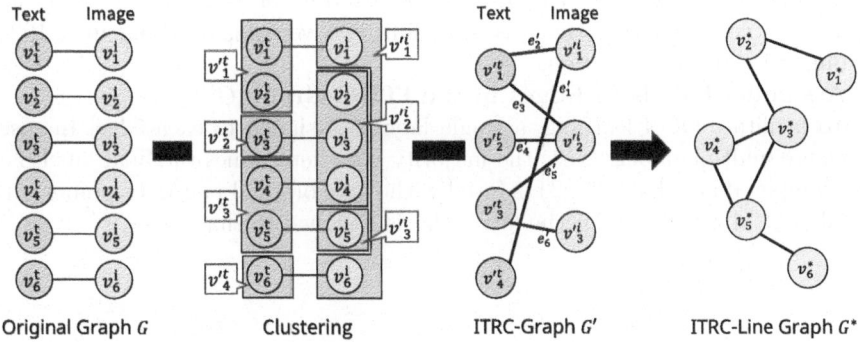

Fig. 3. The process of Clustered Edge Embedding

3.4   Embedding of Clustered Pseudo-Graphs

The model's input consists of the embedding set of image-text pairs V, and the output is the node embeddings in the ITRC-Line Graph G^*, denoted as $\{\mathbf{v}_1^*, \mathbf{v}_2^*, \ldots, \mathbf{v}_L^*\}$, where L represents the number of nodes in G^*.

Construction of ITRC-Graph. First, for the SNS post dataset P, a heterogeneous graph G is constructed, where each image and text is treated as a node, and an edge is formed between them in the paired data. While this edge representation captures relationships within pairs, the graph structure is sparse, making it challenging for GNN-based learning to leverage information across pairs.

To address this issue, clustering is applied, as illustrated in Fig. 3, to construct the ITRC-Graph $G' = \{\mathcal{V}', \mathcal{E}'\}$, where each cluster is treated as a node. This approach transforms the sparse structure into a denser graph, facilitating information propagation across pairs. The construction procedure is as follows:

1. **Applying K-means Clustering**
   The number of clusters K is set, and K-means clustering is applied separately to all images and texts in the entire pair set V, dividing them into K clusters each (a total of $2K$ clusters).
2. **Defining the Node Set $\mathcal{V}'$ of the ITRC-Graph G'**
   Each cluster is treated as a pseudo-node, forming the node set $\mathcal{V}'$. The embedding of each node is computed as the mean vector of the embeddings belonging to the corresponding cluster.
3. **Defining the Edge Set $\mathcal{E}'$ of the ITRC-Graph G'**
   The edge set $\mathcal{E}' = \{e'_1, e'_2, \ldots, e'_L\}$ is constructed by connecting clusters to which the pairs P_n belong. Each edge e'_l connects a node from a text cluster and a node from an image cluster, ensuring that no duplicate edges exist between the same pair of nodes. The embedding of each edge $\mathbf{e}'_l$ is initialized by concatenating the corresponding two vectors, resulting in a 1024-dimensional representation.
4. **Assigning Labels to Edges in the ITRC-Graph G'**
   To facilitate GCN learning, a single label is assigned to each edge. In cases where multiple labels exist, the majority vote determines the final label. For example, if an edge (v_1^{tl}, v_1^{ti}) is fused with three original pair data points with labels $\{\text{"}A\text{"}, \text{"}B\text{"}, \text{"}B\text{"}\}$, the label "$B$" is assigned to that edge.

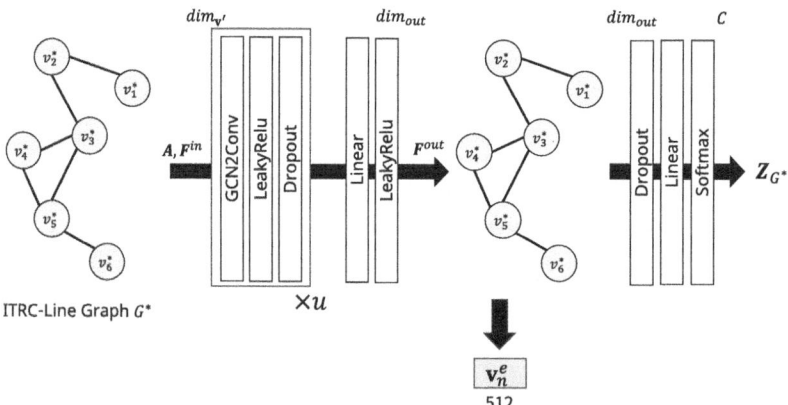

Fig. 4. Learning Model of ITRC-Line Graph

ITRC-Line Graph Construction and Representation Learning. For the ITRC-Graph G' constructed in the previous section, learning is performed using the labels assigned in Step 4. However, to apply GCN, edges must be treated as nodes. To achieve this, the ITRC-Line Graph $G^* = \{\mathcal{V}^*, \mathcal{E}^*\}$ is constructed by swapping the nodes and edges of G', as shown in Fig. 3. Here, $\mathcal{V}^* = \{v_1^*, v_2^*, \ldots, v_L^*\}$, where each node v_l^* corresponds to an edge e_l' in G'. The edge set $\mathcal{E}^*$ is defined by constructing edges between all pairs of nodes in G^* that share an endpoint in G'.

To update the edge representations in ITRC-Graph G', a model based on GCNII [3] is constructed, as illustrated in Fig. 4. Using this GCN model, supervised learning is performed to learn node representations in the ITRC-Line Graph G^*.

The model takes as input the initial node embedding matrix:

$$\mathbf{F}^{in} = \begin{bmatrix} \mathbf{v}_1^* & \cdots & \mathbf{v}_L^* \end{bmatrix}^\top \in \mathbb{R}^{L \times dim_{v'}} \tag{1}$$

and applies u layers of GCNII convolutional layers (GCN2Conv), Leaky ReLU, and Dropout. A fully connected layer then transforms the embeddings into $dim_{out} = 512$ dimensions, followed by a Leaky ReLU activation. The resulting intermediate embedding matrix is denoted as: $\mathbf{F}^{out} \in \mathbb{R}^{L \times dim_{out}}$. Subsequently, Dropout is applied to $\mathbf{F}^{out}$, followed by another fully connected layer that maps the embeddings to C dimensions. Finally, the class probabilities of each node in G^* are obtained via the Softmax function:

$$\mathbf{F}^{out} = f_{GCNmodel}(\mathbf{A}, \mathbf{F}^{in}; \mathbf{\Theta_1}) \tag{2}$$

$$\mathbf{Z}_{G^*} = \text{softmax}(\mathbf{F}^{out} \mathbf{\Theta_2}) \tag{3}$$

where $\mathbf{A} \in \mathbb{R}^{L \times L}$ is the adjacency matrix of G^*, $f_{GCNmodel}(\cdot)$ represents the GCN model, $\mathbf{\Theta_1}$ denotes its trainable parameters, and $\mathbf{\Theta_2} \in \mathbb{R}^{dim_{out} \times C}$ represents the parameters of the output layer.

Since the dimension of $\mathbf{Z}_{G^*}$ is constrained by the number of classes C, the learned intermediate embedding $\mathbf{F}^{out}$ is extracted after training. Specifically, the embedding of the node v_l^* corresponding to the edge between the clusters of the original pair P_n is used as the edge embedding $\mathbf{v}_n^e$, yielding a 512-dimensional representation.

3.5   Fusion of Representations

Through the Node Embedding and Clustered Edge Embedding processes, we obtain three types of 512-dimensional vectors: the text vector $\mathbf{v}_n^t$, the image vector $\mathbf{v}_n^i$, and the node vector of the ITRC-Line Graph (i.e., the edge vector of the ITRC-Graph) $\mathbf{v}_n^e$. These vectors are fused to form the final fused vector $\mathbf{X}_n$ using one of three fusion methods. The proposed fusion methods are listed as follows:

– **Method 1 (Averaging):** Preserves the original dimensionality by calculating the mean of the vectors.

- **Method 2 (Concatenation)**: Increases the dimensionality while retaining all information by stacking the vectors.
- **Method 3 (Averaging + Concatenation)**: Takes an intermediate app-roach between Method 1 and Method 2 by averaging the text and image vectors, then concatenating the result with the edge vector.

These fusion methods combine all three vectors before classification.

3.6   Classification of Fusion Representations

The classification is performed using a three-layer MLP, taking the fused vector $\mathbf{X}_n$ generated in Sect. 3.5 as input. After the linear layers in the first and second layers, ReLU activation and dropout are applied to prevent overfitting. Finally, the predicted label $\hat{y}^n$ is output using Softmax.

The input to the classifier has the same dimensionality as $\mathbf{X}_n$, while the output is a C-dimensional vector corresponding to the number of labels. The class with the highest likelihood is selected as the predicted label $\hat{y}^n$.

4   Experiment

We compare multiple models proposed in Sect. 3 through experiments with var-ious variations and discuss the optimal model.

4.1   Experiment Setup

Dataset. In this study, we use DisRel [15]. The distribution of the three label types is as follows: "Complementary" consists of 1,781 pairs, "Similar" consists of 2,919 pairs, and "Unrelated" consists of 291 pairs, resulting in a total of 4,991 pairs. Since the previous study [15] excluded "Unrelated" from the analysis, we also conducted our experiments using only the two classes, "Com-plementary" and "Similar". The number of data pairs used is $N = 4700^1$, and the number of classification classes is $C = 2$.

Analyzed Models. As summarized in Table 1, we construct nine comparative models based on the fusion of three vectors ($\mathbf{v}_n^t$, $\mathbf{v}_n^i$, $\mathbf{v}_n^e$). Six models fuse two vectors using averaging or concatenation, while three models combine all three vectors using averaging, concatenation, or both. The two models without edge vectors are baselines, while the remaining seven models with edge vectors are proposed models.

In the experiments, we compare these models to assess the effective utilization of multimodal information and evaluate the proposed methods against existing models [15]. Additionally, we also compare models that use Edge2Vec [18] to directly embed the edges of G'. Since the edge vectors embed by Edge2Vec have 64 dimensions, they are concatenated to preserve information.

---

[1] Although the paper [15] states 4,600 pairs, the dataset has been updated to version 2 as of August 12, 2025, containing 4,700 pairs (https://github.com/tsosea2/DisRel.

Table 1. Comparison of fused vector models: "T" represents Text, "I" represents Image, and "E" represents Edge. "A" represents average method, and "C" represents concatenation method.

Model Name	Model Type	T	I	E	Fusion Method
T+I(A)	Baseline	✓	✓		Average
T+I(C)	Baseline	✓	✓		Concatenation
T+E(A)	Proposed	✓		✓	Average
T+E(C)	Proposed	✓		✓	Concatenation
I+E(A)	Proposed		✓	✓	Average
I+E(C)	Proposed		✓	✓	Concatenation
T+I+E(A)	Proposed	✓	✓	✓	Average
T+I+E(C)	Proposed	✓	✓	✓	Concatenation
T+I+E(A+C)	Proposed	✓	✓	✓	Average + Concatenation

Model Implementation

Node Embedding For text and image encoding, CLIP encoders are used. The original study [14] compared ResNet and ViT as image encoders, demonstrating that ViT outperforms ResNet. Therefore, we adopt the ViT model in this study. Specifically, we use the pretrained CLIP model "ViT-B/32"[2], from which we obtain 512-dimensional vectors for both text and image representations.

Clustered Edge Embedding We apply K-means [9] clustering with $K = 100$ and cosine similarity to classify images and texts into 100 clusters using the "Kmean-sClusterer" from the nltk.cluster library [2]. Next, we construct the ITRC-Graph G' with clusters as nodes. To avoid instability and computational complexity, edge reduction is performed by removing edges with the same label between training nodes and connecting only the $J = 5$ nearest nodes based on Euclidean distance.

For edge labeling in G', majority voting is applied: if all edges belong to training data, the label is determined by majority voting; if both training and test data coexist, only training data labels are considered; in the case of a tie, a label is chosen randomly; if all edges belong to test data, the label is determined within the test nodes.

Finally, we construct the ITRC-Line Graph G^* from G' and train the model using PyTorch Geometric [5]. The GCNII layers are set to $u = 64$, learning rate to 0.005, and optimizer to Adam [6]. The loss function is cross-entropy error, and training is done for 100 epochs with learning rate decay. The evaluation of G^* is discussed in Sect. 5.2.

Fusion and Classification. We generate the fused vector $\mathbf{X}_n$ for the 9 models described in Sect. 4.1. Then, we implement and train a three-layer MLP using

---

[2] https://github.com/OpenAI/CLIP.

PyTorch. During training with this model, the batch size is set to 4, and the optimizer used is Adam [6]. The loss function is set to cross-entropy error. We apply learning rate decay using the LambdaLR scheduler, and training is conducted for 100 epochs. If the loss does not improve for a certain period, training is stopped at that point. The model with the lowest validation loss is selected for evaluation.

Evaluation. In this experiment, we compare the classification performance of the models based on the presence or absence of image, text, and ITRC-Line Graph node vectors, as well as the fusion methods described in Sect. 3.5. Specifically, we evaluate the 9 models proposed in Sect. 4.1, along with a comparison against existing models.

For evaluation, the 4,700 pairs from the DisRel dataset are randomly split into training, validation, and test sets in a ratio of 6 : 2 : 2. The evaluation metrics include precision, recall, and F1 score for each label, as well as the macro-F1 score and accuracy, calculated solely on the test data.

4.2   Results

Table 2. Comparison of proposed models by fusion methods and existing models (rounded to three decimal places, average with various of 10 times)

	Similar			Complementary			Macro F1	Accuracy
	P	R	F1	P	R	F1		
ViLBERT-REL-MT (Best Model from [15])	**0.82**	**0.82**	**0.82**	0.63	0.62	0.62	0.72	**0.76**
T+E(Edge2Vec [18])	0.79	**0.82**	0.80	**0.69**	0.65	**0.67**	**0.74**	0.75
I+E(Edge2Vec [18])	0.72	**0.82**	0.76	0.62	0.47	0.53	0.65	0.69
T+I(A)+E(Edge2Vec [18])	0.78	**0.82**	0.80	0.68	0.63	0.66	0.73	0.75
T+I(C)+E(Edge2Vec [18])	0.78	**0.82**	0.80	**0.69**	0.63	0.66	0.73	0.75
T+I(A)	0.78	**0.82**	0.80	0.68	0.63	0.65	0.72	0.74
T+I(C)	0.78	**0.82**	0.80	**0.69**	0.63	0.66	0.73	0.75
T+E(A)	0.79	**0.82**	0.81	**0.69**	0.65	**0.67**	**0.74**	**0.76**
T+E(C)	0.80	**0.82**	0.81	**0.69**	**0.66**	**0.67**	**0.74**	**0.76**
I+E(A)	0.76	**0.82**	0.79	0.66	0.58	0.61	0.70	0.73
I+E(C)	0.76	**0.82**	0.79	0.66	0.59	0.62	0.70	0.73
T+I+E(A)	0.78	0.81	0.80	0.68	0.63	0.65	0.73	0.74
T+I+E(C)	0.79	**0.82**	0.80	0.68	0.64	0.66	0.73	0.75
T+I+E(A+C)	0.78	0.81	0.80	0.67	0.63	0.65	0.72	0.74

We compare the performance of the proposed models described in Sect. 4.1 with the model from previous research [15]. The precision, recall, and F1 score for "Similar" and "Complementary" labels in DisRel, as well as the Macro-F1 score and accuracy, are calculated, and the results are summarized in Table 2. The values for each model represent the average of 10 trials, where P indicates precision and R indicates recall. Additionally, T, I, and E denote the use of text, image, and edge vectors (ITRC-Line Graph node vectors), respectively. For the existing models, we compare our approach with two models: ViLBERT-REL-MT [10] and models using Edge2vec [18]. The former model demonstrated the highest performance by incorporating a help task from a different dataset, as reported in [15]. The result of ViLBERT-REL-MT is quoted from [15]. The latter model is evaluated in the same experiment as the proposed method.

First, comparing the proposed models with the existing model, we find that the method of concatenating text and edge vectors using T+E(C) achieves the highest performance. Notably, compared to the existing best model, ViLBERT-REL-MT, T+E(C) improves precision by 6 points, recall by 4 points, and F1 score by 5 points for the "Complementary" label, along with a 2-point increase in Macro-F1. This significant improvement demonstrates a substantial enhancement in the accuracy of detecting complementary relationships. Moreover, while maintaining the performance for "Similar" relationships, the proposed model also contributes to the overall task performance. Additionally, T+E(C) consistently outperform those using Edge2Vec [18]. However, the overall accuracy remains comparable to existing methods, indicating the need for further improvement in comprehensive performance.

Next, we compare the baseline and proposed models in this study. Among the baseline models, the method of concatenating image and text vectors (T+I(C)) exhibits higher performance than averaging. Both the best baseline and the proposed best models (T+E(C)) outperform the highest-performing existing model in terms of "Complementary" and Macro-F1 scores, suggesting that the use of CLIP encoders effectively contributes to the detection of "Complementary" labels. Furthermore, T+E(C) improves recall for the "Complementary" label by 3 points and Macro-F1 by 1 point compared to T+I(C). This result indicates that leveraging edge information contributes to performance improvement, highlighting that, in addition to the internal relationships of image-text pairs, considering external relationships through clustering and ITRC-Line Graph effectively enhances the accuracy of detecting complementary relationships.

Finally, we analyze the impact of different fusion methods on performance through comparisons within the proposed models. The results show that concatenation methods exhibit slightly better performance than averaging methods, likely due to preserving the original embeddings. However, when fusing all three vectors (text, image, and edge), the performance tends to decline compared to fusing only text and edge vectors. This observation suggests that text features are more critical than image features in identifying complementary relationships.

In summary, our method fusing text and edge vectors significantly outperforms existing methods in detecting "Complementary" relationships, while the overall performance remains comparable to previous approaches.

5 Discussion

First, the distribution of each embedding per label by CLIP of DisRel images and texts is analyzed (Sect. 5.1). Next, the performance of the representation learning of the ITRC-Line Graph is evaluated (Sect. 5.2). Finally, as a case study, a qualitative analysis is performed by comparing the actual outputs (Sect. 5.3).

5.1 Analysis of Dataset

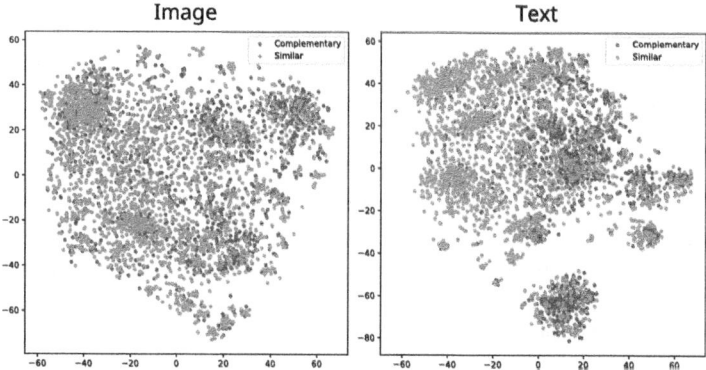

Fig. 5. Distribution of image and text embeddings after dimensionality reduction by the CLIP encoder. "Complementary" and "Similar" class in DisRel are illustrated in blue and orange, respectively.

To visualize and analyze the DisRel dataset, we embedded images and texts into 512-dimensional vectors using the CLIP encoder and then reduced the dimensionality to 2D using t-SNE [11]. We plotted 4,700 pairs labeled as either "Complementary" or "Similar" by coloring them differently according to their labels.

The result is shown in Fig. 5. While no significant difference was observed in the distribution of images, the text vectors revealed a high-density region corresponding to the "Complementary" label. This observation suggests that text features may play a crucial role in relation classification.

5.2 Analysis of ITRC-Line Graph Representation Learning

In Sect. 3.4, we evaluate the performance of representation learning for the ITRC-Line Graph G^*. This evaluation analyzes the sub-tasks related to edge vector learning presented in Sect. 4.2, using the average results of 10 trials. It should be noted that the original pair data P and the labeling method differ.

In this experiment, we utilize the ITRC-Line Graph G^* after edge reduction. Following the procedure described in Sect. 4.1, we split the graph into training

Table 3. Statistics of ITRC-Line Graph

The number of edge		The number of node(L)				
Total	after reduction (Average)	Total	Training (Average)	Test(Average)		
				Total	Sim	Com
48702	35925	1928	1409	519	336	183

Table 4. Classification Results for ITRC-Line Graph (rounded to the third decimal place, average of 10 times).

Similar			Complementary			Macro F1	Accuracy
P	R	F1	P	R	F1		(Acc)
0.72	0.88	0.79	0.63	0.39	0.48	0.64	0.70

and test nodes. The proportion of training data averaged 0.73, and the detailed information is shown in Table 3.

The experimental results are shown in Table 4. The accuracy is 70%, and the Macro-F1 score is 0.64. The F1 score is not high due to the low recall for the "Complementary" label, while the recall and F1 score for "Similar" are higher. On the other hand, by combining the edge vectors with the original pairs, we observed an improvement in the performance for the "Complementary" label. Out of the dataset P (4700 pairs), about 82% (on average, 3853 pairs) of the labels on the ITRC-Line Graph matched. Furthermore, following the labeling rules in Sect. 4.1, some test data were learned as training nodes, even though they were not included in the majority vote, indicating that some of the test data were effectively learned.

From these results, it can be suggested that aggregating image and text embeddings at the nodes of the ITRC-Line Graph, combined with majority voting for labeling, is effective. This approach considers the internal and external interactions between image-text pairs, particularly contributing to the discrimination of the "Complementary" label. Additionally, it is likely that further fusion of the text vector with the edge vectors formed from the image and text is also important.

5.3   Case Study

We analyzed the output of 10 test data samples for T+E(C) (T+E (Concatenation)) and the baseline model (T+I (Concatenation)), comparing five examples. Additionally, we compared the results with ChatGPT's output. The relevance definitions from Sect. 1 were provided as prompts to ChatGPT. The GPT-4o model (as of January 2025) was used for this experiment [12]. The output results are shown in Table 5.

Pair I: The text and image describe the impact of Hurricane Irma across Florida. All models correctly identified this as "Complementary".

Pair II: The text describes the weakened Hurricane Irma, and the image shows its intensity. The baseline model misclassified this as "Similar", while the proposed model and ChatGPT correctly identified it as "Complementary".

Pair III: The text discusses storm surges one week after Irma, and the image shows rising water levels. This relationship should be "Similar", but the baseline model and ChatGPT misclassified it. The proposed model performed better.

Pair IV: The text describes a flood warning, and the image shows the flood. The proposed model misclassified this as "Similar", while ChatGPT correctly identified it as "Complementary" by focusing on the warning aspect.

Pair V: Although expected to be "Complementary" from the perspective of city officials, all models classified it as "Similar". This highlights a potential issue with proper noun entities, as noted in previous research [15], suggesting the need for relabeling in the dataset.

These analyses reveal differences in the focus between the image and text, as well as challenges with labels, emphasizing important considerations when embedding multimodal data into a common space.

Table 5. Results for Each of the Relevance Prediction Labels.

	I	II	III	IV	V
Image					
Text	"Hurricane Irma downgraded to Category 1, but still wreaking havoc across Florida \<URL\> #Irma \<URL\>"	Weakened #Irma lashes much of Fl \<URL\>	King tides rise a week after Irma, highlighting flood risks \<URL\> @entornoi \<URL\>	#HurricaneIrma: Flooding in South Carolina as officials warn residents remain alert \<URL\> \<URL\>	"A tree is down on 17-92 near W 18th St. in Sanford, blocking most of the road. City crew on scene #Irma #WFTV \<URL\>"
Correct	**Com**	**Com**	**Sim**	**Com**	**Com**
T+I(C)	**Com**	Sim	Com	**Com**	Sim
ChatGPT	**Com**	**Com**	Com	**Com**	Sim
T+E(C)	**Com**	**Com**	**Sim**	Sim	Sim

6   Conclusion

In this paper, we highlighted the importance of extracting complementary relationships in the classification of image-text pairs in SNS posts and proposed a graph-based method for automatic classification. Specifically, we introduced a method that integrates GCN-based embeddings obtained from the ITRC-Line Graph with the original image-text pair embeddings and classifies them using an MLP classifier. Experimental results showed that the F1 score for the complementary relationship improved to 0.67, significantly outperforming existing methods. A case study analysis confirmed the effectiveness of the proposed method in distinguishing between complementary and similar relationships. Future challenges include optimizing graph learning and fusion techniques to further improve classification performance. Additionally, we plan to evaluate the generalizability and applicability of our method using different datasets.

Acknowledgments. This work was partly supported by JSPS KAKENHI Grant Numbers JP23K28094, JP25K21275.

References

1. Aggarwal, C.C., He, G., Zhao, P.: Edge classification in networks. In: ICDE2016, pp. 1038–1049. IEEE Computer Society (2016)
2. Bird, S., Klein, E., Loper, E.: Natural Language Processing with Python, Analyzing Text with the Natural Language Toolkit. O'Reilly Media, Sebastopol (2009)
3. Chen, M., Wei, Z., Huang, Z., Ding, B., Li, Y.: Simple and deep graph convolutional networks. In: ICML. Proceedings of Machine Learning Research, vol. 119, pp. 1725–1735. PMLR (2020)
4. Cheng, X., Wang, Y., Liu, Y., Zhao, Y., Aggarwal, C.C., Derr, T.: Edge classification on graphs: new directions in topological imbalance. In: WSDM, pp. 392–400. ACM (2025)
5. Fey, M., Lenssen, J.E.: Fast graph representation learning with PyTorch geometric. arXiv preprint arXiv:1903.02428 (2019)
6. Kingma, D.P., Ba, J.: Adam: a method for stochastic optimization. In: ICLR (Poster) (2015)
7. Kipf, T.N., Welling, M.: Semi-supervised classification with graph convolutional networks. In: ICLR (Poster). OpenReview.net (2017)
8. Lee, K.: What 1 Million Tweets Taught Us About How People Tweet Successfully (2015). https://buffer.com/resources/twitter-data-1-million-tweets/. Accessed 19 Mar 2025
9. Lloyd, S.P.: Least squares quantization in PCM. IEEE Trans. Inf. Theory **28**(2), 129–136 (1982)
10. Lu, J., Batra, D., Parikh, D., Lee, S.: ViLBERT: pretraining task-agnostic visiolinguistic representations for vision-and-language tasks. In: NeurIPS, pp. 13–23 (2019)
11. Maaten, L.V.D., Hinton, G.: Visualizing data using T-SNE. J. Mach. Learn. Res. **9**(86), 2579–2605 (2008)
12. OpenAI: ChatGPT(GPT-4o, Jan 27 version) [Large Language Model] (2025). https://chat.openai.com
13. Otto, C., Springstein, M., Anand, A., Ewerth, R.: Characterization and Classification of Semantic Image-text Relations. Int. J. Multim. Inf. Retr. **9**(1), 31–45 (2020)

14. Radford, A., et al.: learning transferable visual models from natural language supervision. In: ICML. Proceedings of Machine Learning Research, vol. 139, pp. 8748–8763. PMLR (2021)
15. Sosea, T., Sirbu, I., Caragea, C., Caragea, D., Rebedea, T.: Using the image-text relationship to improve multimodal disaster tweet classification. In: ISCRAM, pp. 691–704. ISCRAM Digital Library (2021)
16. Sun, L., Li, Q., Liu, L., Su, Y.: Unsupervised multimodal learning for image-text relation classification in tweets. Pattern Anal. Appl. **26**(4), 1793–1804 (2023)
17. Vempala, A., Preotiuc-Pietro, D.: Categorizing and inferring the relationship between the text and image of twitter posts. In: ACL (1), pp. 2830–2840. Association for Computational Linguistics (2019)
18. Wang, C., Wang, C., Wang, Z., Ye, X., Yu, P.S.: Edge2vec: edge-based social network embedding. ACM Trans. Knowl. Discov. Data **14**(4), 45:1–45:24 (2020)
19. Xu, C., Tan, H., Li, J., Li, P.: Understanding social media cross-modality discourse in linguistic space. In: EMNLP (Findings), pp. 2459–2471. Association for Computational Linguistics (2022)
20. Yuan, Z., et al.: Exploring a fine-grained multiscale method for cross-modal remote sensing image retrieval. IEEE Trans. Geosci. Remote Sens. **60**, 1–19 (2022)

Enhancing Segmentation of Irregular Microstructural Elements Using Extended Channel Information and Transfer Learning

Łukasz Marcjan[1]([✉]) [iD], Sandra Gajoch[1] [iD], Dorota Wilk-Kołodziejczyk[1,2] [iD],
Marcin Małysza[1,2] [iD], Krzysztof Jaśkowiec[2] [iD], and Grzegorz Gumienny[3] [iD]

[1] AGH University of Krakow, Kraków, Poland
lmarcjan@agh.edu.pl
[2] Łukasiewicz Research Network-Kraków Institute of Technology, Kraków, Poland
[3] Lodz University of Tecgnology, Łódź, Poland

Abstract. This work aims to extend the previously proposed method of segmentation of irregular microstructural elements of materials by applying transfer learning and analyzing the approach's effectiveness on other deep neural network architectures. The research enriched input information with additional structural-textural channels obtained from the k-means algorithm, the Sobel operator, and superpixel segmentation (Felzenszwalb). Previous research was limited to U-Net-based architectures; the DeepLabv3 model with ResNet50 encoder was also analyzed in this work. To improve segmentation quality, transfer learning was applied using pretrained weights from two datasets: the universal ImageNet dataset and the specialized MicroNet dataset containing microscopic images. Comparative experiments were carried out for models trained from scratch and using transfer learning, assessing the quality of segmentation based on standard measures such as IoU, Dice score, Precision, Recall, and F1-score. The results show that integrating extended input data with a suitably adapted architecture can significantly improve the quality of segmentation of irregular structures. The best results were achieved for Attention U-Net with pre-training on ImageNet, which confirms the potential of combining nonsupervised methods and transfer learning in tasks of precise segmentation of material microstructures.

Keywords: Machine learning · Neural networks · Segmentation · Microstructure

1 Introduction

Image segmentation is a key stage in the analysis of visual data and is used in many fields, such as medicine [13,18], biology [2,14], industry [1,10], and also satellite image analysis [9]. There are many approaches to segmentation,

R. Wrembel et al. (Eds.): DEXA 2025, LNCS 16047, pp. 19–33, 2026.
https://doi.org/10.1007/978-3-032-02088-8_2

ranging from classical methods such as thresholding and edge detection through clustering techniques to modern solutions based on superpixels and artificial intelligence. Deep learning has revolutionized segmentation tasks by achieving superior performance compared to traditional computer vision techniques [20].

Segmentation methods have been applied to analyze biomedical images [18]. Due to the difficulties in identifying microstructures often characterized by local blurring and irregular pore structures, models can misclassify individual regions. Using edge detection methods allows for the precise separation of regions of interest, which is an important stage in data processing before further analysis using deep learning algorithms [8,13]. Edge detection is a fundamental tool in image processing, particularly in feature detection and extraction tasks [10]. Integrating these techniques allows for more accurate segmentation results and can also form the basis for further improvements in diagnostic imaging. Detecting boundaries and segmenting microstructure images is crucial for reconstructing the 3D microstructure of a material [10]. Due to the limited amount of labeled data in materials science, images are cut from larger sizes (e.g. 1024×1024) to many smaller ones (e.g. 256×256) [10]. This method, combined with augmentation techniques, allows one to increase the data needed to train artificial intelligence algorithms.

Sobel edge detection has been extensively applied in medical image analysis, including the extraction of trabecular patterns in bone imaging [13]. The algorithm's effectiveness in highlighting object boundaries makes it particularly suitable for preprocessing in deep learning pipelines, where edge information can enhance feature representation [8]. A binary image is created based on the edges obtained for later work with the deep learning algorithm.

Metallographic image analysis based on the UNet algorithm, which uses only the RGB color space and its transformation to HSV and YUV, does not provide sufficient effectiveness in image segmentation [1]. The K-means clustering algorithm is widely used in image segmentation due to its computational efficiency and the ability to partition images into homogeneous regions [15,16]. Studies have demonstrated that combining K-means with other techniques significantly improves segmentation accuracy compared to using K-means alone [7], particularly in scenarios with complex texture patterns found in microstructural analysis. However, in cases where the edges are unclear, blurred, or jagged, and the colors of the bordering phases are similar, methods based on edge detection and color intensity may be insufficient. In such situations, it is reasonable to use superpixel algorithms [14]. The use of superpixels improved the effectiveness of classification of plant species by combining texture information with spectral data [2], and has shown promise in biological research for segmentation of phase contrast images in cell microscopy [14]. The use of superpixels in biological research has allowed the segmentation of phase contrast images in cell microscopy, even in the case of densely packed cells [14].

This work continues previous research conducted by our team, in which we proposed a method for segmenting irregular microstructural structures through input data enrichment. This approach integrated classic RGB channels with additional structural and textural information obtained using unsupervised image analysis methods: the k-means algorithm, Sobel operator, and superpixel segmentation according to Felzenszwalb's method. The preliminary results indicated an improvement in segmentation accuracy, especially when using UNet-based architectures.

The dataset used in this study and previous research data supporting the methodological foundation of this work, including preliminary results with U-Net architectures and initial validation of the multi-channel approach, are available from the corresponding author upon reasonable request. These preliminary data are not publicly available as they form part of ongoing research that will be submitted for publication.

Previous research confirmed that models trained exclusively on RGB images can learn to detect carbides, but the highest accuracy was achieved using the Attention U-Net model trained on images augmented with additional channels. This model achieved the best results both in terms of quantitative measures (Recall, Precision, F1-score, Dice, IoU) and in the visual assessment of the correctness of detection of heterogeneous and irregular structures, such as carbides present in the microstructure of gray cast iron.

In this paper, the scope of the analysis was extended to include the application of the proposed method to extend the input channels to other architectures, including DeepLabv3 [4], and the possibility of using transfer learning was also investigated to further improve the quality of segmentation. The encoder used was the ResNet50 architecture [6], and the pretrained weights came from two different sources: the popular ImageNet dataset [19] and the specialized MicroNet dataset of microscopic images [17].

2   Materials and Method

2.1   Dataset and Material Characterization

In this study, irregularly shaped carbides in the microstructure of gray cast iron were chosen as an example of data with high heterogeneity in segmentation. Gray cast iron is widely used in industry because of its low production cost and good mechanical strength. One of the important quality parameters is the presence of carbides, which negatively affect the machining properties of the material, make it difficult to process and deteriorate the characteristics of the final product [5].

The analysis included 52 images of gray cast iron microstructures obtained using an optical microscope at a magnification of 500x and a resolution of 2560×1920 pixels. Experts manually labeled each image using the DjangoLabeller tool [3], and the results were saved in COCO-compatible JSON format.

Data were divided into training, validation and test sets in a proportion of 70%/15%/15%. The images were scaled down to a lower resolution to reduce the demand for computing resources, improve the model's generalizability, and enable more effective detection of more minor details. Each image and its corresponding mask were divided into 20 fragments of 512×512 pixels. This resulted in an extended dataset of 1040 images.

Due to the complex and irregular nature of the studied structures, the process of their manual annotation is associated with a certain degree of subjectivity and risk of errors. However, the prepared masks served as ground truth in subsequent experiments related to automatic segmentation.

2.2   Hybrid Segmentation Methodology

In this study, we rely on a previously proposed hybrid segmentation method for irregular microstructural elements, particularly carbides, in gray cast-iron images. The method's basic idea is to enrich conventional RGB images with additional structural and textural information extracted using unsupervised learning techniques. A deep neural network then processes this enriched input data to improve segmentation accuracy.

To extend the standard three-channel RGB representation, additional channels were generated using three unsupervised algorithms: K-Means clustering, Sobel edge detection, and Felzenszwalb's superpixel segmentation. These algorithms were chosen for their ability to capture complementary aspects of image structure, such as texture homogeneity, object boundaries, and local region similarity. Figure 1 shows a visual overview of the methodology.

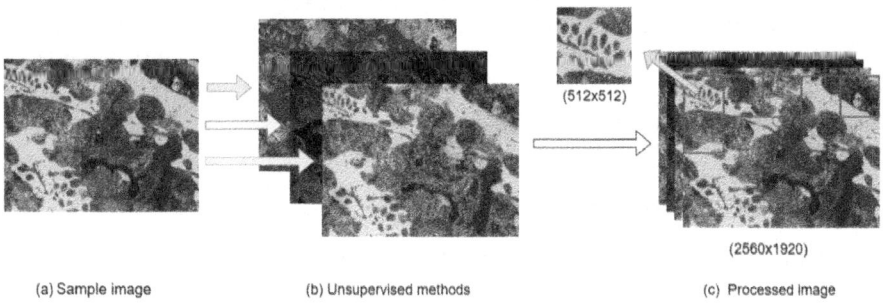

(a) Sample image   (b) Unsupervised methods   (c) Processed image

Fig. 1. Overview of the hybrid segmentation pipeline: (a) input microstructure image, (b) unsupervised processing methods including K-Means, Sobel edge detection, and Felzenszwalb superpixels, (c) composite multi-channel representation.

The method consists of the following key stages:

1. **Structural Feature Extraction:**
   - *K-Means clustering* partitions each image into regions with similar color and texture characteristics. The number of clusters was empirically selected using PCA-based dimensionality reduction and visual validation. The K-means were selected for their ability to partition hyperspherical clusters in intensity space, making them well suited for separating high-contrast carbide phases from the metallic matrix [7].

- *Sobel edge detection* enhances the boundaries of objects by computing intensity gradients. The resulting edge maps serve as additional channels for improved boundary recognition. Provided optimal boundary recognition through gradient magnitude computation, crucial for irregular microstructure delineation.
- *Felzenszwalb superpixel segmentation* generates compact homogeneous regions that reflect localized texture and tone similarities, supporting the network's understanding of context-sensitive structures. Chosen for adaptive region grouping that preserves texture homogeneity while maintaining computational efficiency.

2. **Data Preparation:** The outputs from the above methods are combined with the original RGB images to form a multi-channel input dataset. Each image is further divided into 512×512 pixel patches to manage memory usage and emphasize more minor structural elements.
3. **Model Training:** The final input trains convolutional neural networks adapted to handle high-dimensional input data.

This hybrid technique, initially proposed in our earlier work, combines unsupervised preprocessing with supervised learning to improve segmentation performance in complex, heterogeneous microstructures.

The applied unsupervised methods are commonly adopted in image segmentation workflows and were selected through empirical evaluation. Their combined use was found to complement each other effectively in capturing distinct structural characteristics relevant to the targeted microstructural features.

The preliminary validation of this approach, conducted using U-Net architectures, demonstrated significant improvements in segmentation accuracy compared to standard RGB-only methods. Detailed quantitative results and methodological validation data are available from the corresponding author upon reasonable request.

2.3   Data Augmentation

A series of data augmentation operations were performed to enhance the diversity of the training data and improve model generalization. Initially, all microstructure images and their corresponding segmentation masks were cropped into patches with a resolution of 512×512 pixels.

Each of these patches was subsequently transformed using a set of predefined geometric modifications.

- 90° rotation,
- Horizontal flipping,
- Vertical flipping.

These transformations were applied once to every patch, generating three variants for each original sample. As a result, the size of the training dataset was quadrupled, introducing additional variability while preserving structural consistency with the original images.

2.4   Neural Network Architectures

The study analyzed the effectiveness of different deep learning architectures for microstructure segmentation, focusing on the impact of transfer learning. The ResNet50 encoder was used as a common base element in all cases. Three common segmentation architectures were selected:

- **U-Net with ResNet50 encoder**
- **Attention U-Net with ResNet50 encoder**
- **DeepLabV3 with ResNet50 encoder**

Three training scenarios were considered for each of the above architectures:

1. *From scratch learning* - All layers, including the ResNet50 encoder, were randomly initialized and trained exclusively on the data used in this study.
2. *Transfer learning with MicroNet weights* the encoder was initialized with weights previously trained on the MicroNet [12] microscopic dataset published by NASA. This dataset is particularly similar to the data used in this study, potentially increasing the effectiveness of knowledge transfer.
3. *Transfer learning with ImageNet weights* weights trained on the ImageNet dataset, commonly used in general computer vision tasks, were used.

All models were implemented in the TensorFlow/Keras environment. A typical encoder architecture enabled a direct comparison of the impact of both the network structure itself and the source of the initial weights on the segmentation quality. The analysis aimed to determine whether pretraining on domain-similar data (MicroNet) could yield measurable benefits compared to the standard ImageNet-based approach.

Since the input data in this study contained an extended multi-channel representation (beyond standard RGB images), it was necessary to adapt the encoder's input structure using transfer learning. The MicroNet and ImageNet datasets contain three-channel images, so the pretrained versions of ResNet50 have weights for the R, G, and B channels only.

The first convolutional layer was expanded to support the full number of input channels, enabling the use of additional channels (such as edge maps, superpixel segments, or cluster masks). Alternative initialization strategies were also explored, including initializing additional channel weights with small values. However, preliminary experiments indicated that zero initialization provided the most stable training dynamics for the multi-channel adaptation scenario used in this study. Zero initialization was selected to maintain gradient stability while preserving pretrained RGB knowledge. This approach ensures that additional channels do not introduce random noise.

This initialization method preserves learning stability in the early epochs without introducing noise or unwanted randomness. At the same time, it enables the model to gradually learn appropriate filters based on the data. This allows the network to utilize the knowledge accumulated in the original RGB channels while adapting to the extended input.

2.5   Training Procedure

All segmentation models were implemented and trained using the open-source PyTorch environment. The Adam algorithm with a manually set initial learning rate was used to optimize the neural network weights.

In order to increase the stability of the learning process and improve convergence, an adaptive learning rate reduction mechanism was implemented - if there was no improvement in the loss function value on the validation set for five consecutive epochs, the learning rate was automatically halved.

A maximum of 100 training epochs was used, incorporating early stopping and checkpoint mechanisms to prevent overfitting.

Binary cross-entropy was used as the main loss function, due to its effectiveness in binary segmentation tasks. To complement the evaluation of segmentation quality, the Dice coefficient was also analyzed as an auxiliary metric.

2.6   Evaluation of Segmentation Quality

A set of commonly used evaluation measures was used to comprehensively assess the effectiveness of segmentation models. They allow for the assessment of both the quality of the shape representation of objects and the accuracy of classification at the level of individual pixels.

The Jaccard coefficient, also known as the Intersection over Union (IoU), is a widely-adopted metric in semantic segmentation that measures the degree of overlap between predicted and ground truth masks [11]. IoU is particularly valuable because it penalizes both under-segmentation and over-segmentation more severely than alternative metrics [11]. It is defined in Eq. (1):

$$J(A, B) = \frac{|A \cap B|}{|A \cup B|} \tag{1}$$

In this formula, A represents the mask predicted by the model and B represents the reference mask. The numerator represents the number of common pixels (common part) and the denominator represents the sum of pixels from both masks (common part + differences). A value of 1 indicates full coverage, while a value of 0 indicates no match.

The Dice coefficient, while closely related to IoU, is often preferred in medical image segmentation due to its differentiable nature and suitability as a loss function [18]. Its mathematical form is shown in Eq. (2):

$$Dice(A, B) = \frac{2|A \cap B|}{|A| + |B|} \tag{2}$$

Both metrics   Dice and IoU   take values from 0 to 1, where a higher value means greater similarity between the predicted and reference segmentation masks.

Since the segmentation task was treated as a binary pixel classification problem, the evaluation also used the F1-score measure, which is the harmonic mean of precision and sensitivity (Eq. 3):

$$F1 = \frac{2 \times Precision \times Recall}{Precision + Recall} \tag{3}$$

In order to calculate this measure, a confusion matrix was created, containing the number of correct positive classifications (TP), false positives (FP), false negatives (FN) and correct negatives (TN). Based on these, the following indicators were determined: sensitivity, precision, and overall accuracy (Eqs. 4–6):

$$Recall = \frac{TP}{TP + FN} \tag{4}$$

$$Precision = \frac{TP}{TP + FP} \tag{5}$$

$$Accuracy = \frac{TP + TN}{TP + TN + FP + FN} \tag{6}$$

Sensitivity determines the percentage of the true positive pixels that the model correctly identified as positive. Precision describes what percentage of all pixels classified as positive are actually positive. Accuracy indicates what percentage of all pixels were classified correctly.

In addition to the quantitative analysis, a qualitative evaluation was also performed in the form of a visual analysis. The model predictions were compared with the expert annotations, paying particular attention to the precision of carbide boundary mapping and the identification of systematic errors. This approach provided additional interpretive context for the numerical results obtained and helped identify potential directions for further improvement of the method.

3   Results and Discussion

In this section, the learning process for each of the model versions evaluated was analyzed. Training was carried out using the early stopping mechanism, which interrupted the learning process if there was no improvement in the loss function value in the validation set for 10 consecutive epochs. The aim of this approach was to limit the phenomenon of overfitting and to select the best model based on the course of validation. Checkpoint mechanisms were used to maintain the best model.

The training curves shown in Fig. 2 illustrate the loss function values on the training and validation sets for the U-Net, Attention U-Net and DeepLabv3 models in three configurations: training from scratch, using a pre-trained encoder on ImageNet and on the MicroNet set.

In all cases, the models trained from scratch showed faster and more stable convergence, reaching relatively low final values of the loss function on the validation set. This effect was most pronounced for the DeepLabv3 model, which

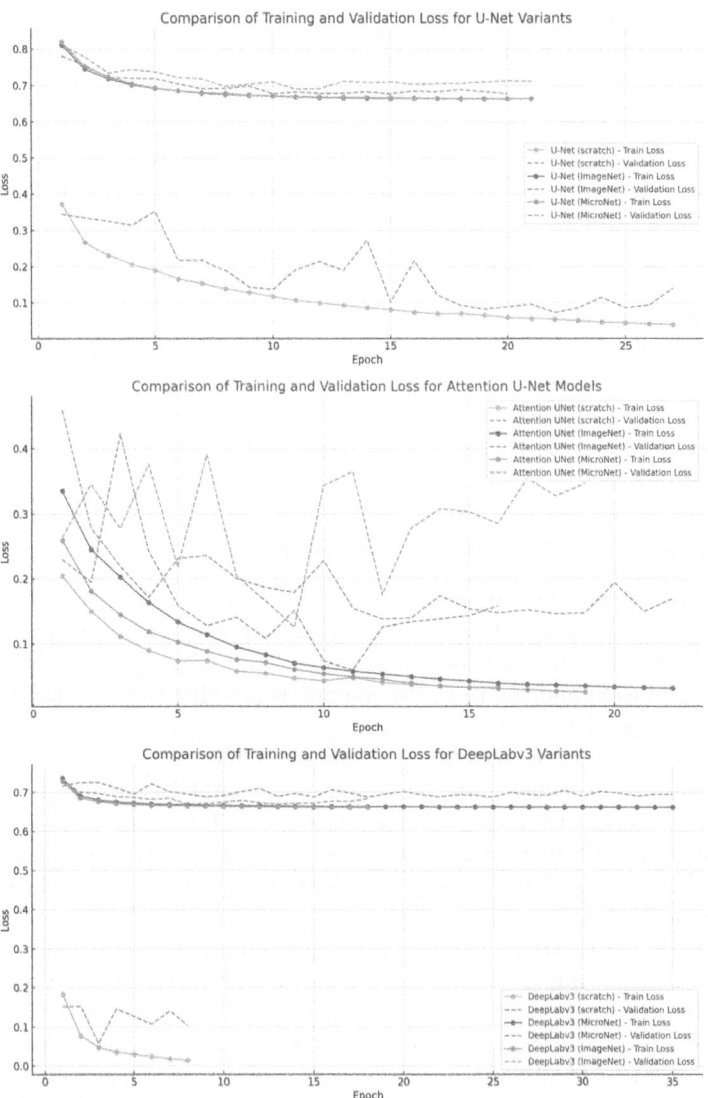

Fig. 2. Training and Validation Loss for different neural network architectures: U-Net (top), Attention U-Net (middle), DeepLabv3 (bottom)

achieved a validation loss of around 0.10. In contrast, the variants with pre-trained encoders on ImageNet and MicroNet converged much more slowly and stopped at much higher loss values (above 0.67), which indicates a mismatch of the weights to the extended input data.

A similar trend was observed with the U-Net architecture. The model learned from scratch reached stable convergence and finished training with a validation loss of around 0.14. Versions with pre-trained weights (both Ima-geNet and MicroNet) did not show significant improvement during training, and the loss values remained high throughout the learning process.

For the Attention U-Net, the model learned from scratch also showed bet-ter learning dynamics, significantly reducing the loss value on all sets. The version with a pre-trained encoder on ImageNet initially learned more slowly, but ultimately achieved a competitive loss level. In contrast, the weight-based variant from MicroNet showed an unstable validation curve with large fluctua-tions, which may suggest overfitting or insufficient adaptation of the weights to the analyzed dataset.

In conclusion, although pre-trained encoders should theoretically acceler-ate convergence and improve model generalization, their effectiveness largely depends on the compliance of the source data with the characteristics of a spe-cific task, in this case the segmentation of microstructure images with extended input channels.

3.1   Quantitative Comparison of Models

A comparison of the performance of the tested segmentation models was carried out based on a set of standard evaluation measures: precision, recall, F1-score, Dice coefficient, and Jaccard coefficient (IoU). The evaluation was performed on both the validation and the test sets, making it possible to assess the generaliz-ability of each model.

Table 1 and Table 2 show detailed results for all models and training scenarios. Table 1 contains the values of the precision, recall, and F1-score metrics, allowing the precision of the classification to be assessed at the pixel level. On the other hand, Table 2 focuses on measures of spatial similarity, Dice, and IoU, commonly used in segmentation tasks to assess the degree of coverage between the predicted and the reference mask.

The Attention U-Net model with pretraining on ImageNet achieved the best overall results, obtaining the highest values of all metrics on the test set: IoU (0.8749), F1-score (0.9316), and Dice (0.9316). This result emphasizes the effec-tiveness of using weights pre-trained on large sets of natural images, especially in combination with attention mechanisms that support selecting relevant features in images with a heterogeneous structure.

The DeepLabv3 model trained from scratch also achieved good results, with a high segmentation accuracy (IoU = 0.8375, F1-score = 0.9083). This demon-strates the effectiveness of the encoder-decoder architecture with attractive spa-tial pyramid pooling, even without prior learning on external data sets.

The exact opposite was observed for models with pretraining on the MicroNet dataset, especially for DeepLabv3 and U-Net, where the IoU values did not exceed 0 13. This suggests that transferring the representation from microscopic data does not always bring benefits if the structure of the input data or its domain differs significantly from the original.

Interestingly, models with pretraining on ImageNet (U-Net, DeepLabv3) also performed poorly. This is most likely because the pretrained weights were only

Table 1. Precision, Recall and F1-score for different models on the validation and test sets

Model	Pretraining	Dataset	Precision	Recall	F1-score
DeepLabv3	None	Val	0.7360	0.8158	0.7263
DeepLabv3	None	Test	0.9293	0.8924	0.9083
U-Net	None	Val	0.5653	0.8926	0.6512
U-Net	None	Test	0.8672	0.9288	0.8950
Attention U-Net	None	Val	0.5570	0.8510	0.6191
Attention U-Net	None	Test	0.7830	0.9352	0.8376
DeepLabv3	MicroNet	Val	0.0839	1.0000	0.1392
DeepLabv3	MicroNet	Test	0.1054	1.0000	0.1761
U-Net	MicroNet	Val	0.0839	1.0000	0.1391
U-Net	MicroNet	Test	0.1359	0.7035	0.2038
Attention U-Net	MicroNet	Val	0.6061	0.7372	0.5921
Attention U-Net	MicroNet	Test	0.8420	0.9069	0.8522
DeepLabv3	ImageNet	Val	0.0839	1.0000	0.1391
DeepLabv3	ImageNet	Test	0.1054	1.0000	0.1761
U-Net	ImageNet	Val	0.0839	1.0000	0.1391
U-Net	ImageNet	Test	0.1054	1.0000	0.1761
Attention U-Net	ImageNet	Val	0.6798	0.8095	0.6990
Attention U-Net	ImageNet	Test	**0.9328**	**0.9318**	**0.9316**

prepared for three RGB channels, while the input data in this study contained more information (additional channels). In such cases, the missing weights were initialized to zeros, which may have limited the network's ability to learn effectively in the initial stages of training.

In conclusion, the results indicate that transfer learning can significantly improve segmentation quality. However, its effectiveness depends on the consistency of the model architecture with the input data and the method of integrating pre-trained weights in the case of an extended number of channels.

3.2   Visual Analysis of Predicted Results

In addition to the quantitative evaluation, a visual analysis of the segmentation results was also carried out to assess the qualitative accuracy of the models masks. The comparison focused on how well the models reproduce the masks annotated by experts and how they deal with visually ambiguous or structurally similar areas.

Models trained from scratch generally fit the reference masks (ground truth) better. Their predictions usually more accurately reproduced the actual carbide boundaries, and the number of false-positive detections was noticeably lower.

Table 2. Wartości IoU i Dice Score dla różnych modeli na zbiorach walidacyjnym i testowym

Model	Pretraining	Dataset	IoU	Dice Score
DeepLabv3	None	Val	0.6507	0.7263
DeepLabv3	None	Test	0.8375	0.9083
U-Net	None	Val	0.5287	0.6512
U-Net	None	Test	0.8139	0.8950
Attention U-Net	None	Val	0.5336	0.6191
Attention U-Net	None	Test	0.7430	0.8376
DeepLabv3	MicroNet	Val	0.0839	0.1392
DeepLabv3	MicroNet	Test	0.1054	0.1761
U-Net	MicroNet	Val	0.0839	0.1391
U-Net	MicroNet	Test	0.1231	0.2038
Attention U-Net	MicroNet	Val	0.4856	0.5921
Attention U-Net	MicroNet	Test	0.7776	0.8522
DeepLabv3	ImageNet	Val	0.0839	0.1391
DeepLabv3	ImageNet	Test	0.1054	0.1761
U-Net	ImageNet	Val	0.0839	0.1391
U-Net	ImageNet	Test	0.1054	0.1761
Attention U-Net	ImageNet	Val	0.6062	0.6990
Attention U-Net	ImageNet	Test	**0.8749**	**0.9316**

This indicates that despite the lack of prior learning, the networks could effectively adapt to the specifics of the analyzed data set during the whole training process.

However, most models using pretrained encoders (both on ImageNet and MicroNet) tended to incorrectly label areas that resembled carbides but were not labeled by experts. False positive predictions appeared most often in regions with colors and textures similar to carbides, which may suggest that these models relied too heavily on visual features learned during pretraining rather than on patterns characteristic of a given type of microstructure.

The Attention U-Net with pretraining on ImageNet showed the most stable and visually consistent results of all the models tested. In this case, the number of misclassifications of objects that were visually similar but did not have carbides was the lowest. This model was also more precise in the location of carbide boundaries. The attention mechanism probably contributed to filtering out irrelevant features and focusing predictions on semantically relevant areas of the image.

4   Conclusion

The study analyzed the impact of enriching the input data with additional structural information on the quality of image segmentation of gray cast iron microstructures. The models achieved better segmentation performance by incorporating additional channels derived from k-means clustering, Sobel edge detection, and superpixel segmentation into standard RGB images.

Among the tested architectures, the Attention U-Net that utilized ImageNet pretraining demonstrated the best overall performance, achieving the highest Dice and Intersection over Union (IoU) scores. This suggests that combining extended input representations with attention mechanisms, along with pre-trained weights, can significantly enhance the effectiveness of segmentation processes.

The study highlighted that the effectiveness of transfer learning depends on the proper adaptation of the model architecture to the number and type of input channels. The analysis showed that models pretrained on MicroNet performed poorly, emphasizing that simply ensuring similarity in the domain is not enough to achieve optimal results.

Future work should concentrate on further automating the segmentation process, improving the stability of learning with limited datasets, and investigating alternative pretraining methods for multichannel input. Furthermore, it is important to ensure the accuracy and consistency of ground-truth labels, as labeling noise can significantly impact evaluation results. Future research will also focus on hyperparameter optimization and a comprehensive analysis of alternative initialization strategies for additional channels.

Although this study employed zero initialization for stability, systematic investigation of other approaches–including constant value initialization, RGB weight averaging, and adaptive Gaussian initialization–warrants detailed exploration to optimize multi-channel transfer learning performance. Such studies could provide valuable information on the theoretical foundations of weight initialization in extended channel scenarios and potentially improve segmentation accuracy across different types of material microstructures.

Acknowledgments. The authors gratefully acknowledge the financial support provided under the University Grant System for Research Work Involving Doctoral Students 5th Edition, implemented within the framework of the Initiative for Excellence Research University (IDUB) at AGH University of Krakow (mini-grant no. 10547). The dataset used in this study is available from the corresponding author upon reasonable request. Previous research data supporting the methodological foundation of this study, including preliminary results with U-Net architectures and initial validation of the multi-channel approach, are also available from the corresponding author upon reasonable request. These preliminary data are not publicly available as they form part of ongoing research that will be submitted for publication elsewhere.

Conflict of Interest. The authors declare no conflict of interest.

References

1. Biswas, M., Pramanik, R., Sen, S., et al.: Microstructural segmentation using a union of attention guided u-net models with different color transformed images. Sci. Rep. **13**(1), 1–14 (2023). https://doi.org/10.1038/s41598-023-32318-9
2. Blanco, S.R., Heras, D.B., Argüello, F.: Texture extraction techniques for the classification of vegetation species in hyperspectral imagery: Bag of words approach based on superpixels. Remote Sens. **12**, 2633 (2020). https://doi.org/10.3390/RS12162633
3. Britefury: Django labeller. https://github.com/Britefury/django-labeller. Accessed Oct 2024
4. Chen, L.C., Papandreou, G., Schroff, F., Adam, H.: Rethinking atrous convolution for semantic image segmentation (2017). https://arxiv.org/abs/1706.05587
5. Gundlach, R.B., Janowak, J.F., Bechet, S., Rohrigtt, K.: On the problems with carbide formation in gray cast iron. MRS Proc. **34**, 251 (1984). https://doi.org/10.1557/PROC-34-251
6. He, K., Zhang, X., Ren, S., Sun, J.: Deep residual learning for image recognition (2015). https://arxiv.org/abs/1512.03385
7. Khrissi, L., El Akkad, N., Satori, H., Satori, K.: Image segmentation based on K-means and genetic algorithms. In: Bhateja, V., Satapathy, S.C., Satori, H. (eds.) Embedded Systems and Artificial Intelligence. AISC, vol. 1076, pp. 489–497. Springer, Singapore (2020). https://doi.org/10.1007/978-981-15-0947-6_46
8. Kong, W., Chen, J., Song, Y., Fang, Z., Yang, X., Zhang, H.: Sobel edge detection algorithm with adaptive threshold based on improved genetic algorithm for image processing. Int. J. Adv. Comput. Sci. Appl. **14** (2023). https://doi.org/10.14569/IJACSA.2023.0140266
9. Li, P., Zhao, W., Fu, C., et al.: Segmentation of backscattered electron images of cement-based materials using lightweight u-net with attention mechanism (lwaunet). J. Build. Eng. **77**, 107547 (2023). https://doi.org/10.1016/J.JOBE.2023.107547
10. Liu, W., Chen, J., Liu, C., et al.: Boundary learning by using weighted propagation in convolution network. J. Comput. Sci. **62**, 101709 (2022). https://doi.org/10.1016/J.JOCS.2022.101709
11. Müller, D., Soto-Rey, I., Kramer, F.: Towards a guideline for evaluation metrics in medical image segmentation. arXiv preprint arXiv:2202.05273 (2022). https://doi.org/10.48550/arXiv.2202.05273
12. NASA: Pretrained microscopy models (2022). https://github.com/nasa/pretrained-microscopy-models. Accessed 04 Apr 2025
13. Nguyen, T.P., Chae, D.S., Park, S.J., Yoon, J.: A novel approach for evaluating bone mineral density of hips based on sobel gradient-based map of radiographs utilizing convolutional neural network. Comput. Biol. Med. **132**, 104298 (2021). https://doi.org/10.1016/J.COMPBIOMED.2021.104298
14. On, V., Zahedi, A., Phandthong, R.: Applications of superpixels on phase contrast microscopy. In: Proceedings of the 2023 5th International Conference on Transdisciplinary AI (TransAI), pp. 262–265 (2023). https://doi.org/10.1109/TRANSAI60598.2023.00035
15. Pambudi, E., Andono, P., Pramunendar, R.: Image segmentation analysis based on k-means pso by using three distance measures. ICTACT J. Image Video Process. **9**, 1821–1826 (2018). https://doi.org/10.21917/ijivp.2018.0256

16. Pugazhenthi, A., Kumar, L.S.: Selection of optimal number of clusters and centroids for k-means and fuzzy c-means clustering: a review. In: 2020 5th International Conference on Computing, Communication and Security (ICCCS), pp. 1–4 (2020). https://doi.org/10.1109/ICCCS49678.2020.9276978
17. Stuckner, J., Harder, B., Smith, T.M.: Microstructure segmentation with deep learning encoders pre-trained on a large microscopy dataset. npj Comput. Mater. **8**, 200 (2022). https://doi.org/10.1038/s41524-022-00878-5
18. Xie, Q., Hao, K., Wei, B., et al.: Adaptive dual-path spatial-frequency network for medical microstructure segmentation. Expert Syst. Appl. **275**, 127032 (2025). https://doi.org/10.1016/J.ESWA.2025.127032
19. Yang, K., Yau, J., Fei-Fei, L., Deng, J., Russakovsky, O.: A study of face obfuscation in imagenet. In: International Conference on Machine Learning (ICML) (2022)
20. Zhuang, F., et al.: A comprehensive survey on transfer learning. Proc. IEEE **109**(1), 43–76 (2021). https://doi.org/10.1109/JPROC.2020.3004555

Deep-RVT: A Residual Vision Transformers for Human Action Recognition

Sayda Elmi[1,2(✉)], Morris Bell[1], and Sai Karthik Navuluru[2]

[1] School of Medicine, Yale University, New Haven, USA
{saida.elmi,morris.bell}@yale.edu
[2] University of New Haven, West Haven, USA
snavu3@unh.newhaven.edu

Abstract. In action recognition, while combining spatio-temporal videos with skeleton features can enhance recognition performance, it necessitates distinct models and balanced feature representations for cross-modal data. Addressing these challenges, we introduce the Deep Residual Vision Transformer (Deep-RVT), a novel architecture that combines the training stability and resilience provided by residual connections with the representational power of Vision Transformers in a seamless manner. The spatio-temporal dynamics required for identifying human actions in video sequences are precisely captured by Deep-RVT. Our model efficiently propagates low-level features across layers by embedding residual pathways into the Transformer blocks, which resolves the vanishing gradient issue and speeds up the training process. Our design leverages the integration of local and global features to encode subtle motion cues and spatial arrangements, which are essential for identifying a broad range of human activities. The proposed network is validated by lots of mainstream benchmarks. Many experimental results, conducted on the Penn-Action, ImageNet-1K and ImageNet-22K, show that the proposed network outperforms most state-of-the-art methods. Our code is available at https://anonymous.4open.science/r/RVT-6D6F/README.md.

Keywords: Action Recognition · Spatio-Temporal Learning · Vision Transformers · Residual Neural Network

1 Introduction

Action recognition, a longstanding research area, entails categorizing human actions based on video frames and finds applications in diverse fields such as human-robot interaction [4], healthcare [28], and video surveillance [14]. With the advent of deep learning, action recognition research has evolved into three distinct approaches. Firstly, the video-based approach [25,33], involves employing deep learning models solely on video frames to discern actions. This approach

R. Wrembel et al. (Eds.): DEXA 2025, LNCS 16047, pp. 34–48, 2026.
https://doi.org/10.1007/978-3-032-02088-8_3

often experiences a notable decline in performance due to various disturbances encountered in real-world scenarios, such as variations in camera angles, sizes of human subjects, and complex backgrounds. The second approach involves using the skeleton structure [6,13], where actions are identified based on human skeletons and their joint trajectories across different time frames fed into deep learning models. However, this method needs an additional deep learning model to extract skeletons from images. Furthermore, the accuracy of the skeleton extractor and the extent of overlap among skeletons significantly influence action recognition in this approach. The third strategy involves leveraging cross-modal data, combining both video and skeletal information [10]. Deep learning models are trained to process both RGB video frames and human skeletal features simultaneously, often yielding high recognition performance. Nonetheless, the fusion of video and skeletal data entails a complex process and typically requires a separate sub-model for cross-modal learning.

As a novel learning paradigm within the deep learning domain, the Vision Transformer (ViT) [12] has garnered significant interest recently due to its remarkable performance across various computer vision tasks, including image classification [15,21], image segmentation [35], object tracking [38], and action recognition [8]. ViT's self-attention mechanism, a cornerstone of its architecture, excels at capturing spatial relationships within images, making it particularly effective for tasks like image classification. However, in action recognition, where both long-range temporal features and dynamic feature variations over time are crucial, ViT, relying on the conventional multi-head attention mechanism, faces challenges due to its high computational demands [23].

To address these challenges, we propose the Deep Residual Vision Transformer (Deep-RVT), a novel architecture that combines the training stability and resilience provided by residual connections with the representational power of Vision Transformers. Our Deep-RVT model efficiently propagates low-level features across layers by embedding residual pathways into the Transformer blocks, resolving the vanishing gradient issue and speeding up the training process. Our design leverages the integration of local and global features to encode subtle motion cues and spatial arrangements, essential for identifying a broad range of human activities.

The proposed network is validated by extensive experiments conducted on several mainstream benchmarks, including the Penn-Action, ImageNet-1K, and ImageNet-22K datasets. These datasets are widely recognized in the research community and serve as standard benchmarks for evaluating the performance of action recognition models. Using these benchmarks allows us to demonstrate the robustness and generalizability of our model across diverse and challenging datasets.

In this study, we compare our proposed Deep-RVT model with several recent state-of-the-art methods to highlight its performance improvements. These comparisons are crucial for demonstrating the advancements our model offers in terms of accuracy and efficiency.

In summary, our contributions are threefold:

- We introduce a novel Deep Residual Vision Transformer (Deep-RVT) architecture that integrates residual connections within Transformer blocks to enhance gradient flow and information propagation.
- We present an innovative adaptive learning module that dynamically adjusts residual path contributions based on input, optimizing the trade-off between feature preservation and transformation.
- We provide extensive experimental validation on multiple benchmarks, showing that our model outperforms state-of-the-art methods in action recognition tasks.

2   Related Work

Image-Based Action Recognition: Image-based action recognition aims to identify actions solely through sequential [25,33] or static images [17,18]. The fundamental process of video-based action recognition entails dissecting the action into smaller semantic elements and grasping the significance of each element in action identification [18]. Utilizing video frames in this approach allows for processing with a straightforward single model. However, with long videos, the efficacy could notably suffer due to diverse background noises.

Skeleton-Based Action Recognition: Skeleton-based action recognition has emerged as a highly effective approach due to its abstraction from raw video data, enabling a focus on the dynamics of human motion. It aims to identify actions by employing a series of spatio-temporal joint coordinates extracted from video frames using a pose estimator and applying them to a graph convolutional network (GCN) [7,9], 3D convolutional neural network (3DCNN) [13], and convolutional neural network (CNN) [6]. Skeleton sequences offer the advantage of being immune to contextual disturbances such as changes in background and lighting [13], yet they suffer from a drawback: the recognition performance heavily relies on the accuracy of the pose extractor and necessitates an additional classifier for recognition.

Video and Skeleton-Based Action Recognition: The fusion of video and skeleton-based features represents a synergistic approach that combines the rich contextual information of video frames with the high-level abstraction of skeleton data. Das et al.'s Video-Pose Network (VPN) [11] was a significant step forward, demonstrating the power of cross-modal learning by distilling pose information into the video recognition process. Multi-modal Mutual Learning (MML) strategy was introduced in [39,42] showcasing the potential of simultaneously learning from both modalities, enabling the model to benefit from the complementary strengths of each data type. Kim et al. explored the use of deep residual networks for joint video and skeleton-based action recognition [22], proposing a model that dynamically adjusts its focus between modalities to improve recognition performance. However, these approaches often require separate models for each data type, making the training process more complex.

Transformer-Based Action Recognition: The adoption of transformer models in action recognition has introduced a paradigm shift, emphasizing the importance of capturing long-range dependencies and interactions within video sequences [2,5,31]. ViViT model [2] demonstrated the feasibility of applying self-attention mechanisms directly to video data, but computational challenges were faced. STAR-Transformer model [1] builds upon these insights, integrating spatial-temporal attention mechanisms that efficiently handle the complexities of action recognition. By embedding these mechanisms within a transformer framework, the STAR-Transformer model addresses the scalability issues and sets a new benchmark for accuracy and efficiency in action recognition models.

Most approaches to action recognition employing transformers use video frames as input tokens [3,37,40], with relatively few incorporating the transformer's skeleton [16,27,29]. However, transformer-based action recognition often faces challenges due to high computational costs associated with self-attention applied to numerous 3D tokens in a video. Additionally, there is still a need for a method to integrate cross-modal information using transformers. Therefore, this study marks the initial attempt to utilize spatiotemporal cross-modal data as input tokens for Vision Transformers (ViTs). Our approach differentiates itself by embedding residual pathways into the Transformer blocks, addressing the vanishing gradient issue and enhancing training stability.

3 Methodology

In this section, we first introduce the overall framework, then we present the architecture of the Deep-RVT model. The different components of our model will be described, i.e., (i) residual token embedding and (ii) residual projection for attention.

Figure 1 shows the overall operational structure of the Deep-RVT model. The video frames are fed into the Deep-RVT structure to give multi-class tokens. Deep-RVT consists of a multiple layer encoderdecoder output separable multiclass feature, which is used as input for the downstream action recognition network.

3.1 Deep-RVT Model

As shown in Fig. 2, a multi-stage hierarchy design is used where each stage consists of two parts. First, the input (or 2D reshaped token maps) is processed by the Residual Token Embedding layer, which is implemented as a residual layer with overlapping patches and tokens reshaped to a 2D spatial grid as input. An additional normalization layer is applied to the tokens. This allows each stage to progressively reduce the number of tokens (i.e., feature resolution)

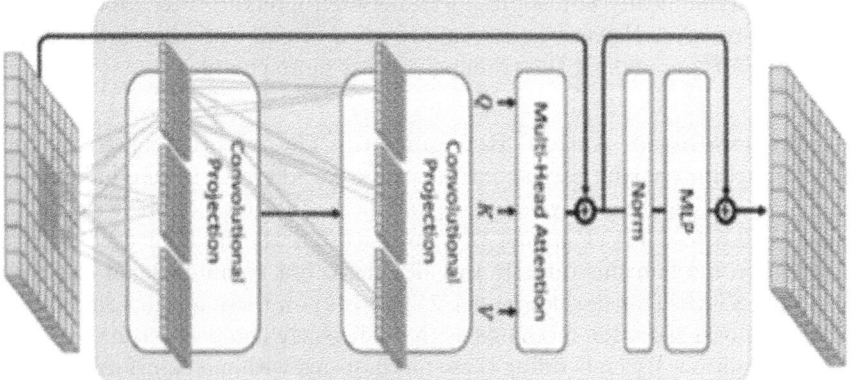

Fig. 1. Framework Architecture. The proposed framework takes as input the video frames and transforms it into a class-separable multi-feature representation. The multi-class token is an aggregation of global grid and joint map tokens obtained by feeding video and pose sequences as showed in Fig. 2

while simultaneously increasing the width of the tokens (i.e., feature dimension), achieving spatial downsampling and increasing the richness of representation.

Different from prior transformer-based architectures, position embeddings are not added to the tokens. Instead, a stack of Residual Transformer Blocks comprises the remainder of each stage. The architecture of the Residual Transformer Block is shown in Fig. 3. A depth-wise separable convolution operation, referred to as Residual Projection, is applied to query, key, and value embeddings instead of the standard position-wise linear projection in ViT. Furthermore, unlike CVT, the classification token is added to each stage, and the output of the final stage is used for classification. An MLP Head is utilized upon the classification token to predict the class. Figure 3 also shows how Residual Projection is performed for the Multi-Head Self-Attention module. The novelty of our Deep-RVT model lies in the integration of residual pathways into the transformer blocks, ensuring efficient gradient flow and better feature propagation across layers. This design addresses the high computational demands of ViTs and enhances the model's ability to capture spatio-temporal dynamics in action recognition tasks.

3.2   Residual Token Embedding

Residual Token Embedding (RTE) is another module used in the RVT model to address the problem of vanishing gradients by enabling the network to better capture long-term dependencies and retain information from earlier layers.

Formally, given an input token sequence $x_i \in \mathbb{R}^{L_i \times D_i}$ at layer i, the residual token embedding layer learns a function $f(\cdot)$ that maps x_i into new tokens $f(x_i) \in \mathbb{R}^{L_i \times D_i}$, where $f(\cdot)$ is the residual operation that adds the original input token sequence to the output of a learnable function $g(\cdot)$ that operates on the input. This is expressed as:

$$f(x_i) = x_i + g(x_i) \tag{1}$$

The residual token embedding layer allows us to adjust the token feature dimension and the number of tokens at each layer by varying the parameters of the learnable function $g(\cdot)$, similar to the Residual Token Embedding layer. The output tokens $f(x_i)$ are then normalized by layer normalization for input into the subsequent layers of the network.

By incorporating residual connections in the network, the residual token embedding layer allows the network to capture long-term dependencies better, retain information from earlier layers, and mitigate the problem of vanishing gradients for deep connections. This technique has been shown to be effective in a wide range of natural language processing tasks, including language modeling, machine translation, and text classification.

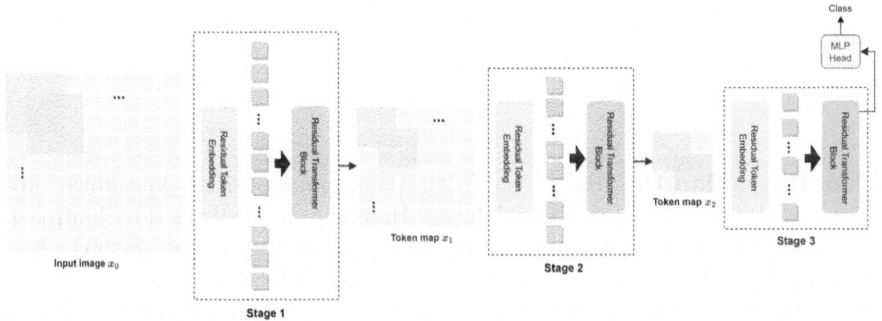

Fig. 2. Deep-RvT Model: Integrating the Residual Networks in the Transformer Architecture.

3.3 Residual Projection for Attention

The Residual Projection layer further represents the local spatial context and improves efficiency by allowing the undersampling of the K and V matrices.

In essence, the Transformer block with Residual Projection is just a broadening of the original Transformer block. Previous works [12] have attempted to improve the Transformer Block for speech recognition and natural language processing by adding more residual modules, but this has resulted in a more complex design and higher computational costs. Instead, the Residual Projection layer, made up of depth-wise separable convolutions, replaces the original position-wise linear projection for Multi-Head Self-Attention (MHSA).

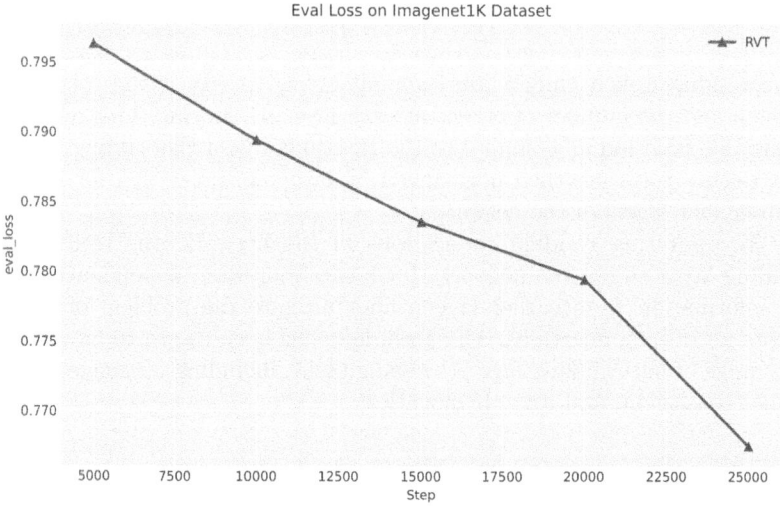

Fig. 3. Residual Projection

Token Projection Details. In CVT [34], the original position-wise linear projection used a 1×1 convolutional layer. However, in RVT, a new approach using $s \times s$ Residual Projection is used, as shown in Fig. 3. This involves first reshaping the tokens into a 2D token map, followed by a Residual Projection layer implemented using a depth-wise separable convolution layer with kernel size s. Finally, the projected tokens are flattened into 1D for further processing. Mathematically, this can be expressed as:

$$x_i^{q/k/v} = \text{Flatten}(\text{Conv2d}(\text{Reshape2D}(x_i), s)) \tag{2}$$

Here, $x_i^{q/k/v}$ represents the token input for $Q = K = V$ matrices at layer i, x_i is the unperturbed token prior to the Residual Projection, Conv2d is a depth-wise separable convolution implemented as Depth-wise Conv2d $\rightarrow$ BatchNorm2d $\rightarrow$ Point-wise Conv2d, and s refers to the convolution kernel size.

The resulting Transformer Block with the Residual Projection layer is a generalization of the original Transformer Block design, which can be trivially implemented using a residual layer with a kernel size of 1×1 for the position-wise linear projection layer.

Residual Projection Efficiency. There are two main efficiency benefits to the Residual Projection layer design in the Residual Vision Transformer. First, instead of using standard $s \times s$ convolutions, efficient depth-wise separable convolution is utilized. This significantly reduces the number of parameters and FLOPs required for each Residual Projection layer compared to the original position-wise linear projection. Second, the Residual Projection layer is leveraged to reduce the computation cost of the MHSA operation. By using a stride larger than 1 for key and value projection, the number of tokens is reduced by a factor of 4, resulting in a corresponding 4x reduction in computational cost. This strategy incurs a minimal performance penalty. Additionally, the Residual Projection layer's local context modeling compensates for the loss of information incurred by resolution reduction. The integration of residual connections within the transformer blocks ensures robust gradient flow, enabling the model to learn more effectively from complex spatio-temporal data. Our extensive experiments demonstrate that the Deep-RVT model significantly outperforms existing state-of-the-art methods in both accuracy and computational efficiency, particularly in action recognition tasks.

4   Experiments

In this section, we describe the implementation details, including the dataset and training hyper-parameters applied.

4.1   Data-Set Description

The experiment was conducted using the representative action recognition datasets, Penn-Action [41], ImageNet-1K and ImageNet-22K.

- **The Penn-Action**: It includes 15 different action classes, such as baseball swings, jumping jacks, and pushups, for a total of 2,326 RGB video sequences.
- **ImageNet-1K**: It contains over 1,000 object categories with approximately 1.2 million high-resolution images. The dataset is structured according to the WordNet hierarchy, where each node of the hierarchy is depicted by hundreds and thousands of images. Commonly used in various computer vision tasks, it serves as a benchmark for image classification models.
- **ImageNet-22K**: An extended version of the ImageNet dataset, comprising approximately 22,000 categories with over 14 million high-resolution images. Unlike ImageNet-1K, this dataset includes a wider variety of categories, not limited to object recognition, and is utilized for tasks requiring a broader and more diverse set of images. The expansive category range of ImageNet-22K offers a challenging environment for more comprehensive image classification and action recognition experiments.

The proposed Deep-RVT was implemented using TensorFlow, and pre-trained ResNet was applied as the backbone network. When training the model, the Penn-Action and ImageNet datasets used 16 fixed frames. For all datasets, we utilized a batch size of 4, 100 epochs, and a stochastic gradient descent (SGD) optimizer. To ensure robustness and consistency in our results, we conducted multiple runs for each dataset and averaged the results to mitigate any anomalies or outliers.

4.2 Benchmarks

The following existing algorithms have been chosen for comparison with our proposed model.

- **Pr-VIPE** [30]: It is a view-invariant probabilistic embedding. It was pre-trained using the Human3.6M dataset [19].
- **UNIK** [36]: A unified framework for skeleton-based action recognition (UNIK). UNIK was pre-trained using the Posetics dataset reconstructed from the Kinect-400 [20] dataset.
- **PoseMap** [24]: It is a long-term localization using 3D LiDARs.
- **CVT** [34]: It has improved efficacy over the Vision Transformers for image processing tasks.
- **DeiT** [32]: Data-efficient image transformers (DeiT) that aim to achieve high performance with less data by leveraging data augmentation and knowledge distillation.
- **Swin Transformer** [26]: Hierarchical Vision Transformer using shifted windows (Swin Transformer) for capturing local and global features efficiently, which is effective for various computer vision tasks.

These benchmarks were chosen for their relevance and state-of-the-art performance in the field of action recognition. Each model represents a distinct approach, allowing for a comprehensive evaluation of our proposed method against diverse methodologies.

4.3 Performance

Comparative Study: In our comprehensive evaluation of the Deep Residual Vision Transformer architecture, we observed remarkable performance enhancements across several challenging benchmarks, demonstrating the effectiveness of integrating residual connections within Transformer blocks. Notably, Deep-RVT achieved superior accuracy metrics on the ImageNet-1K, ImageNet-22K, and Penn Action datasets, outperforming the existing works and establishing new state-of-the-art performance standards (Tables 1 and 2).

The performance of various state-of-the-art methods on action recognition tasks was evaluated across three datasets: Penn Action, ImageNet-1K, and ImageNet-22K. Our model was compared to Pr-VIPE, UNIK, PoseMap, CVT, DeiT, Swin Transformer and Deep-RVT outperformed the existing methods, suggesting its robustness across datasets as shown in Table for a detailed comparison. Notably, Deep-RVT achieved the highest accuracy of 99.7% on the Penn Action dataset and consistently led across the ImageNet datasets. This indicates a higher generalizability of Deep-RVT to diverse datasets, which could be attributed to its advanced architecture. Lower accuracies on the ImageNet datasets suggest these datasets pose more complex challenges, potentially requiring more specialized techniques to achieve similar performance levels as seen with the Penn Action dataset, but it is still out-performing the state-of-the-art existing works (Fig. 4).

Table 1. Features and Pre-Training comparison with the-state-of-the-art methods.

Methods	Pre-Training	Features		
		RGB	Est. Pose	Annot. Pose
Pr-VIPE [30]	✓			✓
UNIK [36]	✓			✓
PoseMap [24]	×	✓	✓	✓
CVT [34]	×	✓		✓
DeiT [32]	×	✓		✓
Swin Transformer [26]	×	✓		✓
Deep-RVT [ours]	×	✓		✓

Table 2. Performance comparison with other state-of-the-art methods on the Penn Action dataset, ImageNet-1K and ImageNet-22K.

Methods	Reference	Datasets		
		Penn Action	ImageNet-1K	ImageNet-22K
Pr-VIPE	[30]	96.9	78.0	80.0
UNIK	[36]	97.4	79.0	81.0
PoseMap	[24]	97.8	80.0	81.5
CVT	[34]	99.4	81.6	83.3
DeiT	[32]	99.5	81.8	83.5
Swin Transformer	[26]	99.6	82.0	83.8
Deep-RVT [ours]	[ours]	99.7	82.4	84.2

On the Penn Action dataset, Deep-RVT achieved a top accuracy of 99.7%, outperforming DeiT and Swin Transformer. Its innovative design, including adaptive learning modules and efficient attention mechanisms, optimizes feature preservation and transformation. This enhances Deep-RVT's generalization capabilities, making it a powerful architecture for applications requiring nuanced understanding of temporal dynamics and action recognition

Analysis on the Attention Layers Impact: In Fig. 5, the x-axis of the graph represents the number of attention layers in the Deep-RVT model, ranging from 1 to 5. The y-axis represents the accuracy of the model in percentage. The graph shows a clear peak at 3 layers, where the accuracy reaches the highest point, marked by a red dot and a dashed blue vertical line for emphasis. This suggests that for the Deep-RVT model, 3 attention layers result in the optimal performance using the ImageNet-1K dataset.

The accuracy increases from just over 78% with 1 layer to an accuracy above 82% with 3 layers, before declining with 4 and 5 layers, where the accuracy drops back down to slightly under 80% and over 79%, respectively. This trend indi-

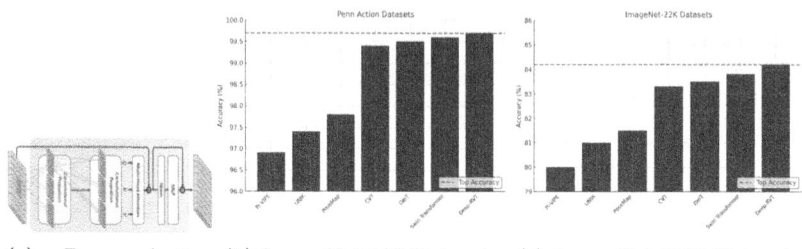

(a)   Penn   Action (b) ImageNet-1K Datasets (c) ImageNet-22K Datasets
Datasets

Fig. 4. Performances of the Existing works on: (a) Penn Action Datasets, (b) ImageNet-1K Datasets, and (c) ImageNet-22K Datasets.

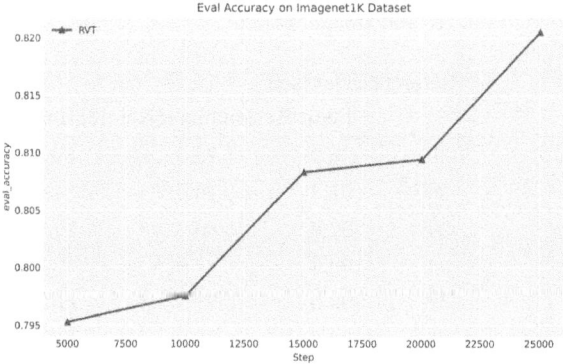

Fig. 5. Evaluation of the Attention Layers

cates that adding more attention layers up to a point can improve the model's performance; however, beyond this optimal point, additional layers may increase the model complexity and decrease its performance, which could be an indication of overfitting or inefficiencies in the model architecture for this particular task. Overall, the graph could be interpreted to support the idea that there is an optimal number of attention layers for the Deep-RVT model when it comes to maximizing accuracy on the ImageNet-1K dataset, and that number appears to be 3 based on our experiments. Furthermore, the RVT's innovative design, which includes adaptive learning modules and efficient attention mechanisms, not only addresses the issue of quadratic complexity in self-attention but also optimizes the balance between feature preservation and transformation. This enables the RVT to adapt dynamically to different inputs, enhancing its generalization capabilities across a wide range of visual recognition tasks. These results underscore the potential of RVT as a versatile and powerful architecture for advancing the field of computer vision, particularly in applications requiring nuanced understanding of temporal dynamics and action recognition. By leveraging the strengths of residual connections within transformer blocks, our model demonstrates a significant reduction in training time and computational resources while achieving superior accuracy. This efficiency makes the Deep-RVT model highly suitable for deployment in real-world applications where both performance and resource utilization are critical.

5   Ablation Study

In our ablation study, we conducted a series of controlled experiments where we removed or altered specific components of the Deep-RVT model: (i) **Residual connections**, (ii) **Adaptive learning module** and (iii) **Efficient attention mechanisms**. We then measured the impact of these changes on key performance metrics such as accuracy, loss, and computational efficiency across various datasets, including Penn-Action, ImageNet-1K, and ImageNet-22K.

The results of the ablation study are presented in Table 3. Each row/column of the table corresponds to a different variant of the Deep-RVT model, where a specific component has been ablated. The performance metrics clearly show a decline when essential components are removed, indicating their importance. For instance, removing the residual connections resulted in a significant drop in accuracy across all datasets, underscoring their role in improving gradient flow and information propagation (Fig. 6).

Table 3. Ablation Study Results: Impact of Removing Different Components on Model Performance

Component	Penn Action Accuracy	ImageNet-1K Accuracy	ImageNet-22K Accuracy	Remarks
Full Model (Deep-RVT)	**99.7%**	**82.4%**	**84.2%**	Baseline
Without Residual Connections	97.2%	79.8%	81.0%	Significant drop
Without Adaptive Learning Module	98.0%	80.2%	82.0%	Moderate drop
Without Efficient Attention Mechanisms	97.5%	79.5%	81.5%	Increased complexity

Removing residual connections led to a decrease in accuracy by 2.5% on the Penn-Action dataset. Indeed, residual connections help in alleviating the vanishing gradient problem, allowing for deeper network architectures and better gradient flow. This results in improved learning and higher accuracy. Disabling the adaptive learning module reduced the model's ability to generalize, as evidenced by a 2.2% drop in accuracy on the ImageNet-1K dataset. This is because, the adaptive learning module dynamically adjusts the residual paths based on the input, optimizing the trade-off between feature preservation and transformation. This enhances the model's adaptability to varying inputs. In addition, replacing efficient attention mechanisms with standard ones increased computational complexity and reduced accuracy by 2.9%. This can be explained by the fact that efficient attention mechanisms reduce the quadratic complexity of self-attention, making the model more computationally efficient while preserving performance. Overall, the consistent decline in performance metrics upon

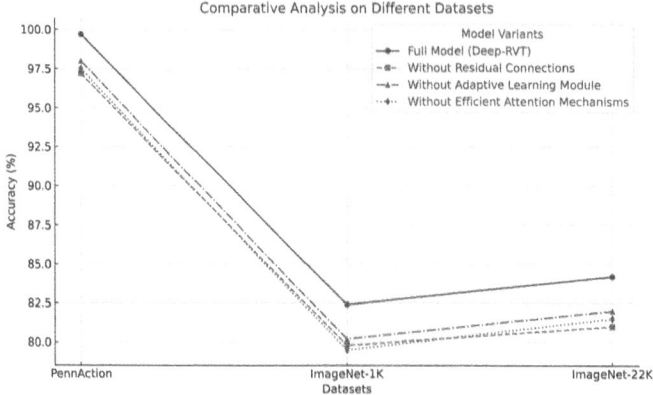

Fig. 6. Comparative analysis on PennAction dataset, ImageNet-1K, and ImageNet-22K datasets.

component removal across multiple datasets strengthens the validity of our conclusions. These results are not dataset-specific but generalizable across diverse data. To ensure a rigorous experimental setup, we kept all other variables constant while ablating components. This isolates the effect of each component on performance.

6   Conclusion

In conclusion, this study introduces the Residual Vision Transformer (RVT), a pioneering architecture that melds the dynamic receptive fields of Vision Transformers with the robustness and training stability of residual learning. Through extensive experiments across challenging benchmarks, including ImageNet-1K, ImageNet-22K, and the Penn Action dataset, RVT has demonstrated superior performance, setting new state-of-the-art metrics in action recognition.

References

1. Ahn, D., Kim, S., Hong, H., Ko, B.C.: Star-transformer: a spatio-temporal cross attention transformer for human action recognition. In: Proceedings of the IEEE/CVF Winter Conference on Applications of Computer Vision, pp. 3330–3339 (2023)
2. Arnab, A., Dehghani, M., Heigold, G., Sun, C., Lučić, M., Schmid, C.: Vivit: a video vision transformer. arXiv preprint arXiv:2103.15691 (2021)
3. Yu, J., et al.: Co-training transformer with videos and images improves action recognition, pp. 1–15. arXiv preprint arXiv:2112.07175 (2021)
4. Bandi, C., Thomas, U.: Skeleton-based action recognition for human-robot interaction using self-attention mechanism. In: The International Conference on Automatic Face and Gesture Recognition (FG), pp. 1–8 (2021)

5. Bertasius, G., Wang, H., Torresani, L.: Is space-time attention all you need for video understanding? arXiv preprint arXiv:2102.05095 (2021)
6. Caetano, C., Sena, J., Brémond, F., Dos, S.J.A., Schwartz, W.R.: Skelemotion: a new representation of skeleton joint sequences based on motion information for 3d action recognition. In: The Conference on Advanced Video and Signal based Surveillance (AVSS), pp. 1–8 (2019)
7. Cai, J., Jiang, N., Han, X., Jia, K., Lu, J.: Jolo-gcn: mining joint-centered light-weight information for skeleton-based action recognition. In: The Winter Conference on Applications of Computer Vision (WACV), pp. 2735–2744 (2021)
8. Chen, J., Ho, C.M.: Mm-vit: multi-modal video transformer for compressed video action recognition. In: The Winter Conference on Applications of Computer Vision (WACV), pp. 1910–1921 (2022)
9. Cheng, K., Zhang, Y., He, X., Chen, W., Cheng, J., Lu, H.: Skeleton-based action recognition with shift graph convolutional network. In: CVPR, pp. 183–192 (2020)
10. Das, S., Sharma, S., Dai, R., Bremond, F., Thonnat, M.: Vpn: learning video-pose embedding for activities of daily living. In: The European Conference on Computer Vision (ECCV), pp. 72–90 (2020)
11. Das, S., et al.: Video-pose network: a novel approach for cross-modal action recognition. In: Proceedings of a Conference (2019)
12. Dosovitskiy, A., et al.: An image is worth 16x16 words: transformers for image recognition at scale. arxiv 2020. arXiv preprint arXiv:2010.11929 (2010)
13. Duan, H., Chen, K., Zhao, Y., Lin, D., Dai, B.: Revisiting skeleton-based action recognition. In: CVPR, pp. 2969–2978 (2022)
14. Elharrouss, O., Almaadeed, N., Al-Maadeed, S., Bourida, A., Beghdadi, A.: A combined multiple action recognition and summarization for surveillance video sequences. Appl. Intell. **51**(2), 690–712 (2021)
15. Elmi, S., Bell, M.: Res-vit: residual vision transformers for image recognition tasks. In: 35th IEEE International Conference on Tools with Artificial Intelligence, ICTAI, pp. 309–316. IEEE (2023)
16. Girdhar, R., Carreira, J., Doersch, C., Zisserman, A.: Action transformer: a self-attention model for short-term action understanding. In: Proceedings of the IEEE/CVF Conference on Computer Vision and Pattern Recognition (2019)
17. Girish, D., Singh, V., Ralescu, A.: A survey on still image based human action recognition. Pattern Recogn. **47**(10), 3343–3361 (2014)
18. Girish, D., Singh, V., Ralescu, A.: Understanding action recognition in still images. In: CVPRW, pp. 370–371 (2020)
19. Ionescu, C., Papava, D., Olaru, V., Sminchisescu, C.: Human3. 6m: large scale datasets and predictive methods for 3d human sensing in natural environments. IEEE Trans. Pattern Anal. Mach. Intell. (TPAMI) **36**(7), 1325–1339 (2013)
20. Kay, W., et al.: The kinetics human action video dataset. arXiv preprint arXiv:1705.0695 (2017)
21. Kim, S., Nam, J., Ko, B.C.: Vit-net: interpretable vision transformers with neural tree decoder. In: The International Conference on Machine Learning (ICML), pp. 1–13 (2022)
22. Kim, Y.A., et al.: Deep residual networks for joint video and skeleton action recognition. In: Proceedings of Some Conference (2017)
23. Liang, Y., Zhou, P., Zimmermann, R., Yan, S.: Dualformer: local-global stratified transformer for efficient video recognition. arXiv preprint arXiv:2112.04674 (2021)
24. Liu, M., Yuan, J.: Recognizing human actions as the evolution of pose estimation maps. In: Proceedings of the IEEE Conference on Computer Vision and Pattern Recognition, pp. 1159–1168 (2018)

25. Liu, X., Pintea, S.L., Nejadasl, F.K., Booij, O., van Gemart, J.C.: No frame left behind: Full video action recognition. In: CVPR, pp. 14892–14901 (2021)
26. Liu, Z., et al.: Swin transformer: Hierarchical vision transformer using shifted windows. In: Proceedings of the IEEE/CVF International Conference on Computer Vision, pp. 10012–10022 (2021)
27. Plizzari, C., Cannici, M., Matteucci, M.: Skeleton-based action recognition via spatial and temporal transformer networks. In: The Computer Vision and Image Understanding (CVIU), vol. 208, p. 103219 (2021)
28. Serpush, F., Menhaj, M.B., Masoumi, B., Karasfi, B.: Wearable sensor-based human activity recognition in the smart healthcare system. Comput. Intell. Neurosci. **2022**(1), 1391906 (2022)
29. Shi, L., Zhang, Y., Cheng, J., Lu, H.: Decoupled spatialtemporal attention network for skeleton-based action-gesture recognition. In: The Asian Conference on Computer Vision (ACCV), vol. 208, p. 103219 (2020)
30. Sun, J.J., Zhao, J., Chen, L.-C., Schroff, F., Adam, H., Liu, T.: View-invariant probabilistic embedding for human pose. In: Vedaldi, A., Bischof, H., Brox, T., Frahm, J.-M. (eds.) ECCV 2020. LNCS, vol. 12350, pp. 53–70. Springer, Cham (2020). https://doi.org/10.1007/978-3-030-58558-7_4
31. Tong, Z., Song, Y., Wang, J., Wang, L.: Videomae: masked autoencoders are data-efficient learners for self-supervised video pre-training, pp. 183–192. arXiv preprint arXiv:2203.12602 (2022)
32. Touvron, H., Cord, M., Douze, M., Massa, F., Sablayrolles, A., Jégou, H.: Training data-efficient image transformers & distillation through attention. In: International Conference on Machine Learning, pp. 10347–10357. PMLR (2021)
33. Wang, Z., She, Q., Smolic, A.: Action-net: multipath excitation for action recognition. In: CVPR, pp. 13214– 13223 (2021)
34. Wu, H., et al.: Cvt: introducing convolutions to vision transformers. In: Proceedings of the IEEE/CVF International Conference on Computer Vision, pp. 22–31 (2021)
35. Yan, X., Tang, H., Ma, H., Sun, S., Kong, D., Xie, X.: After-unet: axial fusion transformer unet for medical image segmentation. In: The Winter Conference on Applications of Computer Vision (WACV), pp. 3971–3981 (2022)
36. Yang, D., Wang, Y., Dantcheva, A., Garattoni, L., Francesca, G., Bremond, F.F.: Unik: a unified framework for real-world skeleton-based action recognition. In: BMVC 2021-The British Machine Vision Conference (2021)
37. Yi, F., Wen, H., Jiang, T.: Asformer: transformer for action segmentation. In: The British Machine Vision Conference (BMVC), pp. 1–15 (2021)
38. Zeng, F., Dong, B., Wang, T., Zhang, X., Wei, Y.: Motr: end-to-end multiple-object tracking with transformer. arXiv preprint arXiv:2105.03247 (2021)
39. Zhang, A., et al.: Multi-modal mutual learning for joint video and skeleton action recognition. In: Proceedings of Another Conference (2020)
40. Wang, X., et al.: Oadtr: online action detection with transformers. In: ICCV, pp. 7565–7575 (2021)
41. Zhang, W., Zhu, M., Derpanis, K.G.: From actemes to action: a strongly-supervised representation for detailed action understanding. In: The International Conference on Computer Vision (ICCV), pp. 2248–2255 (2013)
42. Zolfaghari, S., et al.: Multi-modal fusion for video action recognition. In: Proceedings of Yet Another Conference (2018)

Recommender Techniques

Food Recommendation With Balancing Comfort and Curiosity

Yuto Sakai[1]([✉]) [ID] and Qiang Ma[2] [ID]

[1] Kyoto University, Kyoto, Japan
sakai.yuto.b66@kyoto-u.jp
[2] Kyoto Institute of Technology, Kyoto, Japan
qiang@kit.ac.jp

Abstract. Food is a key pleasure of traveling, but travelers face a trade-off between exploring curious new local food and choosing comfortable, familiar options. This creates demand for personalized recommendation systems that balance these competing factors. To the best of our knowledge, conventional recommendation methods cannot provide recommendations that offer both curiosity and comfort for food unknown to the user at a travel destination. In this study, we propose new quantitative methods for estimating comfort and curiosity: Kernel Density Scoring (KDS) and Mahalanobis Distance Scoring (MDS). KDS probabilistically estimates food history distribution using kernel density estimation, while MDS uses Mahalanobis distances between foods. These methods score food based on how their representation vectors fit the estimated distributions. We also propose a ranking method measuring the balance between comfort and curiosity based on taste and ingredients. This balance is defined as curiosity (return) gained per unit of comfort (risk) in choosing a food. For evaluation the proposed method, we newly collected a dataset containing user surveys on Japanese food and assessments of foreign food regarding comfort and curiosity. Comparing our methods against the existing method, the Wilcoxon signed-rank test showed that when estimating comfort from taste and curiosity from ingredients, the MDS-based method outperformed the Baseline, while the KDS-based method showed no significant differences. When estimating curiosity from taste and comfort from ingredients, both methods outperformed the Baseline. The MDS-based method consistently outperformed KDS in ROC-AUC values.

Keywords: Food Recommendation · Travel · Comfort · Curiosity

1 Introduction

Trying new food or local specialties that one has never eaten before is one of the most enjoyable aspects of traveling, satisfying our curiosity, which is one of the most important motivations to travel. It is desirable that the food that are eaten are made up of ingredients that one can eat and they taste good according

© The Author(s), under exclusive license to Springer Nature Switzerland AG 2026
R. Wrembel et al. (Eds.): DEXA 2025, LNCS 16047, pp. 51–65, 2026.
https://doi.org/10.1007/978-3-032-02088-8_4

to one's own taste. Especially when it comes to food during travel, there is often no second chance to make a different choice if the food does not meet one's expectations.

Research on food recommendation has been actively conducted. Gao et al. [4] and Thongsri et al. [11] proposed food recommendation systems that satisfy users' food preferences. In the context of meals during travel, if recommendations are made using users' food preferences as in conventional methods, the recommended food are likely to be close to the user's preferred taste, ensuring a sense of comfort in enjoying the meal. However, since the taste is similar to the foods they usually enjoy, it does not accompany curiosity.

Correia et al. [1] concluded that the uniqueness and tradition of food are important factors in recommending food to tourists visiting a travel destination. By recommending food based on the local food or popular food in that area, it is possible to recommend food that satisfy users' curiosity. However, the recommendation results may include food made with ingredients that the user cannot eat or food with strong flavors that do not suit their palate. Therefore, comfort is not guaranteed.

To the best of our knowledge, there is no method yet that recommends food at travel destinations that provide both comfort and curiosity for unknown food. In our previous research [10], we proposed a neural network model that takes recipe data composed of food image data, ingredients, and cooking procedures as input and outputs the taste representation vector and ingredient representation vector. We demonstrated through case studies that it is possible to discover food that ensure comfort and curiosity by comparing the representation vectors of the target food's taste and ingredients with the user's past food history using interaction data from recipe sites. However, how to recommend food with balancing comfort and curiosity is still a challenge.

In this study, we propose the concepts of comfort and curiosity in food recommendation and their quantification methods, KDS (Kernel Density Scoring) and MDS (Mahalanobis Distance Scoring), to recommend food that balance comfort and curiosity. The user's food history distribution is probabilistically estimated using kernel density estimation in KDS and distance-based estimation using Mahalanobis distances between foods in MDS. We evaluate how well the representation vectors of the recommended food fit the estimated distribution. The comfort-curiosity score indicates greater comfort with smaller values and greater curiosity with larger values. Additionally, we propose a ranking measure that estimates the balance of comfort and curiosity from the taste and ingredients of the food. By ranking candidate foods based on this measure, we realize food recommendation. The balance of comfort and curiosity is defined by associating the taste and ingredients of the food with the concepts of risk and return on a one-to-one basis, formulating curiosity as the return obtained per unit of risk in choosing a food.

Figure 1 shows an example of the proposed method and recommended foods. It assumes a user who regularly eats familiar food in Japan and visits China for the first time. In this case, conventional preference-based recommendation methods would recommend food like fried rice or soup dumpling, which are commonly eaten in daily life. Also, recommending food like century eggs, which are challenging in both taste and ingredients for the user, is not desirable. In this study, we achieve a balance of comfort and curiosity by recommending food like stinky tofu, which arouses curiosity in taste but is comforting in familiar ingredients, or conversely, food like fried frog, which are close in taste to food the user has eaten before but have unknown ingredients that arouse curiosity[1,2,3,4,5]

.

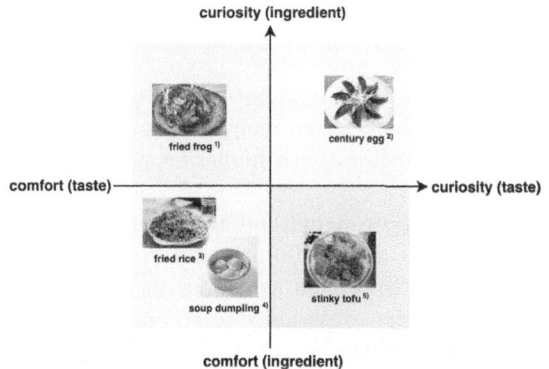

Fig. 1. Example of the recommended food

The major contributions of this study are as follows:

- We propose a novel recommendation method considering both comfort and curiosity. At first, we define and quantifies comfort and curiosity from aspects of taste and ingredients. Then, we propose an integrated measure to estimate the balance of comfort and curiosity to food recommendation (Sect. 3).
- We construct a new dataset that records user emotions towards food, evaluated from the perspectives of taste and ingredients, based on interaction data between users and food collected through crowdsourcing (Sect. 4).
- We demonstrate through experiments that the proposed recommendation method can achieve food recommendations that ensure both comfort and curiosity (Sect. 4).

[1] https://x.gd/ASuZB.

[2] https://x.gd/yGA7a.

[3] https://x.gd/krZJr.

[4] https://x.gd/euQxs.

[5] https://x.gd/8UesA.

2   Related Work

2.1   Research on Relevance-Oriented Recommendations

Conventional food recommendation systems has focused on recommending food based on the relevance of historical data, specifically the user's food preferences. Gao et al. [4] define food recommendation as the problem of predicting a user's preference for a recipe. They propose a recommendation system based on a neural network model called Hierarchical Attention based Food Recommendation (HAFR) to estimate food preferences. HAFR estimates the probability that a user accepts a particular recipe using three inputs: historical data indicating whether the user has eaten a specific recipe, ingredient data, and image data of the food. Ueda et al. [12] also propose a recipe recommendation method based on food preferences. Their research estimates a user's food preferences, including liked and disliked ingredients, from recipes the user has viewed or cooked in the past, and recommends recipes with higher scores when they contain more preferred ingredients.

Thongsri et al. [11] propose a method that not only estimates food that match a user's preferences using collaborative filtering but also calculates basal metabolic rate based on features such as gender, height, and weight, solving a knapsack problem with calorie constraints to recommend food that align with user preferences while maintaining health.

Zhang et al. [13] propose a personalized restaurant recommendation system that extracts visual features from images using deep convolutional neural networks and combines them with collaborative filtering methods.

2.2   Research on Serendipity-Oriented Recommendations

Research on recommendation systems focus on modeling user preferences and recommending items that match those preferences. However, merely recommending items related to the user's past behavior can lead to feelings of boredom or dissatisfaction. There is a risk of missing out on items that offer unexpected value. In recent trends, research on serendipity-oriented recommendations has become active to address these issues. The concept of serendipity in recommendation systems has not yet reached a consensus on its definition, but Fu et al. [3] argue that serendipity consists of four elements: unexpectedness, novelty, diversity, and relevance.

Li et al. [7] propose a recommendation system that incorporates unexpectedness as an element to provide surprise and satisfaction. Their proposed method embeds items experienced by the user in the past into a latent space and forms interest clusters using the mean shift clustering method. They model unexpectedness by calculating the weighted distance between recommended items and each cluster. Lo et al. [8] incorporate the concept of novelty into recommendations by focusing on items that have existed for a long time but are unpopular. Cui et al. [2] introduce diversity into recommendations by defining the similarity between items using Pearson's correlation coefficient and determining the diversity of an item as the reciprocal of the sum of similarities with all items in the user's history set. Zhang et al. [14] demonstrate that it is possible to simultaneously improve unexpectedness, novelty, and diversity in music recommendations. Their research proposes entropy-based and graph-based algorithms in addition to conventional accuracy-focused recommendation frameworks, achieving serendipity-oriented recommendations by linearly complementing recommendation rankings output by multiple recommendation frameworks.

Recommendations based on user food preferences are useful in everyday situations, but they may not suffice for food at travel destinations in unfamiliar lands, where there is a desire to try local and unusual food. Therefore, research is being conducted on recommendation factors beyond relevance. Kauppinen et al. [6] conducted interviews to understand memorable dining experiences. They identified four main themes, including tourism, and emphasized tourists' willingness to try new meals as a key characteristic of their dining experiences. Jiménez et al. [5] conducted research analyzing the relationship between gastronomy, culture, and tourism, revealing that tourists consider food an important part of the cultural identity of tourist destinations. Correia et al. [1] analyzed data from tourists in Hong Kong, highlighting food quality, uniqueness, tradition, and service as key recommendation factors, with a focus on uniqueness and tradition.

In short, There is no agreed-on definition of serendipity in recommendation system research, and each study independently defines it while associating it with concepts such as unexpectedness, novelty, and diversity. Similarly, the concepts of comfort and curiosity addressed in this study have not yet been defined in the context of recommendation systems. To the best of our knowledge, recommendation methods that balance these two concepts does not yet exist.

3   Proposed Method

The proposed recommendation system is illustrated in Fig. 2. First, it receives the user's food history and information on candidate food for recommendation, and outputs a set of representation vectors for each food using the taste network model proposed in our previous research [10]. Next, the comfort and curiosity scoring module inputs the set of representation vectors of the history and candidate foods, and calculates the comfort and curiosity scores for each candidate food. Finally, the ranking module measures the balance of comfort and curiosity of candidate foods and ranks. This section details the newly defined quantitative analysis methods for comfort and curiosity, the measure for estimating the balance of comfort and curiosity, and the recommendation method for food that incorporate both comfort and curiosity. The definitions of the terms used in the explanation are shown in Table 1.

3.1   Definition of Comfort and Curiosity

In this study, "comfort" refers to the feeling of reassurance when selecting a food that aligns with one's taste and ingredient preferences. On the other hand,

Table 1. Definition of Terms in the Proposed Method

F	Set of candidate foods for recommendation
H	Set of foods eaten in the past
$p(\cdot)$	Probability density function for the distribution of history H
$Score_f$	Score for food f
$Score'_f$	Scaled score
$Score^{taste}$	Score for taste representation vector
$Score^{ing}$	Score for ingredient representation vector
$d_{i,j}$	Distance between $h_i, h_j \in H$
d_i	Distance between $h_i \in H, f \in F$
d_{in}	Average distance between all pairs of foods within H
d_{out}	Average distance between candidate food and all foods within H

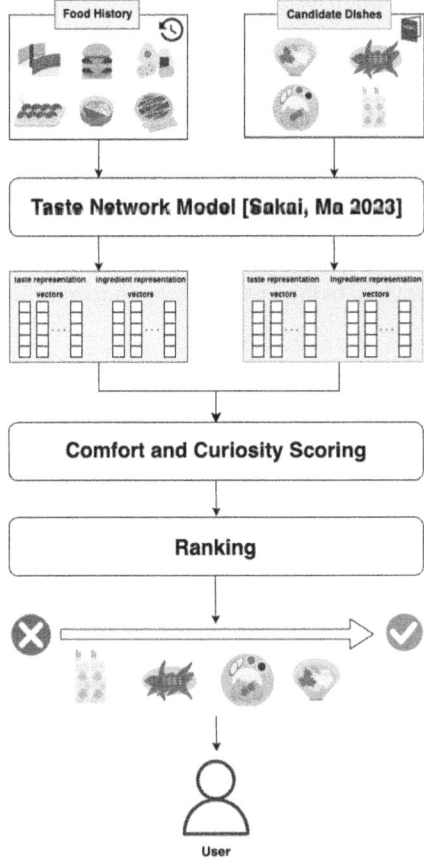

Fig. 2. Overview of Our Recommendation System

"curiosity" refers to the desire to try new or region-specific food that one does not usually eat. Therefore, a food that possesses comfort is defined as one that belongs to the domain formed by the taste and ingredient trends of foods previously eaten. Additionally, a food that possesses curiosity is defined as one that differs from the taste and ingredient trends of foods previously eaten.

3.2   Quantification of Comfort and Curiosity

This section describes the methods for quantifying comfort and curiosity as defined in Sect. 3.1. As mentioned before, we propose two methods: Kernel Density Scoring (KDS) and Mahalanobis Distance Scoring (MDS).

Kernel Density Scoring (KDS). The score $Score_f$ for a food f is evaluated using a probabilistic approach based on how close the representation vector of food f is to the distribution formed by the representation vectors of food h included in the set of foods H previously eaten. Since the distribution of the food history may not fit a known distribution, a non-parametric method, Kernel Density Estimation (KDE) [9], is used to estimate the probability density function of the distribution. The procedure is as follows:

1. Perform principal component analysis on the set of representation vectors of food $h_1, h_2, ..., h_n \in H$ and the candidate food f, reducing each representation vector to a two-dimensional vector.
2. Estimate the probability density function $p(h)$ using KDE on the reduced-dimensional representation vectors of food $h_1, h_2, ..., h_n \in H$.
3. Calculate the probability density $p(f)$ that food f belongs to the distribution of history H.
4. Take the negative logarithm of the probability density obtained in step 3, and set $Score_f = -\log p(f)$.

Mahalanobis Distance Scoring (MDS). The score $Score_f$ for a food f is evaluated using a distance-based approach based on how close the representation vector of food f is to the distribution formed by the representation vectors of food h included in the set of foods H previously eaten. The spatial relationship between the candidate food and the history distribution is captured while considering the internal structure formed by the history. The Mahalanobis distance is adopted as the distance measure to consider the spread of the distribution. The procedure is as follows:

1. Perform principal component analysis on the set of representation vectors of foods $h_1, h_2, ..., h_n \in H$ and the candidate food f, reducing each representation vector to a two-dimensional vector.
2. Calculate the average Mahalanobis distance d_{in} between all pairs of reduced-dimensional representation vectors of food $h_1, h_2, ..., h_n \in H$.

$$d_{in} = \frac{1}{n(n-1)} \sum_{i=1}^{n} \sum_{j=1, j \neq i}^{n} d_{i,j}$$

3. Calculate the average Mahalanobis distance d_{out} between the representation vector of food f and each representation vector of food h_i included in the history.

$$d_{out} = \frac{1}{n}\sum_{i=1}^{n} d_i$$

4. Take the ratio of the values obtained in steps 2 and 3, and set $Score_f = \dfrac{d_{out}}{d_{in}}$.

3.3   Ranking Method

This section describes the method for ranking the set of recommended foods using the score $Score_f$ defined in Sect. 3.2.

First, the scaling of $Score_f$ is performed. When the score for a food h_i included in the history H is $Score_{h_i}$, the scaled value $Score'_{h_i}$ is defined as follows:

$$Score_{max} = \max_{h \in H} Score_h$$

$$Score'_{h_i} = \frac{Score_{h_i}}{Score_{max}}$$

Similarly, the scaled score $Score'_{f_i}$ for a food f_i included in the set of recommended foods F is defined as follows:

$$Score'_{f_i} = \frac{Score_{f_i}}{Score_{max}}$$

Next, the degree of balance between comfort and curiosity is calculated. Comfort and curiosity are obtained from the taste and ingredients of the food, respectively, and the balance $Score_{f_i}^{total}$ for each food $f_i \in F$ is defined as follows. Here, the formula is shown for the case where comfort is obtained from the taste and curiosity from the ingredients.

$$Score_f^{total} = \frac{Score_f'^{ing} - r_f}{Score_f'^{taste}} \tag{1}$$

r_f is defined as following procedures. First, when calculating the score $Score^{taste}$, we identify the food h_{min} with the smallest value of $Score_{h_i}^{taste}$. The food h_{min}, which has the smallest value of $Score^{taste}$, is considered to be a food with assured comfort as it best fits the distribution of foods previously eaten. Then, we calculate the curiosity score $Score_{h_{min}}^{ing}$ for food h_{min}, scale it, and set $r_f = Score_{h_{min}}'^{ing}$.

Finally, for all foods $f_i \in F$ included in the set of recommended food candidates, we calculate $Score_{f_i}^{total}$ using the formula (1), and rank the foods by sorting the values in descending order to achieve a balance between comfort and curiosity.

4   Experiment

4.1   Dataset

In this study, the task is to recommend food that balance comfort and curiosity. To the best of out knowledge, there is no existing research addressing a similar task, nor is there a dataset that records evaluations of comfort and curiosity. Therefore, we conducted a survey using crowdsourcing to construct a dataset that records real users' food histories and their evaluations of comfort and curiosity towards food. The platform used for crowdsourcing was "Lancers"[6]. Workers were informed that their response data would be used for research and analysis, and consent was obtained before collecting responses.

The crowdsourcing task consisted of questions regarding a set of foods, including 12 types of Japanese food and 9 or 10 types of foods from either Southeast Asia, China, or Europe. Workers answered the following four questions for each food:

i. Whether they have eaten this food before.
ii. Whether they feel this food suits their taste and whether they would like to try it.
iii. Whether they feel this food suits their preference for ingredients and whether they would like to try it.
iv. List foods they usually eat in their daily life, other than those listed in the survey.

For question i, workers answered with either "yes" or "no." For questions ii and iii, they selected from the following four options:

1. Feels suitable + Would like to try
2. Feels suitable + Would not like to try
3. Feels unsuitable and anxious + Would like to try
4. Feels unsuitable and anxious + Would not like to try

For question iv, workers provided as many foods as possible that they usually eat in their daily life in a free-text format.

These tasks were created for each set of foods composed of Japanese and Southeast Asian food, Japanese and Chinese food, and Japanese and European food, with 100 different workers responding to each task.

From the collected response data, the following four types of datasets were created:

all_food A dataset that records the recipe data of 40 types of foods prepared as questions in the crowdsourcing task. The breakdown of the 40 foods is 12 types of Japanese food, 10 types of Southeast Asian food, 9 types of Chinese food, and 9 types of European food. Each recipe contains information such as food ID, food name, ingredients, cooking instructions, and food image.

---

[6] https://www.lancers.jp/.

extended_food A dataset created by extracting food names included in the food history answered in question iv of the crowdsourcing and collecting recipes that can be made. It records 170 types of food recipes, and the information each recipe contains is the same as in the *all_food* dataset.

user_food_interaction A dataset that records the interaction between workers and foods, as well as the evaluation values of comfort and curiosity, answered in questions i, ii, and iii of the crowdsourcing. The minimum components of the dataset include worker ID, food ID from *all_food*, experience (answer to question i), taste evaluation (answer to question ii), and ingredient evaluation (answer to question iii).

user_food_extended_interaction A dataset that records the interaction between workers and foods answered in question iv of the crowdsourcing. The minimum components of the dataset include worker ID and food ID from *extended_food*.

4.2   Experiment Settings

We evaluate the performance of the proposed recommendation method using the dataset described in Sect. 4.1. The experimental procedure is as follows:

1. For users recorded in the dataset, set all foods from the region targeted in the crowdsourcing task as candidate foods for recommendation, and obtain a ranking based on the balance of comfort and curiosity defined in Sect. 3.3. The user's food history includes all foods they have eaten, as recorded in the *user_food_interaction* and *user_food_extended_interaction* datasets.
2. Define the relevant items for recommendation to users in Experiment 1, where comfort is derived from taste and curiosity from ingredients, and Experiment 2, where curiosity is derived from taste and comfort from ingredients.
   - The relevant items for Experiment 1 are the set of foods with a taste evaluation of "feels suitable" and an ingredient evaluation of "would like to try." This corresponds to the green area in the second quadrant of Fig. 1.
   - The relevant items for Experiment 2 are the set of foods with a taste evaluation of "would like to try" and an ingredient evaluation of "feels suitable." This corresponds to the red area in the fourth quadrant of Fig. 1.
3. Compare the output ranking with the set of correct items to evaluate the performance of the recommendation.

The ranking performance is evaluated by using $Precision@K, Recall@K$, and $NDCG@K (K = 1, 3, 5)$. The recommendation methods evaluated include two methods using MDS and KDS for quantifying comfort and curiosity, and the Baseline method that generates rankings randomly, making a total of three methods. The evaluation value of the Baseline method is obtained by repeating the evaluation of the ranking output by randomly arranging the candidate food set 10^6 times and using the average value.

Finally, based on the evaluation values of each method, we statistically test the effectiveness of the proposed recommendation method. In this test, considering that the set of values for each evaluation metric does not guarantee a normal distribution, we perform a two-sided test using the Wilcoxon signed-rank test. The data used for the test includes a total of 27 values based on the combination of three evaluation metrics ($Precision, Recall, NDCG$), three K values, and three regions (Southeast Asia, China, Europe) for each method. Then, with a significance level of 0.05, we test whether there is a significant difference between the evaluation results of the Baseline and the proposed method.

4.3   Experiment Results

Experiment 1. The recommendation performance when obtaining comfort from taste and curiosity from ingredients is shown in Tables 2, 3, 4, and 5. The Wilcoxon signed-rank test results showed that the p-value between the evaluation values of the MDS-based method and the Baseline was $0.01523 < 0.05$, confirming that the MDS-based method is superior to the Baseline. On the other hand, the p-value between the evaluation values of the KDS-based method and the Baseline was $0.1023 > 0.05$, indicating no significant difference in performance.

Figure 3 shows the ROC curve for taste-comfort. Looking at the average ROC curve across all regions, the ROC curve of the proposed method based on MDS and KDS is positioned above the Baseline, indicating that the proposed method has higher recommendation performance. In the comparison between the MDS-based and KDS-based methods, the AUC values were 0.5690 for the former and 0.5619 for the latter, demonstrating that the MDS-based method is superior to the KDS-based method.

Looking at the ROC curves for each region, Table 2 shows that in the recommendation case for Southeast Asian food, both proposed methods perform worse than the Baseline. The ROC curves of the KDS-based and MDS-based methods are positioned below that of the Baseline, indicating that the proposed methods are making recommendations that reverse positive and negative examples. In the recommendation cases for Chinese and European food, the proposed methods show higher performance than the Baseline. In particular, the recommendation for Chinese food is shown to be successful with high accuracy.

Table 2. Recommendation Performance of Southeast Asian Food

Method	P@1	P@3	P@5	R@1	R@3	R@5	NDCG@1	NDCG@3	NDCG@5
Baseline	0.4753	0.4753	0.4753	0.1000	0.3000	0.5000	0.4753	0.4963	0.5406
MDS	0.3951	0.3868	0.4123	0.0751	0.2220	0.3916	0.3951	0.3976	0.4418
KDS	0.3210	0.3128	0.4049	0.0474	0.1567	0.3737	0.3210	0.3156	0.4023

Table 3. Recommendation Performance of Chinese Food

Method	P@1	P@3	P@5	R@1	R@3	R@5	NDCG@1	NDCG@3	NDCG@5
Baseline	0.5040	0.5040	0.5040	0.1111	0.3333	0.5555	0.5040	0.5202	0.5619
MDS	**0.7216**	**0.7491**	0.6619	**0.1656**	**0.5378**	0.7503	**0.7216**	**0.7784**	0.7782
KDS	**0.7216**	0.7388	**0.6887**	0.1623	0.5152	**0.7878**	0.7216	0.7647	**0.7967**

Table 4. Recommendation Performance of European Food

Method	P@1	P@3	P@5	R@1	R@3	R@5	NDCG@1	NDCG@3	NDCG@5
Baseline	0.6756	0.6756	0.6756	0.1111	0.3333	0.5556	0.6756	0.6846	0.7067
MDS	**0.8506**	**0.7969**	0.7195	**0.1507**	**0.3994**	0.6069	**0.8506**	**0.8157**	**0.7929**
KDS	0.8161	0.7663	**0.7241**	0.1399	0.3801	**0.6077**	0.8161	0.7842	0.7842

Table 5. Average Recommendation Performance Across Three Regions

Method	P@1	P@3	P@5	R@1	R@3	R@5	NDCG@1	NDCG@3	NDCG@5
Baseline	0.5516	0.5516	0.5516	0.1074	0.3222	0.5371	0.5516	0.5670	0.6031
MDS	**0.6558**	**0.6443**	0.5979	**0.1305**	**0.3864**	0.5829	**0.6558**	**0.6639**	**0.6710**
KDS	0.6196	0.6060	**0.6059**	0.1165	0.3506	**0.5897**	0.6196	0.6215	0.6610

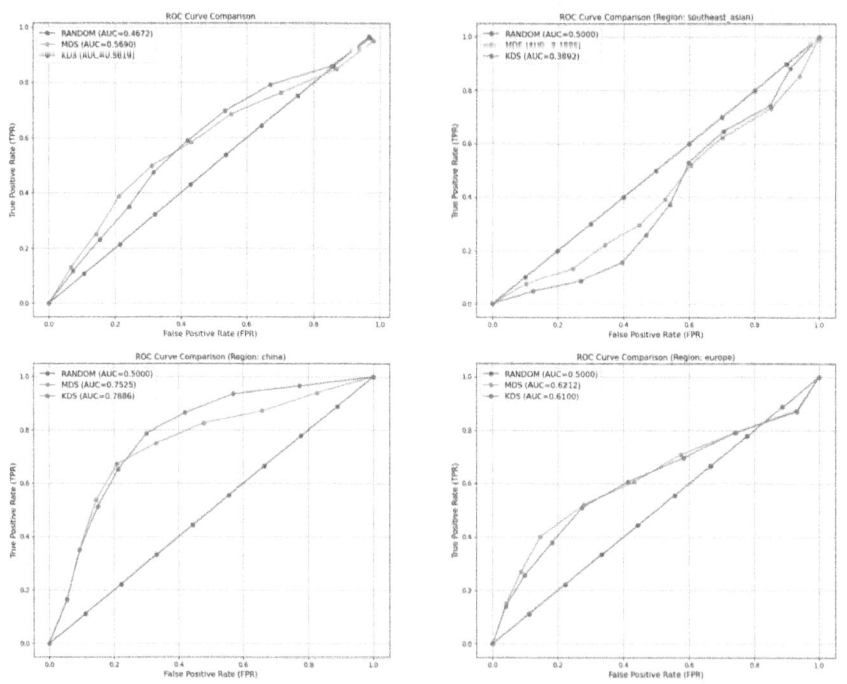

Fig. 3. ROC Curve: Taste-Comfort (Top Left: Average Across All Regions, Top Right: Southeast Asian Food, Bottom Left: Chinese Food, Bottom Right: European Food)

Experiment 2. The recommendation performance when obtaining curiosity from taste and comfort from ingredients is shown in Tables 6, 7, 8, and 9. The Wilcoxon signed-rank test results showed that the p-value between the evaluation values of the MDS-based method and the Baseline was $0.00003978 < 0.05$, and the p-value between the evaluation values of the KDS-based method and the Baseline was $0.001938 < 0.05$. This confirms that the proposed methods are superior to the Baseline.

Figure 4 shows the ROC curve for taste-curiosity. Looking at the average ROC curve across all regions, the ROC curve of the proposed method based on MDS and KDS is positioned above the Baseline, indicating that the proposed method has higher recommendation performance. In the comparison between the MDS-based and KDS-based methods, the AUC values were 0.5384 for the former and 0.5085 for the latter, demonstrating that, as in Experiment 1, the MDS-based method is superior to the KDS-based method.

Table 6. Recommendation Performance of Southeast Asian Food

Method	P@1	P@3	P@5	R@1	R@3	R@5	NDCG@1	NDCG@3	NDCG@5
Baseline	0.1615	**0.1615**	**0.1615**	0.1000	0.2999	0.5000	0.1615	0.2476	0.3317
MDS	**0.1923**	0.1282	**0.1615**	**0.1615**	0.2865	**0.5667**	**0.1923**	0.2561	**0.3766**
KDS	0.1538	0.1538	0.1462	0.1038	**0.3635**	0.5058	0.1538	**0.2799**	0.3415

Table 7. Recommendation Performance of Chinese Food

Method	P@1	P@3	P@5	R@1	R@3	R@5	NDCG@1	NDCG@3	NDCG@5
Baseline	0.1547	0.1547	0.1546	0.1111	0.3335	0.5556	0.1547	0.2565	0.3548
MDS	**0.2609**	**0.2464**	**0.1913**	**0.1594**	**0.5000**	**0.6739**	**0.2609**	**0.3710**	**0.4463**
KDS	**0.2609**	0.1739	0.1826	**0.1594**	0.3116	0.6087	**0.2609**	0.2738	0.4037

Table 8. Recommendation Performance of European Food

Method	P@1	P@3	P@5	R@1	R@3	R@5	NDCG@1	NDCG@3	NDCG@5
Baseline	0.1277	0.1278	0.1278	**0.1111**	0.3333	0.5555	0.1277	0.2443	0.3380
MDS	**0.1500**	**0.2000**	**0.1500**	0.0917	**0.4833**	**0.6333**	**0.1500**	**0.3193**	**0.3795**
KDS	**0.1500**	0.1333	0.1400	0.0917	0.3417	0.5833	**0.1500**	0.2488	0.3560

Table 9. Average Recommendation Performance Across Three Regions

Method	P@1	P@3	P@5	R@1	R@3	R@5	NDCG@1	NDCG@3	NDCG@5
Baseline	0.1480	0.1480	0.1480	0.1074	0.3223	0.5370	0.1480	0.2494	0.3415
MDS	**0.2011**	**0.1915**	**0.1676**	**0.1375**	**0.4233**	**0.6246**	**0.2011**	**0.3155**	**0.4008**
KDS	0.1882	0.1537	0.1563	0.1183	0.3389	0.5659	0.1882	0.2675	0.3670

Looking at the ROC curves for each region, unlike Experiment 1, the proposed methods outperform the Baseline in all three regions. However, there was no significantly higher performance than the Baseline, as seen in the ROC curve for Chinese food in Experiment 1. Additionally, in some metrics such as Precision@3 and Precision@5 in Table 6 and Recall@1 in Table 8, the proposed methods performed worse than the Baseline, indicating room for improvement.

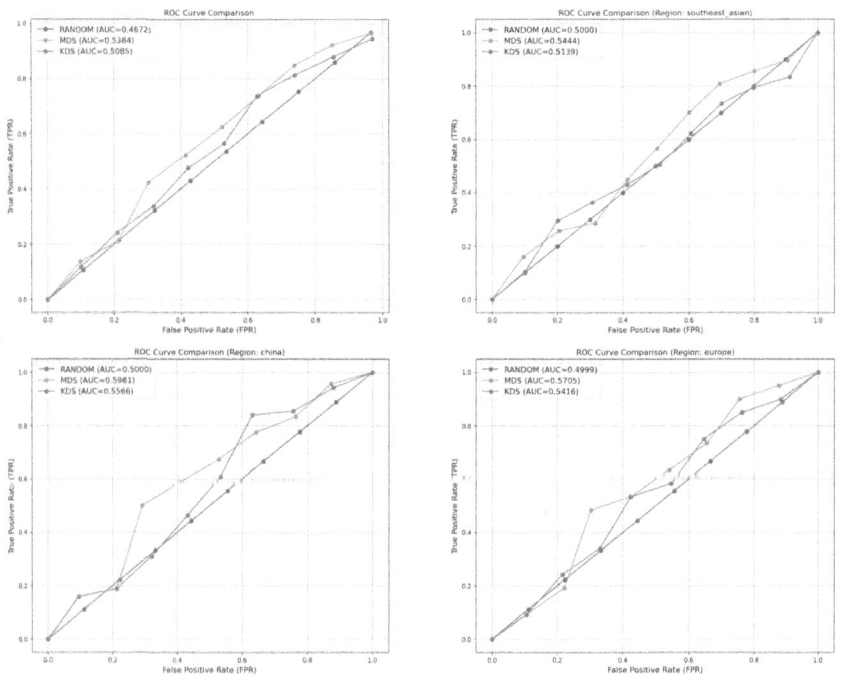

Fig. 4. ROC Curve: Taste-Curiosity (Top Left: Average Across All Regions, Top Right: Southeast Asian Food, Bottom Left: Chinese Food, Bottom Right: European Food)

5   Conclusion

In this study, we addressed the challenge to recommend food that balance comfort and curiosity.

First, we quantified the qualitative concepts of "comfort" and "curiosity" that food possess, taking into account past food history data. Next, we formulated curiosity (return) per unit risk obtained by eating food using these concepts and proposed a ranking method for food recommendations that considers the balance between comfort and curiosity.

Finally, we used a newly constructed dataset obtained through crowdsourcing, comparing the performance of the proposed methods using KDS and MDS against a baseline of random recommendation. The results indicated that the MDS-based method outperformed the Baseline when taste was associated with comfort and ingredients with curiosity. In contrast, no significant difference was observed between the KDS-based method and the Baseline. Additionally, when taste was associated with curiosity and ingredients with comfort, both the MDS-based and KDS-based methods showed improved performance compared to the Baseline. Overall, the MDS-based method demonstrated superiority across all cases.

Acknowledgments. This work was partly supported by JSPS KAKENHI (23K28094)

References

1. Correia, A., Kim, S., Kozak, M.: Gastronomy experiential traits and their effects on intentions for recommendation: a fuzzy set approach. Int. J. Tour. Res. **22** (2019)
2. Cui, Z., et al.: Personalized recommendation system based on collaborative filtering for iot scenarios. IEEE Trans. Serv. Comput. **13**(4), 685–695 (2020)
3. Fu, Z., Niu, X., Maher, M.L.: Deep learning models for serendipity recommendations: a survey and new perspectives. ACM Comput. Surv. **56**(1) (2023)
4. Gao, X., et al.: Hierarchical attention network for visually-aware food recommendation. IEEE Trans. Multimedia **22**(6), 1647–1659 (2020)
5. Jiménez-Beltrán, F.J., López-Guzmán, T., González Santa Cruz, F.: Analysis of the relationship between tourism and food culture. Sustainability **8**(5) (2016)
6. Kauppinen-Räisänen, H., Gummerus, J., Lehtola, K.: Remembered eating experiences described by the self, place, food, context and time. Brit. Food J. **115** (2013)
7. Li, P., Que, M., Jiang, Z., Hu, Y., Tuzhilin, A.: Purs: personalized unexpected recommender system for improving user satisfaction, pp. 279–288. Association for Computing Machinery (2020)
8. Lo, K., Ishigaki, T.: PPNW: personalized pairwise novelty loss weighting for novel recommendation. Knowl. Inf. Syst. **63**(5), 1117–1148 (2021)
9. Parzen, E.: On estimation of a probability density function and mode. Ann. Math. Stat. **33**(3), 1065–1076 (1962)
10. Sakai, Y., Ma, Q.: Taste representation learning toward food recommendation balancing curiosity and comfort. In: Delir Haghighi, P., Pardede, E., Dobbie, G., Yogarajan, V., ER, N.A.S., Kotsis, G., Khalil, I. (eds.) Information Integration and Web Intelligence, pp. 382–397. Springer, Cham (2023). https://doi.org/10.1007/978-3-031-48316-5_36
11. Thongsri, N., Warintarawej, P., Chotkaew, S., Saetang, W.: Implementation of a personalized food recommendation system based on collaborative filtering and knapsack method. Int. J. Electr. Comput. Eng. **12**(1) (2022)
12. Ueda, M., Asanuma, S., Miyawaki, Y., Nakajima, S.: Recipe recommendation method by considering the user's preference and ingredient quantity of target recipe. Lect. Notes Eng. Comput. Sci. **2209**, 519–523 (2014)
13. Zhang, X., Luo, H., Chen, B., Guo, G.: Multi-view visual bayesian personalized ranking for restaurant recommendation. Appl. Intell. **50**, 2901–2915 (2020)
14. Zhang, Y.C., Séaghdha, D.O., Quercia, D., Jambor, T.: Auralist: introducing serendipity into music recommendation, pp. 13–22 (2012)

ONFOODS: A Substitute Recommendation System in Food Recipes

Maryam Mozaffari$^{(\boxtimes)}$, Anton Dignös , Oswald Lanz , Dominik Matt ,
Gabriele Pasetti Monizza , Matthias Gauly , and Johann Gamper

Free University of Bozen-Bolzano, Bozen-Bolzano, Italy
{maryam.mozaffari,anton.dignoes,oswald.Lanz,dominik.Matt
gabriele.pasettimonizza,matthias.Gauly,johann.Gamper}@unibz.it

Abstract. Food waste is a serious problem in modern society. A specific aspect of food waste concerns meat consumption in gastronomy, where typically only prime cuts of meat are used in the kitchen. To facilitate the usage of all parts of animals and thereby reducing food waste, we present ONFOODS, a system that recommends alternative meat cuts in recipes and integrates inventory data to help with the creation of menus. ONFOODS uses an ontology and a knowledge graph to model recipes, meat cuts and the relationships between the two, similarity measures to find candidates for alternative meat cuts, and inventory data to track the availability of different meat cuts. An intuitive user interface allows the user on one hand to update the knowledge graph and inventory data, and on the other hand to navigate through recipes and choose alternative meat cuts.

Keywords: Recommender System · Ingredient Substitution · Ontologies · Knowledge Graphs

1 Introduction

Substituting ingredients in cooking recipes has great potential to improve customer satisfaction and sustainability. For instance, when substituting an ingredient due to dietary restrictions, recipes can be personalized to customer needs. Similarly, when substituting an ingredient in a recipe with an alternative ingredient that may be about to expire or is abundant, food waste can be reduced. Recent studies [1,4,6,17] on ingredient substitution leverage ontologies and knowledge graphs to model and integrate recipe and ingredient data, enabling more accurate substitution recommendations. While these works are motivated by specific goals, such as satisfying user's dietary restrictions, health considerations, or nutritional needs, our motivation is to reduce meat waste. Since recipes require specific meat cuts that may not be at hand, our approach is to combine substitute recommendations with the inventory that indicates which meat cuts are available at which quantities.

© The Author(s), 2026
R. Wrembel et al. (Eds.): DEXA 2025, LNCS 16047, pp. 66–79, 2026.
https://doi.org/10.1007/978-3-032-02088-8_5

Food waste is mentioned in the United Nations' sustainable development goals as a serious problem in modern society. It is estimated that about half of the food grown is wasted before and after it reaches the consumer [14]. A specific instance of this problem is manifested in meat consumption in gastronomy. Instead of using all parts of animals, restaurants often use only the fine cuts of meat, partially because cooks are not familiar with other parts and do not know how to prepare them, and partially because guests consume only fine cuts. One way to help mitigating meat waste in the gastronomy is to provide an information system, which on one hand stores information about the origin of the animal, expiration date of the meat, etc., and on the other hand provides information about how to use alternative meat cuts for the preparation of meals. To develop such an information system, ontologies and knowledge graphs can be used to provide a standardized conceptual terminology [9]. While an ontology represents metadata or a schema, which captures more complex structures with relationships between a set of concepts, the focus of knowledge graphs is data instances. Ontologies and knowledge graphs can be used for various real-world applications such as recommender systems, decision making, and information retrieval.

In this paper, we focus on meat substitution for recipes due to its perishable nature and high cost, making accurate substitution particularly valuable for chefs and other consumers. Currently, most cooks in restaurants and hotels only rely on particular parts of an animal, which are ordered daily from butcher shops. There are several reasons for this: preference for a particular cut, recipes usually only mention one meat cut for a specific recipe, and using an entire animal requires more knowledge about how to use the different parts. To find meat cut alternatives, we use scoring metrics, such as cosine similarity and the Word2Vec model, based on nutritional composition and sensory attributes.

To tackle these problems, we present a recommendation system for alternative meat cuts to be used by chefs and/or decision-makers in restaurants for finding the most sustainable available meat cut for recipes, which can help in considering the use of entire animals. Our system relies on an ontology to organize meat cuts and recipes. The ontology specifies a hierarchy of meat cuts and a hierarchy of recipes together with relationships between the two hierarchies, which allow to find and rank alternative meat cuts for each recipe. Moreover, our system integrates inventory data that helps to manage the inventory and usage of meat.

The technical contributions of this paper can be summarized as follows:

- We design an ontology for meat substitutes in food recipes and generate a knowledge graph based on the ontology.
- We integrate inventory data into the recommendation system, allowing users to create menus based on available meat cuts.
- We investigate how to apply word embeddings and similarity measures based on nutritional composition and sensory attributes to identify suitable alternative meat cuts.

– We provide an intuitive user interface that facilitates easy navigation through the knowledge graph, allowing users to update data and choose alternative meat cuts.

The rest of the paper is structured as follows. In Sect. 2 we discuss and compare related work. Section 3 describes our approach to score alternative meet cuts. In Sect. 4, the architecture of the ONFOODS system is described, including the ontology, knowledge graph, and the inventory database, followed by a discussion of several use cases in Sect. 5.

2   Related Work

Food ontologies specify the shared terminology for types, properties, and relations among food concepts and, hence, can aid in the resolution of data harmonization issues that span food-related domains. Food ontologies are usually used as the schema and play an important role in the development of food knowledge graphs. There exist several food-related ontologies and knowledge graphs: specialized ones in specific domains, e.g., health and nutrition [2,5], cooking and recipes [7,19], and food safety [15,16], as well as more general food ontologies and knowledge graphs [1,4]. A large-scale and comprehensive knowledge graph is FoodKG [4]. It adopts the food product ontology from FoodOn [1] and covers recipes, ingredients, and nutrition information, which can be used in many applications, such as recipe recommendation and ingredient substitutions.

Several studies have addressed the problem of ingredient substitution in food recipes, employing a variety of techniques, such as heuristics [17], AI techniques [10,13] semantic models [6], and filtering [12].

The work in [17] presents a heuristic using explicit semantic information and word embeddings of ingredients to rank plausible ingredient substitutions in FoodKG. To rank ingredient substitutions, similarities in the properties of the ingredients and the recipes are used.

The authors of [13] investigated how to use natural language processing techniques such as word embedding to find alternative ingredients or recipes. The proposed system replaces an ingredient with similar ingredients or a recipe with similar recipes using the cosine similarity of the word embeddings of the ingredients. The AdaptaFood system [10] focuses on adapting recipes to dietary needs using AI-driven ingredient extraction and substitution. It employs BERT-based language models and image-captioning techniques to extract and analyze ingredient descriptions, and find semantically potential substitutes. The system identifies substitutes based on how similar two ingredients are in terms of their textual descriptions and contextual usage in recipes.

In [6], the authors introduced an ontology design pattern for food recipe ingredient substitution, aiming to enhance the interoperability and knowledge representation in culinary and nutritional domains. Their approach facilitates the systematic modeling of ingredient alternatives based on various criteria such as nutritional value, flavor compatibility, and dietary restrictions.

The work in [12] proposed an ingredient substitute recommendation system based on a filtering process that takes into consideration the original recipe context, the relationship among sets of ingredients and user preferences. According to a category of user food restrictions, the proposed system replaces forbidden ingredients in recipes with safe ingredients.

While previous approaches focus on technical aspects and offer general ingredient substitution in recipes, in this paper we concentrate on an easy-to-use end-to-end system that suggests alternative meat cuts in recipes with the aim to use entire animals, thereby mitigating food waste. In addition to suggest alternative meat cuts, our system stores inventory data, such as expiry date, the origin of the animal, and the amount of different cuts, which are helpful information to support the optimal use of entire animals.

3 Scoring Meat Cut Alternatives

To support chefs in using all parts of an animal and, hence, reducing food waste, it is important to propose suitable alternative meat cuts for each recipe. Since recipes are typically prepared with specific meat cuts that may not be at hand, our recommendation of alternative meat cuts considers both the quality of the alternatives as well as the inventory that indicates which meat cuts are available at which quantities.

To find and score meat cut alternatives, our system offers two options: First, chefs can add alternative meat cuts for each recipe, based on their experience and preference for flavor and texture. Second, alternative meat cuts are determined based on the similarity of the nutritional compositions (protein, lipid, and water content) and the sensory attributes (tenderness, texture, and flavor) that influence culinary performance. These factors can be helpful in determining the suitability of a substitution since meat cuts with similar nutritional composition may differ significantly in their cooking behavior and sensory attributes. To incorporate these aspects, we use a weighted cosine similarity measure, where different attributes contribute differently to the overall similarity score. The weights w_i are empirically determined based on expert opinions to reflect the relative importance of each factor in substitution decisions. Since some attributes (e.g., protein, lipid, water content) are continuous, while others (e.g., tenderness, texture, flavor) are categorical, the categorical attributes are first mapped into numerical vectors by applying the Word2Vec embedding method [8]. This method captures the semantic similarity between sensory attributes by representing them as dense numerical vectors, which can be directly compared in the similarity computation. Additionally, Min-Max normalization is applied to all attributes, mapping them to a normalized [0,1] range. This ensures a balanced contribution of both nutritional and sensory attributes in the similarity computation. Then, both sensory and nutritional attributes are concatenated into a single feature vector for each meat cut, which is used to compute the cosine similarity. More specifically, a substitution score S between two meat cuts A and B with normalized feature vectors $\langle A_i \rangle$ and $\langle B_i \rangle$ for $i = 1 \ldots n$ (representing the nutritional and sensory attributes) is calculated using the weighted cosine similarity as follows:

$$S(A, B) = \frac{\sum_{i=1}^{n} w_i A_i B_i}{\sqrt{\sum_{i=1}^{n} w_i A_i^2} \cdot \sqrt{\sum_{i=1}^{n} w_i B_i^2}} \tag{1}$$

The weights w_i for each attribute reflect its importance in determining the similarity between two meat cuts. These weights are empirically determined based on expert opinions, ensuring that attributes such as tenderness and flavor are appropriately emphasized alongside nutritional composition.

If alternative meat cuts are used, different quantities might be required for the same recipe, depending mostly on the amount of water in the meat. Thus, to adjust the quantities needed for a recipe, we consider the quantity of dry matter in each meat cut and compute the quantity of meat cut A with respect to the quantity of meat cut B as:

$$Q_A = Q_B \times \frac{100 - Water_B}{100 - Water_A} \tag{2}$$

4   System Design and Architecture

The overall system design and architecture of ONFOODS is shown in Fig. 1. It consists of four parts: *substitution module, ontology, knowledge graph*, and *user interface*. We present in detail the parts of our system in Sects. 4.1 to 4.4. Afterwards, in Sect. 4.5 we describe the implementation of ONFOODS.

4.1   Substitution Process

To find and rank meat cut alternatives, our system offers two options: First, chefs can add alternative meat cuts for each recipe, based on their experience and preference for flavor and texture. Second, alternative meat cuts are determined based on the similarity of the nutritional compositions (protein, lipid, and water content) and the sensory attributes (tenderness, texture, and flavor) that influence culinary performance. As shown in Fig. 1, the substitution process consists of two phases:

1. *Scoring alternatives:* This phase uses scoring metrics, including cosine similarity and the Word2Vec model to automatically suggests alternative meat cuts based on the nutritional compositions and the sensory attributes as discussed in Sect. 3.
2. *Ranking alternatives:* This phase allows the user to refine and customize the suggested alternatives. Users can propose new alternatives beyond those automatically generated by the system, reorder the suggested options, and adjust substitution quantities. This interactive process enhances flexibility by integrating user preferences into the system, improving the adaptability and personalization of substitutions.

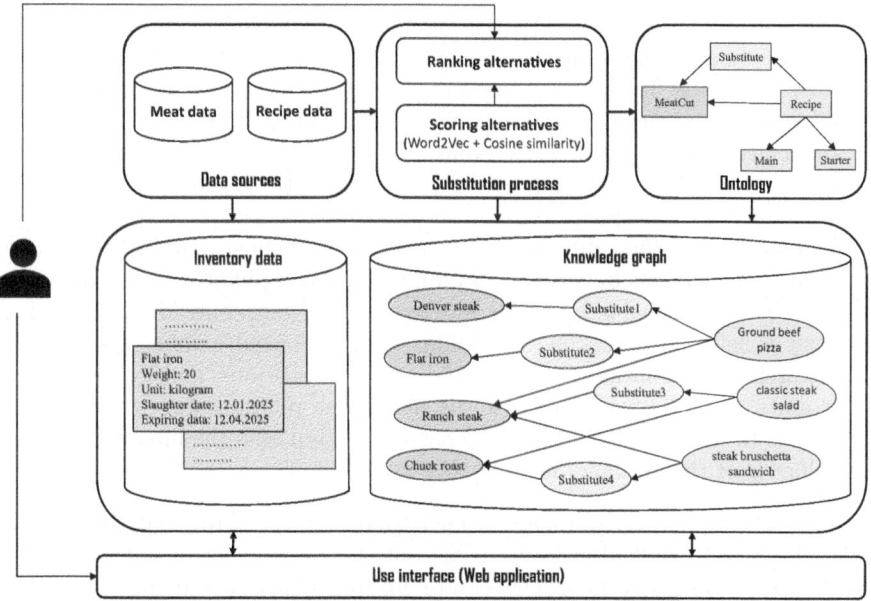

Fig. 1. System architecture.

4.2 Ontology

The ontology specifies the key concepts/classes, relationships and attributes in our specific domain and serves as schema for the knowledge graph. Existing ontologies, such as FoodOn and FoodKG, provide comprehensive models for food items and nutritional information, which only offer generic conceptualizations of substitutions. However, they do not capture the specific substitution relationships required for practical meat cut recommendations, including nutritional and sensory attributes as well as explicit substitution quantities. These features are critical for our use case, where accurate substitutions and the integration with inventory-aware recommendation systems are essential. Thus, the development of a tailored ontology is required.

For the ontology construction, we adopt the method described in [3], which consists of the following five steps: specification, knowledge acquisition, conceptualization, implementation and validation.

– *Specification:* This step entails identifying the purpose of the ontology, defining the scope and objectives of the ontology, and understanding the domain requirements. Our ontology aims at modeling meat cuts and food recipes, and identifying alternative meat cuts to reduce waste.
– *Knowledge acquisition:* This step involves analyzing recipes and meat nutrition information for identifying concepts and relationships, forming an early version of the ontology definition.

- *Conceptualization:* In this step the acquired knowledge is organized into a conceptual model. The identified concepts have been translated into classes, and their attributes have been represented using data properties. The relationships between concepts have been modeled through several object properties.
- *Implementation:* This step uses ontology languages and tools, such as OWL and Protégé [11], to organize classes and properties, define constraints, and implement the ontology.
- *Validation.* The final step ensures correctness and consistency of the ontology using reasoners, such as Pellet [18], an OWL 2 compliant reasoner integrated in Protégé, to evaluate logical integrity and avoid contradictions. Beyond basic consistency checking, we define a set of logical axioms, such as class hierarchies, disjointness, domain and range restrictions, and existential constraints, to formally characterize the classes and relationships in our ontology. This is to ensure not only logical consistency, but also the intended semantic behavior of the ontology for practical applications.

Figure 2 shows the ontology, which models meat data and food recipes with the aim to find alternative meat options for recipes in order to reduce meat waste. The three main classes are *Recipe*, *MeatCut* and *Substitute*. Each recipe is described by several properties, such as quantity, cooking time, and number of servings, and it is recommended for different courses, where we distinguish between starters and main courses. The class *MeatCut* is characterized by nutritional information and sensory attributes. The relationship *hasMeatCut* between the classes *Recipe* and *MeatCut* represents the main cut of meat to be used for a specific recipe. The *Substitute* class represents alternative meat cuts for a recipe, capturing important details, such as the required substitution quantity and a ranking to indicate preferences. The *hasSubstitute* relationship links a recipe to one or more substitute instances, and each substitute instance is further connected to an alternative meat cut through the *isMeatCut* relationship, specifying which meat cut can replace the original. The ontology is publicly available at https://w3id.org/onfoods/ontology.

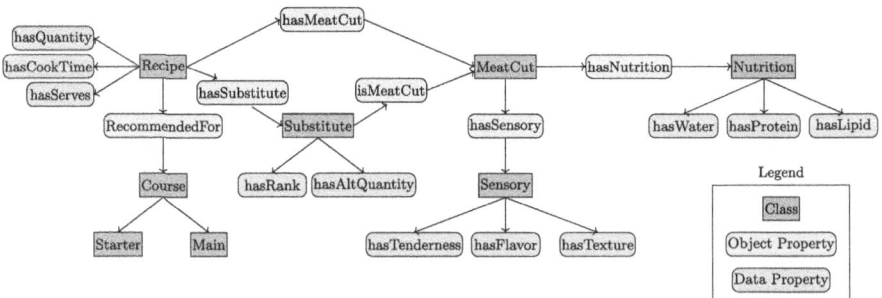

Fig. 2. Ontology

4.3 Knowledge Graph and Inventory Database

The knowledge graph instantiates the schema defined by the ontology with specific instances of the entities and relationships from the domain, i.e., individual recipes, beef cuts, and relationships between these two classes. For the instantiation of the knowledge graph we used two main sources of data:

- The USDA ingredient nutrient database[1], which provides nutritional information for beef cuts, including energy, water content, and macro (carbohydrates, proteins, and lipids) and micro nutrients (vitamins and minerals).
- The second source of data are various online recipe sites.

We chose to collect our recipe data from a variety of sites that provide beef-based recipes, extracting details such as the type of beef cut and its quantity, cooking time, and number of servings. Additional, users can add individual recipes or adapt existing ones.

As a data storage system for the knowledge graph and inventory data about the availability of different meat cuts, we employ the Open Source database management system PostgreSQL. PostgreSQL does not support graph data structures natively, as Neo4j or other dedicated graph databases do. However, we can achieve this by using PostgreSQL tables to represent nodes and relationships.

4.4 User Interface

To make the system more accessible to a broader audience such as chefs and restaurants managers, we developed a user-friendly web interface for ONFOODS. The main pages are shown in Figs. 3, 4, 5, 6 and 7. The graphical user interface is designed with a focus on usability and accessibility, particularly for non-expert users to interact with the substitution recommendation system without requiring technical knowledge. To ensure ease of use, our design decisions emphasize visual clarity, minimal complexity, and intuitive interactions. Users can input recipes, browse beef cuts, receive substitute recommendations based on our metrics, and rank substitutions based on their experience and preference. The web interface design is based on simplicity with clear input fields, easy-to-navigate options, and results that are easy to interpret.

4.5 Implementation

ONFOODS (http://onfoods.projects.unibz.it/) is implemented as a web application, which is built using the ASP.NET MVC framework, leveraging Razor for dynamic web pages and PostgreSQL as the database. The system follows the Model-View-Controller (MVC) architecture, ensuring a clear separation between data management, user interface, and business logic. The backend retrieves available Beef cuts, manages the availability of beef cuts, calculates similarity scores for finding alternative suggestions, and presents them through a responsive

---

[1] https://fdc.nal.usda.gov/.

Razor-based UI. The database schema follows the ontology structure, ensuring semantic organization of the data and making it easier to query and relate different concepts. The frontend of the application is developed using Razor within the ASP.NET MVC framework, ensuring a dynamic and responsive user interface and allowing easy access to functionalities like browsing beef cuts, exploring recipes, and managing inventory.

To address scalability requirements in practical settings, our recommendation system is specifically designed for real-world restaurant environments, where the number of relevant recipes and the granularity of meat cut hierarchies are inherently limited. Specifically, the system supports a three-level hierarchy for meat cuts, reflecting the granularity commonly used in restaurants. During testing we incorporated 200 recipes, which is representative of real-world use cases in restaurants. We observed that the system maintained robust performance and responsiveness as the number of recipes and meat cut variations increased. The system allows for straightforward expansion, making it possible to add new recipes, meat cuts, or substitution options without requiring significant modifications. These features demonstrate that our system is both scalable and adaptable to evolving menu requirements.

5   Use Cases

In our prototype implementation we use beef cuts for a concrete case study. However, the system can easily be applied to other types of meat, such as pork or lamb. In this prototype, due to the availability of data, we primarily rely on nutritional attributes with equal weighting to determine similarity, as sensory attributes for beef cuts were not readily available. In the following we showcase various scenarios that illustrate how users can interact with the ONFOODS system (cf. Figs. 3, 4, 5, 6 and 7).

Scenario 1 – Browsing Beef Cuts. Figure 3 shows how users can explore a variety of beef cuts through the hierarchical structure of beef parts, check their availability in the inventory, and browse through various recipes that use this particular beef cut. This functionality is particularly useful when creating menus because the availability is considered to reduce food waste.

Scenario 2 – Alternative Beef Cuts. This scenario highlights the application's ability to suggest alternative beef cuts based on similar characteristics and the chef's preferences. In Fig. 4, users see the main beef cut for a selected recipe, i.e., a ranked list of recommended alternatives together with cooking time, course, and serves. Each beef cut has assigned two values, the required quantity for the recipe and the selected servings (adjusted based on Sect. 3) and its availability in the inventory. Users can select a beef cut based on the availability. If a beef cut is selected, the inventory is updated prioritizing cuts with earlier expiration date. Consider Fig. 4, which shows two alternative beef cuts for 1.30 kg of "Back ribs": "Large end" is the most similar beef cut, and 1.50 kg are needed. "Short

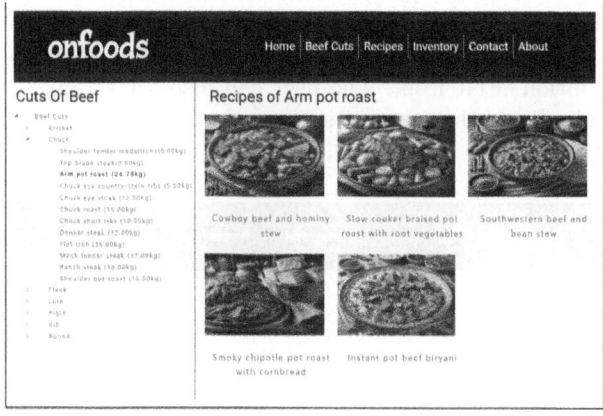

Fig. 3. Hierarchy of beef cuts

ribs" is the second most similar beef cut, for which 1kg are needed for the same recipe.

Scenario 3 – Managing Recipes. The recipes page in Fig. 5a allows users to edit and manage recipes by browsing the list, viewing recipe details, deleting existing recipes, or adding new ones. Figure 5b shows how to add a new recipe. After adding a recipe, the user is redirected to a page that suggests alternative beef cuts based on the similarity score discussed in Sect. 3 (Fig. 6). Additionally, the system allows users to manually add alternative beef cuts and specify their quantities based on their experience, as well as preferences for flavor and texture. Users can also set a ranked order for alternative options and adjust the quantities suggested by the system. Consider Fig. 6, which presents three

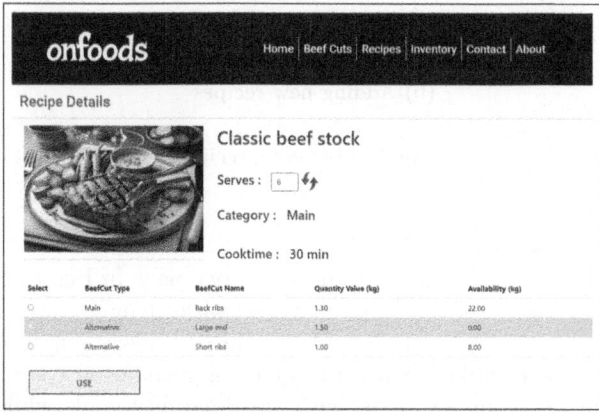

Fig. 4. Details of recipe

alternative beef cuts for the recipe "Ground beef pizza". In addition to the system-suggested alternatives "Large end" and "Short ribs", the user proposes an additional option "Flat iron". The user can also set a ranked order for the alternatives and adjusts their quantities.

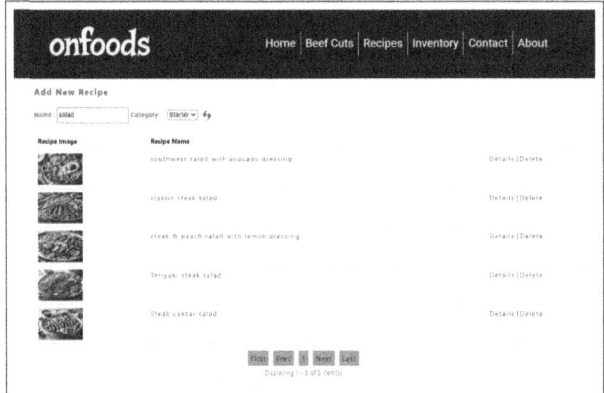

(a) Recipes

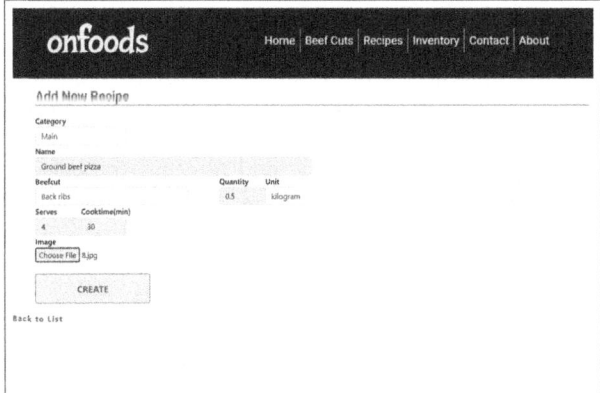

(b) Adding new recipes

Fig. 5. Managing recipes

Scenario 4 – Managing Inventory. The Inventory page in Fig. 7 allows users to manage the availability of beef cuts. This scenario demonstrates how chefs or decision-makers can monitor the inventory and add newly purchased items. By integrating inventory management with the recommendation system, the system ensures that menu planning aligns with sustainability goals while prioritizing the use of soon-to-expire beef cuts, thereby minimizing food waste and ensuring efficient stock usage.

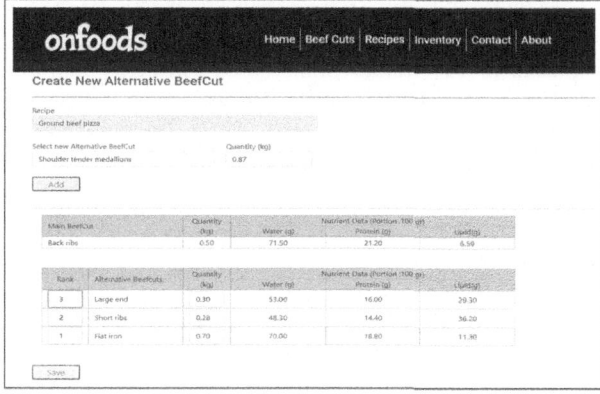

Fig. 6. Adding alternatives

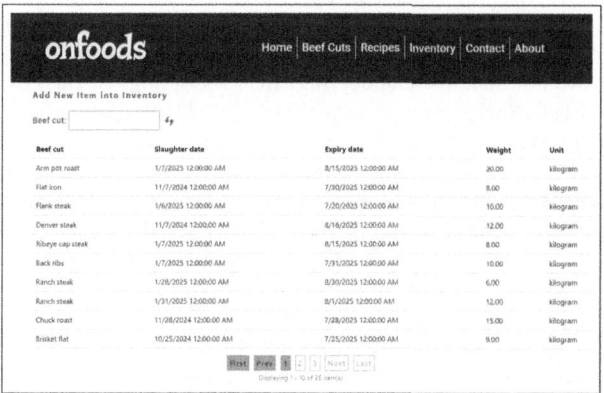

Fig. 7. Inventory data

6   Conclusion and Future Work

Our paper presents ONFOODS, a recommendation system for alternative meat cuts that assists chefs and/or decision-makers in restaurants to find the most sustainable alternative meat cuts for recipes and manage the inventory and usage of meat. ONFOODS relies on an ontology to model recipes, meat cuts and the relationships between the two. To provide an efficient way for users to interact with the system, we developed a user-friendly web application for ONFOODS with a focus on visual clarity, minimal complexity, and intuitive interactions.

In the future, we plan to enhance the system by integrating machine learning algorithms to improve the accuracy of meat cut recommendations based on historical usage patterns and chef preferences. Additionally, collaborations with local chefs are planned, which will help us to suggest region-specific alternative meat cuts that align with local preferences and sustainability goals.

Acknowledgment. This work was supported in part by the National Recovery and Resilience Plan (NRRP), Mission 4 Component 2 Investment 1.3–Call for tender No. 341 of 15 March 2022 of the Italian Ministry of University and Research funded by the European Union-NextGenerationEU; Project code PE00000003, Concession Decree No. 1550 of 11 October 2022 adopted by the Italian Ministry of University and Research, CUP I33C22006890001, project title "ON Foods–Research and innovation network on food and nutrition Sustainability, Safety and Security–Working ON Foods".

References

1. Dooley, D.M., et al.: Foodon: a harmonized food ontology to increase global food traceability, quality control and data integration. NPJ Sci. Food **2**(1), 23 (2018)
2. Dragoni, M., Bailoni, T., Maimone, R., Eccher, C.: HeLiS: an ontology for supporting healthy lifestyles. In: Vrandečić, D., et al. (eds.) ISWC 2018. LNCS, vol. 11137, pp. 53–69. Springer, Cham (2018). https://doi.org/10.1007/978-3-030-00668-6_4
3. Fernández-López, M., Gómez-Pérez, A., Juristo, N.: Methontology: from ontological art towards ontological engineering. In: Proceedings of the Ontological Engineering AAAI-97 Spring Symposium Series, pp. 33–40. American Association for Artificial Intelligence (1997)
4. Haussmann, S., et al.: FoodKG: a semantics-driven knowledge graph for food recommendation. In: Ghidini, C., et al. (eds.) ISWC 2019. LNCS, vol. 11779, pp. 146–162. Springer, Cham (2019). https://doi.org/10.1007/978-3-030-30796-7_10
5. Huang, L., Yu, C., Chi, Y., Qi, X., Xu, H.: Towards smart healthcare management based on knowledge graph technology. In: ICSCA, pp. 330–337. ACM (2019)
6. Lawrynowicz, A., Wróblewska, A., Adrian, W.T., Kulczynski, B., Gramza-Michalowska, A.: Food recipe ingredient substitution ontology design pattern. Sensors **22**(3), 1095 (2022)
7. Lei, Z., Haq, A.U., Zeb, A., Suzauddola, M., Zhang, D.: Is the suggested food your desired?: multi-modal recipe recommendation with demand-based knowledge graph. Expert Syst. Appl. **186**, 115708 (2021)
8. Mikolov, T., Chen, K., Corrado, G., Dean, J.: Efficient estimation of word representations in vector space. In: ICLR (Workshop Poster) (2013)
9. Min, W., Liu, C., Xu, L., Jiang, S.: Applications of knowledge graphs for food science and industry. Patterns **3**(5), 100484 (2022)
10. Morales-Garzón, A., Gutiérrez-Batista, K., Martin-Bautista, M.J.: Adaptafood: an intelligent system to adapt recipes to specialised diets and healthy lifestyles. Multimedia Syst. **31**(1), 1–24 (2025)
11. Musen, M.A.: The protégé project: a look back and a look forward. AI Matters **1**(4), 4–12 (2015)
12. Pacífico, L.D.S., Britto, L.F.S., Ludermir, T.B.: Ingredient substitute recommendation based on collaborative filtering and recipe context for automatic allergy-safe recipe generation. In: WebMedia 2021, pp. 97–104. ACM (2021)
13. Pan, Y., Xu, Q., Li, Y.: Food recipe alternation and generation with natural language processing techniques. In: ICDE Workshops, pp. 94–97. IEEE (2020)
14. Parfitt, J., Barthel, M., Macnaughton, S.: Food waste within food supply chains: quantification and potential for change to 2050. Philos. Trans. Roy. Soc. B Biol. Sci. **365**(1554), 3065–3081 (2010)

15. Pizzuti, T., Mirabelli, G., Grasso, G., Paldino, G.: MESCO (meat supply chain ontology): an ontology for supporting traceability in the meat supply chain. Food Control **72**, 123–133 (2017)
16. Pizzuti, T., Mirabelli, G., Sanz-Bobi, M.A., Goméz-Gonzaléz, F.: Food track & trace ontology for helping the food traceability control. J. Food Eng. **120**, 17–30 (2014)
17. Shirai, S.S., Seneviratne, O., Gordon, M.E., Chen, C.H., McGuinness, D.L.: Identifying ingredient substitutions using a knowledge graph of food. Front. Artif. Intell. **3**, 621766 (2021)
18. Sirin, E., Parsia, B.: Pellet: an OWL DL reasoner. In: Description Logics. CEUR Workshop Proceedings, vol. 104. CEUR-WS.org (2004)
19. Zulaika, U., Gutiérrez, A., López-de-Ipiña, D.: Enhancing profile and context aware relevant food search through knowledge graphs. In: UCAmI. MDPI Proceedings, vol. 2, p. 1228. MDPI (2018)

Inspire Me with Your Questions: Repurposing Historical Questions for New Documents

Yifan Liu[1,2], Yixuan Cao[1,2(✉)], and Ping Luo[1,2,3(✉)]

[1] State Key Laboratory of AI Safety, Institute of Computing Technology, Chinese Academy of Sciences (CAS), Beijing, China
{caoyixuan,luop}@ict.ac.cn
[2] University of Chinese Academy of Sciences, CAS, Beijing, China
[3] Peng Cheng Laboratory, Shenzhen, China

Abstract. Questioning is an effective method of extracting information from documents. While Large Language Models excel at answering user queries in document question-answering systems, users often struggle to formulate effective questions when encountering unfamiliar documents. Fortunately, expert-formulated questions embodying professional knowledge can be transferred to new documents to assist ordinary users. Therefore, we propose a question recommendation approach that transfers expert questions on historical documents to new ones, enabling users to "stand on the shoulders of giants" for enhanced document comprehension. Our approach comprises two modules: 1) A "question reusability classification" module identifies *domain-general* questions applicable across similar documents; 2) A "document-bridged question ranking" module selects semantically appropriate questions for new documents. Experiments on our self-constructed expert question dataset demonstrate that both components significantly impact recommendation accuracy, and performance improves as historical data volume increases.

Keywords: Questions Reusing · Document's Information Accessing · Data Mining

1 Introduction

Asking questions is an effective method for humans to acquire information from complex environments, facilitating cognitive development [6,26] and playing a significant role in education [22,27]. By asking questions, people can rapidly extract information from lengthy contexts or access information that is not directly observable. Today, based on Retrieval-Augmented Generation (RAG) techniques [8], some AI assistants (e.g., GPT-4 [3], Gemini [31]) and programs (e.g., LangChain [2]) with Large Language Models (LLMs) as their backbone allow users to ask questions about long documents, enabling them to quickly grasp the content or find key information. However, without relevant knowledge, people can only pose low-quality questions [28], which fail to enhance

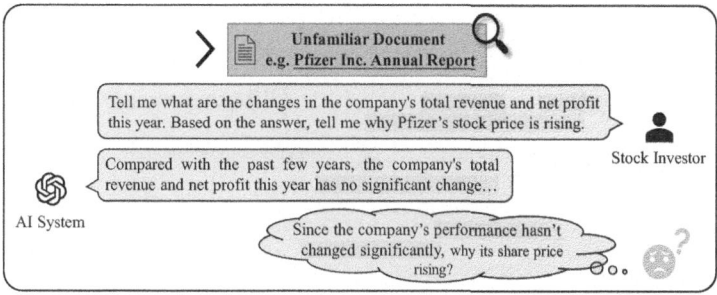

Fig. 1. Questioning on a new document directly. Novice users may be unable to ask the right questions on an unfamiliar document to fulfill their intent.

productivity [11,12]. This indicates that users need assistance to utilize these AI question-answering tools better.

Fortunately, researches show that people excel at recognizing and applying professional questions in new scenarios. Cognitive scientists designed a question-and-answer game inspired by BATTLESHIP[1] to evaluate people's ability to pose and assess questions [28]. They found that participants without gaming experience struggled to create high-quality questions but selected the best ones when choosing from existing options. Furthermore, a subsequent study [17] revealed that people can effectively reuse historical questions in similar new contexts.

Obtaining information through questioning on unfamiliar documents is a similar scenario to the aforementioned study. For example, when a novice investor with financial knowledge but unfamiliar with the pharmaceutical domain seeks to understand why Pfizer's stock price is rising based on its annual report, the proposed question, as shown in Fig. 1, might focus on basic financial metrics like "total revenue" and "net profit". Although the AI system found the correct information, it did not fulfill the user's intention. Meanwhile, expert users, with knowledge in both finance and pharmaceuticals, can pose the correct questions regarding factors like R&D expenditure, which contains domain knowledge that the long-term development of pharmaceutical companies is largely determined by R&D investment. If this expert question (Q3 in Fig. 2) is applied to Pfizer's report, the novice user will obtain more satisfactory information.

Through filtering and clustering, questions from experts about different aspects of pharmaceutical companies can be shared with users to obtain insight into annual reports within the same domain. The expert knowledge contained in questions can be transferred to users with similar interests through repurposing, helping them obtain desired answers more easily. Therefore, when a user uploads a new document, we aim to select and recommend appropriate questions from those previously posed by experts on historical documents.

Some studies [1,5,15] have utilized LLMs to generate new questions for current user-uploaded documents. However, these generative methods necessitate

---

[1] https://en.wikipedia.org/wiki/Battleship_(game).

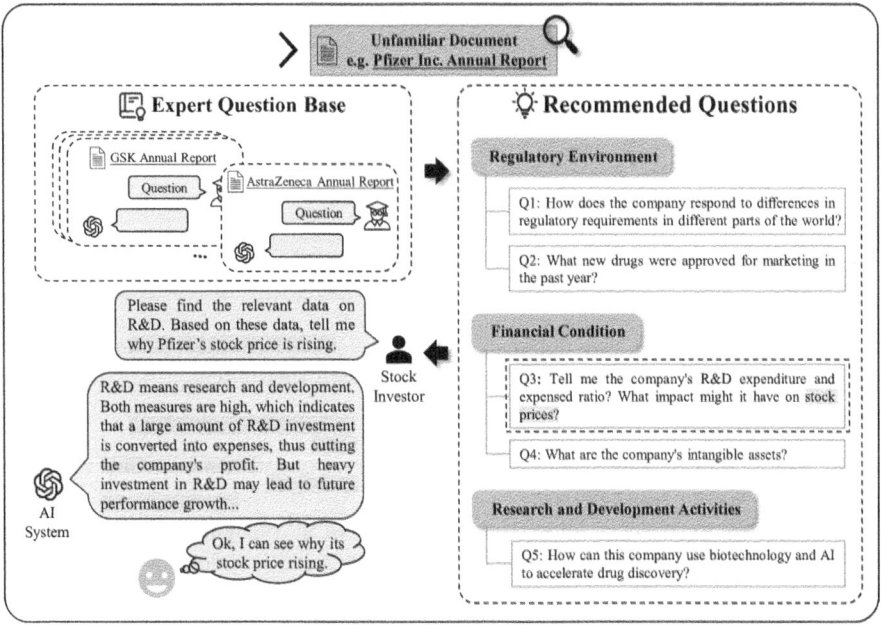

Fig. 2. Questioning on a new document with expert questions. Expert questions can be transferred from historical documents to new similar documents and clustered based on semantics. Therefore, users can reuse interesting questions based on each category's tag, allowing them to obtain the desired answers.

context selection, which may result in unsatisfactory questions if inappropriate content is selected. In contrast, reusing existing questions not only circumvents the challenges of context selection but leverages the expert knowledge contained in historical questions to assist users in better extracting key information from documents (as illustrated in Fig. 2).

We find two challenges when recommending historical questions to users. One is that some historical questions are strongly tied to certain documents, which are difficult to reuse on other documents. For example, the question "Why does Santiago fight the big fish?" can only be used in "The Old Man and the Sea", making it a *document-specific* question. The other is how to select those most suitable for recommendation to the current document from historical questions.

To address these challenges, we propose a method composed of two modules: Question Reusability Classification and Document-Bridged Question Ranking. This two-stage design allows us to systematically address each challenge. The first module assesses questions' reusability and classifies them accordingly, removing *document-specific* questions while retaining reusable ones to form a new knowledge base $\mathcal{K}'$ (details in Sect. 3.2). The second module, following the assumption that "users tend to ask similar questions on similar documents" (as shown in Fig. 2), finds the documents in $\mathcal{K}'$ most similar to the new document

d^*, then rank their corresponding questions based on semantic similarity with d^*. Furthermore, we cluster these questions into several groups to allow users to select reference questions according to their interests (details in Sect. 3.3).

Note that the effectiveness of our proposed method depends on the quality of the question knowledge base. Specifically, knowledge bases containing numerous expert-generated questions that can extract key information from documents provide greater assistance to users (as shown in Fig. 2). However, collecting high-quality expert questions involves issues such as user feature identification and privacy protection, which will be discussed in Section 6.

To validate our approach, we collected a dataset **DQA-UA** (**DQA** between User and **AI** assistant) containing 2,197 documents from finance and computer science (CS), along with professional questions posed about these documents (more details in Sect. 4). We conducted experiments to recommend questions for new documents. We evaluate the quality based on two metrics: 1. Reusability, which refers to whether the recommended questions can be answered by the content of new documents; and 2. Diversity, which indicates whether the recommended questions are semantically rich. The results show that our method can quickly and cost-effectively select suitable recommended questions.

2 Related Works

This paper proposes a method for recommending questions to users on a DQA platform. We introduce related work in: 1. DQA and RAG; 2. Question Reformulation and Question Generation; 3. Relevant datasets.

Document Question-Answering and Retrieval-Augmented Generation. DQA represents an open-book QA task that answers questions based on a provided document. A widely adopted approach for DQA is RAG. In a typical RAG workflow, a retriever (e.g., BM25 [13], ADA2 [21]) identifies document fragments pertinent to the question. Subsequently, LLMs generate answers based on these retrieved fragments [14,18].

Query Reformulation and Question Generation. Since question quality significantly influences the likelihood of obtaining satisfactory answers, researchers have developed various methods categorized into Query Reformulation (QR) and Question Generation (QG). QR aids LLMs in retrieving desired answers by reconstructing the user's input question through techniques like breaking down complex questions [37], restructuring questions [19], or enriching questions with knowledge [29]. However, these methods focus on enhancing initial questions rather than generating new ones. In contrast, QG directly generates questions based on a given context [20]. A common QG technique involves manually collecting or designing seed instructions and using LLMs to generate new instructions [35]. This approach is primarily employed to augment the training datasets of QA models. For instance, Alpaca [30] uses generated instruction data to fine-tune the LLaMA model [32], thereby improving its QA capabilities. However, this technique aims to increase the diversity of questions, which differs

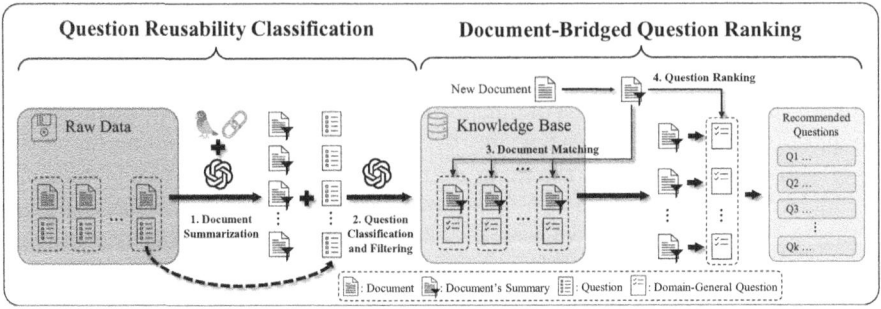

Fig. 3. The framework of the proposed approach

from our objective of providing useful questions to users. Another QG technique involves prompting the LLM to generate contextually relevant questions [7,10,34]. A significant challenge faced by these methods is that when the input is a long document, the input length limitations of LLMs necessitate context selection.

Dataset. Most existing public datasets that include documents and questions are primarily designed to train and evaluate the question-answering capabilities of models. Given our goal to assess the model's ability to generate questions, these datasets are unsuitable for the following reasons:

- Limited variety of question types. Datasets such as SQuAD [25] and HotPotQA [9] are constructed to evaluate reading comprehension and reasoning abilities rather than question quality. Questions in these datasets tend to be similar in form and intent, whereas user queries in real-world scenarios are more diverse. This creates a gap between dataset questions and those in real-world applications.
- Short document length. The contexts provided in existing datasets (e.g., News QA [33] and MS MARCO [23]) are significantly shorter than the input length limits of LLMs, thereby reducing the task's difficulty. Considering that documents uploaded by users often far exceed these input length limits, these datasets are inadequate for our research needs.

3   Methodology

The task in this paper can be formally defined as follows: Given a knowledge base $\mathcal{K} = \{\mathcal{K}_i\}_{i=1}^n$ consisting of n samples, where each sample $\mathcal{K}_i = (d_i, Q_i = \{q_{i,j}\}_{j=1}^{m_i})$ includes one document d_i and m_i questions $q_{i,j}$ (with m_i representing the number of questions associated with the document d_i, and m_i varying for each sample $\mathcal{K}_i$), we aim to select a candidate question set $\mathcal{Q}_{d^*} = \{q_{i,j}^*\}$ that can be recommended for a new document d^*.

3.1 Overview

The framework of our method is illustrated in Fig. 3, comprising two modules. The Question Reusability Classification module eliminates *document-specific* questions that lack applicability across diverse documents (Sect. 3.2). The Document-Bridged Question Ranking module employs document similarity metrics to rank and cluster questions for recommendation (Sect. 3.3). Notably, the first module only needs to run once when constructing the knowledge base, enabling fast and low-cost question recommendations.

3.2 Question Reusability Classification

We introduce some definitions before presenting the methodology. The reusability is determined by whether a question posed on one document can be applied to query other documents within the same **domain**. Reusable questions are termed *domain-general* questions, while others are *document-specific* questions.

We define the document domains and question categories as follows. The **domain** of a document is determined by two attributes: **type** and **topic**. The document type refers to the format or intended purpose of the document, with each type exhibiting a distinctive structural composition. For instance, financial reports often contain sections like financial statements and management discussion; academic papers generally include abstract, introduction, method, and result sections. The document topic pertains to the specific subject matter or disciplinary focus addressed within the document. This may range from quantum mechanics to market analysis or artificial intelligence in the context of academic literature. Therefore, documents sharing both type and topic will likely exhibit substantial content similarity, increasing the probability of question transferability between such documents.

The **categories** of questions include:

- *Domain-general* questions: These questions pertain to common properties that are present in most documents within a specific domain.
- *Document-specific* questions: These questions can only be applied to a limited number of documents containing specific content, as the information required to answer them is unavailable in similar documents.

For example, in annual reports, the question, "What was the company's net profit for last year?" is *domain-general* since net profit is a common attribute of annual reports. Conversely, "What was the net profit of Tesla in 2023?" is *document-specific* because the answer can only be found in Tesla's 2023 annual report. The primary distinction is that *document-specific* questions contain specific entities, limiting their applicability to documents lacking related content.

Question Classification Method. Initially, we utilize an LLM to ascertain a document's type and topic. We generate a summary S_i for document d_i and use S_i as input to prompt the LLM to identify the document's domain D_i. The

> \# Task Description
> In the [Input] section below, I will provide you with a summary of a document. You need to refer to the [Definition] section and think step by step according to the [Think Step] to generate the type and topic of the document.
> \#\# Think Steps
> 1. Infer the content of the document based on its summary;
> 2. Based on your understanding of the document content, generate the type and topic of the document
> \#\# Definition
> -   Document Type: Refers to the format or purpose of the document, which describes the structure of the document and the overall form of the content.
> -   Document Topic: Refers to the content topic of a document, which describes the specific topic or discipline to which the document is concerned.
> \# Constraints
> 1. The output answer must be as concise as possible;
> 2. The output should be in English only.
> \# Input
> The document summary is: {{Document Summary}}
> \# Output

Fig. 4. Document type and topic identification prompt.

prompt template is shown in Fig. 4. The domain identification process can be formulated as:

$$S_i = \text{LLM}_{\text{Summarization}}(d_i), D_i = \text{LLM}_{\text{DomainIdentification}}(S_i). \quad (1)$$

Subsequently, we classify each question $q_{i,j}$ with an LLM, which is provided with both $q_{i,j}$ and D_i. We prompt the LLM to ascertain whether $q_{i,j}$ exhibits broad applicability across most documents within D_i. Considering the complexity of this classification task, we adopt the Chain of Thought (CoT) [36] to decompose the task into two simpler steps. To mitigate potential adverse effects from ambiguous cases, we ask the LLM to label difficult-to-classify questions as *other*. Figure 5 illustrates the classification prompt template. The process can be mathematically expressed as:

$$C_{q_{i,j}} = \text{LLM}_{\text{QuestionClassification}}(q_{i,j}, D_i),$$
$$C_{q_{i,j}} \in \{domain\text{-}general, document\text{-}specific, other\},$$

where $C_{q_{i,j}}$ is the category of question $q_{i,j}$.

Upon obtaining the category labels, we isolate the *domain-general* questions. These questions, along with their corresponding documents (and document summaries), form a new knowledge base $\mathcal{K}' = \{\mathcal{K}'_i\}_{i=1}^{n'}$, where each sample $\mathcal{K}'_i = (d_i, S_i, D_i, Q'_i = \{q_{i,j} \mid C_{q_{i,j}} = domain\text{-}general\}_{j=1}^{m'_i})$. The results in Section 5.5 demonstrate that our method can achieve acceptable classification accuracy.

Note that document type and topic are challenging to define precisely. This implies that LLM may generate varied results for the same document. To address this issue, we prompt the LLM (Fig. 4) to generate concise outputs, preventing overly detailed topics that could affect classification results.

Task Description
In general, we can use document type and topic to describe the domain of documents, which usually have some similar content.
In the [Input] section below, I'll give you a question and a document category and domain that describes a certain class of documents. You need to think step by step according to the [Think Step] to determine whether the question is suitable for questioning most documents in this domain, and classify the questions based on the [Question Category Definition]. If you believe this question does not belong to any category or is difficult to determine its category, please output "Other Question".
Think Steps
1. Think what content should be included in the document based on its type and topic.
2. Determine whether the question is suitable for most documents of this type and topic.
Question Category Definition
- Domain General Question: This type of question can be used on other documents with the same domain; their answer mostly refers to properties common to a class of documents.
- Document Specific Question: This type of question can only be used on a small number of documents that contain specific content, as the information needed to answer these questions does not exist in other similar documents
Constraints
1. The output should be in English only.
Input
Document Type: {{Document Type}}
Document Topic: {{Document Topic}}
Question: {{Question}}
Output

Fig. 5. Question classification prompt.

3.3   Document-Bridged Question Ranking

In this module, we leverage historical documents as a bridge to connect d^* with historical questions. Initially, we identify the set of documents $\mathcal{D}_{d^*} = \{d'_i\}$ within the knowledge base $\mathcal{K}'$ that are most similar to d^*, and aggregate their corresponding questions $\mathcal{Q}_{\mathcal{D}_{d^*}} = \{q_{i,j} \mid d_i \in \mathcal{D}_{d^*}, q_{i,j} \in Q'_i\}$. Then, we rank these questions based on their similarity to the new document d^* and cluster them to obtain the final recommended questions.

Specifically, we firstly utilize an embedding model to generate semantic vector $\mathbf{v}_{S_i}, \mathbf{v}_{D_i}, \mathbf{v}_{q_{i,j}}$ for each document summary S_i, domain D_i, and question $q_{i,j}$, respectively. Next, to identify $\mathcal{D}_{d^*}$, we consider both S_i and D_i as representing different aspects of a document d_i, thus using the weighted sum of their embedding vectors $\mathbf{v}_{S_i}, \mathbf{v}_{D_i}$ to form the document vector $\mathbf{v}_{d_i}$:

$$\mathbf{v}_{d_i} = w_1 \cdot \mathbf{v}_{S_i} + w_2 \cdot \mathbf{v}_{D_i},$$

where w_1, w_2 are the weights of the summary and domain vectors and $w_1 + w_2 = 1$. Then, we use the cosine similarity to measure the similarity between d^* and d_i, which is formulated as:

$$\text{cos-sim}(\mathbf{v}_{d^*}, \mathbf{v}_{d_i}) = \frac{\mathbf{v}_{d^*}\mathbf{v}_{d_i}}{||\mathbf{v}_{d^*}||^2||\mathbf{v}_{d_i}||^2},$$

where $\mathbf{v}_{d^*}$ and $\mathbf{v}_{d_i}$ denote the embedding vector of d^* and d_i, respectively.

Subsequently, we rank the samples in descending order based on cos-sim and add questions of these samples to the candidate set $\mathcal{Q}_{\mathcal{D}_{d^*}}$ until the number of questions exceeds the threshold $a \cdot k$, where k is the number of recommendations in the end and $a \in \{x \in \mathbb{Z} \mid x \geq 1\}$ is a multiplier used to control the number of candidate questions. Then, we remove questions from $\mathcal{Q}_{\mathcal{D}_{d^*}}$ whose cos-sim fall below a threshold τ_s. This process may result in the number of candidate questions dropping below $a \cdot k$. During the deployment of this method, the parameters τ_s or a can be dynamically adjusted to ensure that the number of candidate questions meets the requirements.

Finally, we calculate the cosine similarity between the embedding vector of d^* and each question $q_{i,j}$ in $\mathcal{Q}_{\mathcal{D}_{d^*}}$ and sort them in descending order. To improve the diversity of the recommended questions, we cluster these questions into k classes using K-means and select the top-ranked questions from each class. These k questions form the recommended question set $\mathcal{Q}_{d^*}$.

4   Dataset

As discussed in Sect. 2, existing public datasets are unsuitable for evaluating our method. Since our approach focuses on transferring historical expert questions to new documents, we require questions that are: 1. clearly articulated; 2. from questioners with profound domain understanding, aiming to extract key document information. Therefore, we developed a DQA platform to collect questions from graduate students.

We created an online DQA platform integrating GPT-3.5 with RAG, then collaborated with university professors to assign coursework to finance (126 students) and CS (155 students) graduate students. Finance students were required to submit industry research reports, while CS students were tasked with conducting academic surveys over six weeks. Students used our platform to efficiently extract information by asking questions about documents they collected. On average, each student uploaded 16 documents and asked approximately 9 questions per document. To ensure high-quality questions, we selected students ranking in the top 60% based on coursework scores and chose documents with above-average question counts.

Ensuring user data privacy was critical. Before assigning coursework, we surveyed students regarding their willingness to share their questions, and those who declined were excluded from data collection. Additionally, student-submitted reports (including their questions) were provided to us by course professors after anonymization. For collected data, we excluded irrelevant questions (identified by responses containing phrases like "sorry" or "no relevant content") and employed GPT-4 and manual screening to filter out unpublished documents and inappropriate content. The resulting DQA–UA dataset comprises 18,346 questions and 2,197 documents from the finance and CS fields, with questions in English or Chinese. Finance documents include annual reports, industry research, and economic analyses, while CS documents mainly encompass academic research. Table 1 summarizes the data distribution.

Table 1. Data distribution of the DQA-UA dataset

Field	Documents	English Questions	Chinese Questions
Finance	879	2678	4889
CS	1318	7704	3075

5 Experiments

In this section, we assess our method on our dataset. The evaluation primarily focuses on whether historical professional questions can be accurately transferred to new documents. It mainly consists of two parts: 1. The accuracy of question ranking, which measures the validity of the recommendation; 2. The semantic diversity of the recommended questions measures their richness. We first describe the experimental implementation, then delineate baseline configurations, discuss evaluation metrics, and finally present results and analysis.

5.1 Implementation Details

For backbone models, we employ GPT-3.5 and GPT-4 for summary generation and question classification in the Question Reusability Classification module, respectively. GPT-4 is chosen for classification due to its superior reasoning capabilities (Sect. 5.5), while GPT-3.5 provides cost-effective summarization. We employ the Map-Reduce Chain in LangChain for summarizing. In the Document-Bridged Question Ranking module, we use "text-embedding-ada-002" as the embedding model. We set the similarity threshold τ_s to 0.2 and selected over 20 candidate questions for each document in the test set, choosing 5 questions as recommendations (i.e., $a = 4, k = 5$) to balance document similarity and question diversity. The dataset was split into a training set and a test set in a ratio of 2:1, with the training set serving as the knowledge base. After the Question Reusability Classification, approximately half of the document-specific questions were filtered out, with details in Table 2.

To determine the optimal weight parameter settings, we randomly selected 50 samples each from the financial and CS test datasets to form a validation set. We tested the following 6 parameter settings: w1 was set to 0.4, 0.5, ..., 0.9, and w2 ranged from 0.6 to 0.1. For each document in the validation set, we generated 5 recommended questions using different parameter settings and calculated their Average QAA (details in Sect. 5.3). The results are illustrated in Fig. 6, showing that the optimal values are $w_1 = 0.8, w_2 = 0.2$, as the summary contains more information than the domain label.

Table 2. Question distribution of the training set

Field	Domain General Questions	Document Specific Questions	Other Questions	Total
Finance	2561	2308	175	5044
CS	3804	3195	187	7186

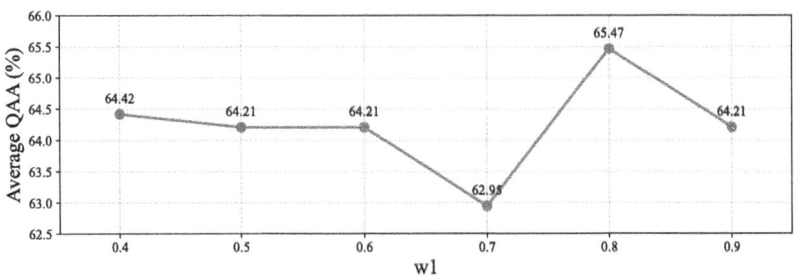

Fig. 6. Performance for different weights

5.2 Baselines

We compared our method with the following baselines:

- **w/o Question Reusability Classification.** All questions from the training set were used as the knowledge base, and Document-Bridged Question Ranking was employed to recommend questions.
- **w/o Document Matching.** Questions were recommended by directly calculating the similarity between *domain-general* questions and the new document without document matching.
- **w/o Question Clustering.** After selecting $a \cdot k$ candidate questions, directly choose those with the highest cosine similarity as recommendations without clustering.

5.3 Evaluation Metrics

Given the challenge of quantifying recommendation quality, we evaluate from two perspectives: recommended questions should be answerable using document content and should be semantically diverse.

Question Answerability (QAA) evaluates whether the answers to recommended questions can be found in the document. Specifically, we use the method mentioned in Sect. 3.2 to generate a summary for the document and employ LangChain [2] as a pipeline to retrieve fragments related to the questions from the document. We use the document summary and relevant fragments as context and utilize LLM to determine whether the question can be answered. Initially, we attempted to prompt the LLM to score the questions based on their answerability, but we found that most questions were either fully answerable or completely unanswerable. For simplicity, we only required the LLM to perform a binary classification, meaning that $QAA_i = 1$ if d^* can answer the ith question in $\mathcal{Q}_{d^*}$, and $QAA_i = 0$ otherwise. In the experiment, we calculate the Average QAA of questions recommended on each document as follows:

$$\text{Average QAA} = \frac{1}{k} \sum_{i=1}^{k} QAA_i.$$

Additionally, we refer to the Mean Reciprocal Rank (MRR) to construct the QA-MRR to check the rationality of the order of recommended questions, which can be formulated as follows:

$$\text{QA-MRR} = \frac{1/i}{\sum_{i=1}^{k} 1/i} \sum_{i=1}^{k} QAA_i.$$

The QA-MRR increases when more relevant questions are in the top position of the recommendation sequence.

Question Diversity is used to evaluate the richness of the recommended questions. We measure the diversity of a set of recommendation questions from two different granularities, following the same principle: the more similar the questions are to each other, the lower the diversity. Initially, we measure diversity from the overall perspective of the questions, using embedding vectors as representations of the overall semantics of each question in the recommendation set, and calculate the average distances between them. The formula is as follows:

$$\text{Div-E} = \frac{1}{k} \sum_{i=1}^{k} \frac{1}{k} \sum_{j=1, j \neq i}^{k} [1 - \text{cos-sim}(\mathbf{v}_{q_i}, \mathbf{v}_{q_j})],$$

where $\mathbf{v}_{q_i}$ represents the embedding vector of the ith question, and $1 - \text{cos-sim}(\cdot)$ represents the distance.

Additionally, we use BERTScore [38] to measure the similarity between questions to construct a more fine-grained diversity metric. The formula is as follows:

$$\text{Div-B} = \frac{1}{k} \sum_{i=1}^{k} \frac{1}{k} \sum_{j=1, j \neq i}^{k} [1 - \text{BERTScore}(q_i, q_j)],$$

where $\text{BERTScore}(q_i, q_j)$ represents the BERTScore-F1 between two questions.

Other widely used automatic evaluation metrics, such as BLEU [24], METEOR [4], and ROUGE [16], which are typically used to evaluate generated results, are not suitable for evaluating recommendation results. Additionally, there exist more sophisticated evaluation approaches that can provide deeper insights into system performance, which will be discussed in Sect. 6.

5.4   Results

As shown in Table 3 and 4, our approach achieved the best or second-best scores across most metrics in both finance and CS fields, indicating a good trade-off between accuracy and diversity. Our method consistently obtained the highest performance on QA-MRR, suggesting the correct ranking of candidate questions.

Compared to w/o Question Reusability Classification, our method's recommendations showed slightly reduced semantic diversity but significantly higher answerability on new documents, indicating accurate filtering of non-reusable questions. Compared to w/o Document Matching, our method achieved lower Average QAA in finance, likely because financial documents share more commonalities, making historical questions more answerable in new documents. However, in CS, the w/o Document Matching method scored lower than ours, suggesting that document filtering is crucial for finding relevant questions. The results in the last row of Table 3 and 4 show that clustering candidate questions improves semantic diversity without compromising answerability.

Table 3. Results on Question Recommendation (Finance).

Method	Average QAA	QA-MRR	Div-E	Div-B
Ours	61.67%	**66.03%**	0.2137	0.5811
w/o Question Reusability Classification	48.74%	50.95%	**0.2174**	0.5870
w/o Document Matching	**63.50%**	65.75%	0.1582	**0.6348**
w/o Question Clustering	60.62%	61.52%	0.2062	0.4885

Table 4. Results on Question Recommendation (CS).

Method	Average QAA	QA-MRR	Div-E	Div-B
Ours	**70.70%**	**71.47%**	0.2394	**0.5992**
w/o Question Reusability Classification	56.64%	56.31%	**0.2397**	0.5969
w/o Document Matching	49.67%	49.00%	0.1852	0.5065
w/o Question Clustering	70.53%	71.16%	0.2168	0.5268

5.5   Analysis

We further analyzed factors affecting recommendation accuracy. Considering the knowledge base's impact on method robustness, we validated the Question Reusability Classification module's effectiveness. We randomly selected

Table 5. Results of Question Reusability Classification

Model	Recall	Precision
GPT-3.5$_{\text{two-stage}}$	83.95%	72.34%
GPT-4$_{\text{one-stage}}$	71.16%	68.88%
GPT-4$_{\text{two-stage}}$	77.69%	88.03%

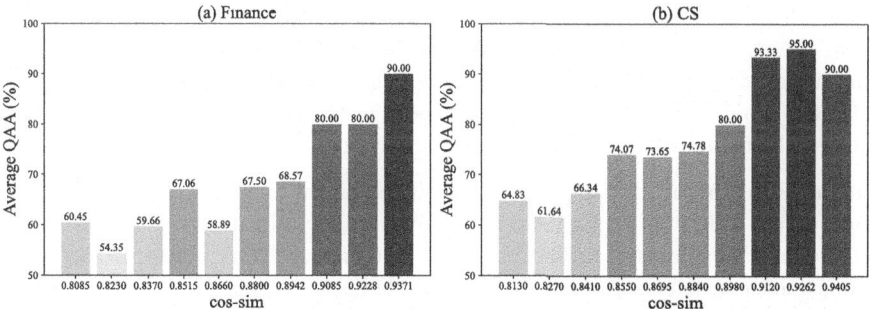

Fig. 7. Performance for different cosine distance values.

1,000 financial questions from DQA-UA and had human experts annotate categories based on answers and relevant documents. These human-annotated labels were considered the gold standard. We compared our two-stage classification method with direct classification based on document summary and question (one-stage) and compared GPT-3.5/GPT-4 as backbone models. Considering *domain-general* questions as positive examples, we calculated precision and recall. The results in Table 5 showed that our approach achieved the highest classification precision.

Moreover, the similarity between historical and new documents is also a key factor. Therefore, we explored the relationship between the cos-sim and the Average QAA. Specifically, each sample i in the test set consists of a document d_i^*, k recommended questions, and their corresponding k' historical documents ($k' \leq k$). We calculated each sample's average cos-sim between d_i^* and k' historical documents. Next, we divided all test samples into ten equal intervals based on their average cos-sim. Finally, we calculated the Average QAA for all test samples and got their mean value in each interval. As shown in Fig. 7, it is evident that QAA is positively correlated with cosine similarity in both the finance and CS fields. This result shows that questions from similar documents are more likely to be suitable for d^*. It also implies that the performance of our approach may improve as the number of documents in the database increases.

6 Discussion and Future Work

This paper proposes a method for repurposing historical expert questions on new documents to help users efficiently read documents with the DQA system. This section summarizes limitations and future directions.

LLM for Questioning. About half of the users' questions were filtered out as *document-specific* (Table 2), containing rich explorable information. Since users tend to learn cross-document patterns and extract key content representations. Using these representations, a language model generates recommended questions that integrate expert knowledge for document-relevant, insightful, and diverse outputs. When deployed, identifying expert users through user features could address the challenge of non-expert participation.

Online Evaluation. In this study, we utilized analytical metrics to assess question quality from specific aspects like answerability and diversity. Upon formal deployment, we plan to conduct an A/B test on our DQA platform to further evaluate user satisfaction, providing a more holistic perspective. This test would involve recommending questions to only a subset of users and comparing interaction frequency or user retention rates between groups.

Privacy Policy. When deploying our method, we will implement stricter privacy policies to protect user privacy. Online users may be more concerned about copyright than students. Therefore, beyond establishing strict data usage guidelines, we will implement incentive measures to encourage question sharing, such as awarding points for frequently clicked questions or building online communities to promote discussions and information sharing.

7 Conclusion

This paper introduces an approach that reuses historical questions on new documents. Through Question Reusability Classification and Document-Bridged Question Ranking, our approach efficiently selects questions from the knowledge base for user-uploaded documents, helping users reuse high-quality questions to obtain key information. In the experiments, we designed multiple evaluation metrics to assess different aspects of question quality. The results demonstrate that our method achieves a good trade-off between accuracy and diversity. Moreover, our method is expected to perform better as the database expands.

Acknowledgment. This work has been supported by National Natural Science Foundation of China (No. 62206265, 62076231) and National Key Research and Development Program of China (No. 2022YFB2702502).

References

1. LlamaIndex (2022)
2. Langchain: Building applications with LLMs through composability (2023)
3. Achiam, J., et al.: GPT-4 technical report. arXiv:2303.08774 (2023)
4. Banerjee, S., Lavie, A.: Meteor: an automatic metric for MT evaluation with improved correlation with human judgments. In: ACL Workshop on Evaluation Measures for MT and Summarization, pp. 65–72 (2005)
5. Bulathwela, S., et al.: Scalable educational question generation with pre-trained language models. In: AIED, pp. 327–339 (2023)
6. Chouinard, M.M., Harris, P.L., Maratsos, M.P.: Children's questions: a mechanism for cognitive development. Monogr. Soc. Res. Child Dev. i–129 (2007)
7. Fukuma, T., et al.: Qana: LLM-based question generation and network analysis for zero-shot key point analysis and beyond. arXiv:2404.18371 (2024)
8. Gao, Y., et al.: Retrieval-augmented generation for large language models: a survey. arXiv:2312.10997 (2023)
9. Geva, M., et al.: Did aristotle use a laptop? A question answering benchmark with implicit reasoning strategies. TACL **9**, 346–361 (2021)
10. Gonzalez, H., et al.: Enhancing human summaries for question-answer generation in education. In: BEA, pp. 108–118 (2023)
11. Graesser, A.C., Langston, M.C., Baggett, W.B.: Exploring information about concepts by asking questions. In: Psychology of Learning and Motivation, vol. 29, pp. 411–436 (1993)
12. Graesser, A.C., Person, N.K.: Question asking during tutoring. Am. Educ. Res. J. **31**(1), 104–137 (1994)
13. Jones, K.S., Walker, S., Robertson, S.E.: A probabilistic model of information retrieval: development and comparative experiments: part 2. IPM **36**(6), 809–840 (2000)
14. Khattab, O., et al.: Demonstrate-search-predict: composing retrieval and language models for knowledge-intensive NLP. arXiv:2212.14024 (2022)
15. Lee, U., et al.: Few-shot is enough: exploring chatgpt prompt engineering method for automatic question generation in english education. Educ. Inf. Technol. 1–33 (2023)
16. Lin, C.Y.: Rouge: a package for automatic evaluation of summaries. In: Text Summarization Branches Out, pp. 74–81 (2004)
17. Liquin, E.G., Gureckis, T.M.: Where questions come from: reusing old questions in new situations. In: CogSci, vol. 44 (2022)
18. Liu, Y., et al.: Exploring the integration strategies of retriever and large language models. arXiv:2308.12574 (2023)
19. Ma, X., et al.: Query rewriting for retrieval-augmented large language models. arXiv:2305.14283 (2023)
20. Mulla, N., Gharpure, P.: Automatic question generation: a review of methodologies, datasets, evaluation metrics, and applications. Prog. Artif. Intell. **12**(1), 1–32 (2023)
21. Neelakantan, A., et al.: Text and code embeddings by contrastive pre-training. arXiv:2201.10005 (2022)
22. Nelson, K.: Structure and strategy in learning to talk. Monogr. Soc. Res. Child Dev. 1–135 (1973)
23. Nguyen, T., et al.: MS MARCO: a human-generated machine reading comprehension dataset (2016)

24. Papineni, K., et al.: Bleu: a method for automatic evaluation of machine translation. In: ACL, pp. 311–318 (2002)
25. Rajpurkar, P., et al.: Squad: 100,000+ questions for machine comprehension of text. arXiv:1606.05250 (2016)
26. Ronfard, S., et al.: Question-asking in childhood: a review of the literature and a framework for understanding its development. Dev. Rev. **49**, 101–120 (2018)
27. Rosenshine, B., Meister, C., Chapman, S.: Teaching students to generate questions: a review of the intervention studies. Rev. Educ. Res. **66**(2), 181–221 (1996)
28. Rothe, A., Lake, B.M., Gureckis, T.M.: Do people ask good questions? Comput. Brain Behav. **1**, 69–89 (2018)
29. Shao, Z., et al.: Enhancing retrieval-augmented large language models with iterative retrieval-generation synergy. arXiv:2305.15294 (2023)
30. Taori, R., et al.: Alpaca: a strong, replicable instruction-following model. Stanford Center Res. Found. Models **3**(6), 7 (2023)
31. Team, G., et al.: Gemini: a family of highly capable multimodal models. arXiv:2312.11805 (2023)
32. Touvron, H., et al.: Llama: open and efficient foundation language models. arXiv:2302.13971 (2023)
33. Trischler, A., et al.: Newsqa: a machine comprehension dataset. arXiv:1611.09830 (2016)
34. Wan, Y., et al.: Sciqag: a framework for auto-generated scientific question answering dataset with fine-grained evaluation (2024)
35. Wang, Y., et al.: Self-instruct: aligning language models with self-generated instructions. arXiv:2212.10560 (2022)
36. Wei, J., et al.: Chain-of-thought prompting elicits reasoning in large language models. In: NeurIPS, vol. 35, pp. 24824–24837 (2022)
37. Yao, S., et al.: React: synergizing reasoning and acting in language models. arXiv:2210.03629 (2022)
38. Zhang, T., et al.: Bertscore: evaluating text generation with bert. arXiv:1904.09675 (2019)

Data Integration

MRF-JOIN: Differentially Private Vertical Data Synthesis via Federated Marginal Join on Shared Attributes

Marin Matsumoto[1(✉)], Tsubasa Takahashi[2], Shun Takagi[3], and Masato Oguchi[1]

[1] Ochanomizu University, Tokyo, Japan
marin@ogl.is.ocha.ac.jp, oguchi@is.ocha.ac.jp
[2] Turing Inc., Tokyo, Japan
tsubasa.takahashi@turing-motors.com
[3] LY Corporation, Tokyo, Japan
shutakag@lycorp.co.jp

Abstract. How can we effectively train a synthetic data generation model for vertically distributed data while preserving privacy? The key challenge in this scenario is accurately reconstructing correlations between attributes held by different parties without compromising privacy. In this study, we assume the existence of shared attributes across vertically distributed tables and explore how these shared attributes can be leveraged for improved performance. To address this, we introduce a differentially private data synthesis method called MRF-JOIN. This method joins privatized Markov Random Fields (MRFs) from different parties on the shared attributes and recovers consistency by leveraging the duplication of shared attributes across parties. Our experiments demonstrate that, compared to existing methods that do not assume shared attributes, the synthetic data generated by MRF-JOIN more effectively preserves the correlations present in the integrated data along with computational and communication efficiency, in vertically federated settings while preserving differential privacy.

Keywords: Data Synthesis · Vertical Federated Learning

1 Introduction

In vertical federated learning (VFL), multiple parties collaborate to leverage the relationships between vertically distributed attributes of the same individuals without revealing their raw data (e.g., medical institutions' diagnosis data and supermarkets' purchase history) [13]. How should we train a synthetic data generation model for integrated data while considering privacy in vertically distributed data? In this study, we consider a VFL algorithm for privacy-preserving synthetic data generation with theoretical guarantees provided by differential privacy (DP) [5].

T. Takahashi—Work done at LY Corporation.

R. Wrembel et al. (Eds.): DEXA 2025, LNCS 16047, pp. 99–114, 2026.
https://doi.org/10.1007/978-3-032-02088-8_7

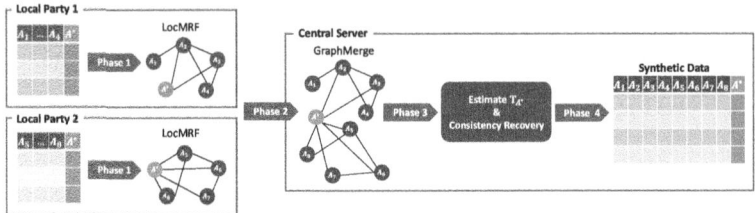

Fig. 1. Overview of MRF-JOIN

The main challenge in solving the VFL task of data synthesis under DP constraints is to efficiently and accurately reconstruct the correlation between attributes held by different parties. Xue et al. [10] proposed VertiGAN, the first generative adversarial network-based model for vertically partitioned data publication. In VertiGAN, each local party sends the local gradients of the generator backbone to the central server. The server summarizes the local gradients and sends them to the local parties. Yuan et al. [20] proposed VFLGAN, which improves the quality of synthetic data compared to VertiGAN. They designed a more efficient and effective mechanism to provide differential privacy guarantees. In GAN-based methods in the vertical setting, communication occurs between local parties and the server every time gradients are updated, resulting in a higher communication volume compared to marginal-based methods. VertiMRF [25] is the first study to adapt the marginal-based approach to VFL settings and shows state-of-the-art performance. VertiMRF introduced techniques based on DP Flajolet-Martin (FM) sketches [16] to estimate cross-party marginals while preserving differential privacy. However, adopting sketch-based cardinality estimation leads to scalability issues with increasing attribute numbers, resulting in lower-quality synthetic data.

On the other hand, typical vertical federated learning assumes no overlapped attributes among different parties. This is not practical when we would like to leverage user-related distributed tables. In most cases, these distributed tables share the same attributes, such as gender, age, and salary. Can we discover and leverage any good properties when such distributed tables share the same attributes to solve the data synthesis in the VFL setting more efficiently? To the best of our knowledge, this study is the first to take advantage of this commonly encountered situation in practice for data synthesis in the VFL setting. Our target use case is privacy-preserving data synthesis in the vertically distributed setting, for which the private data synthesis algorithm should have the following three properties:

R1: Noise resistance against differentially private perturbations,
R2: Computation efficient on local and central entities and
R3: Communication efficient.

To achieve the state-of-the-art results for the aforementioned three aspects, we propose MRF-JOIN, based on the assumption of the existence of shared attribute A^* in Fig. 1. MRF-JOIN incorporates two novel techniques: `GraphMerge` and `ConsistencyRecovery` are both enabled by the presence of shared attributes. In MRF-JOIN, each party initially constructs a local MRF while ensuring DP. The central server then combines local MRFs to form a global MRF, which is used to generate the joint table.

`GraphMerge` is the specific method used to combine local MRFs, which achieves R1 without incurring communication costs or consuming the privacy budget. The key insight behind this significant advantage is based on the following observation. A local MRF contains the correlation information between the shared attribute A^* and other attributes. Consequently, two local MRFs that share the attribute A^* inherently include the correlation information between distributed attributes via A^*. Therefore, we can generate a global MRF that preserves the correlation among distributed attributes without incurring additional privacy or communication costs beyond those required for local MRFs.

`ConsistencyRecovery` is applied to refine the global MRF following `GraphMerge`, which improves R1. Since two local MRFs sharing attribute A^* are independently created, they contain redundant information (i.e., identical counts with independent noise) about A^*. `ConsistencyRecovery` integrates this redundant information to enhance accuracy. Moreover, calibrating A^* also refines the other attributes in the global MRF by maintaining consistency within each local MRF, thereby improving the overall fidelity of the synthetic data.

Finally, we conducted experiments using real-world datasets and compared our method with existing approaches. The experimental results demonstrate that our proposed method generates higher-fidelity synthetic data with lower computational and communication costs, particularly in high-privacy scenarios.

Related Work. We review the existing methods that are not mentioned above. Approaches for generating DP-guaranteed synthetic data can be categorized into GAN-based [3,7,11,22], game-based [8,9,18], and marginal-based [2,21,24]. Among these, marginal-based approaches tend to perform best by approximating high-dimensional joint distributions with multiple low-dimensional marginals, which helps to circumvent the curse of dimensionality, i.e., the exponential growth of contingency histogram sizes with increasing attribute numbers. For example, PrivBayes [21] selects low-dimensional marginals using a Bayesian network to approximate high-dimensional distributions. PrivMRF [2] applies an MRF to model data distribution, allowing flexible low-dimensional marginal selection. PrivSyn [24] greedily searches for numerous low-dimensional marginals to directly represent and synthesize the original dataset. These approaches demonstrate high utility in centralized settings but are challenging to extend directly to VFL settings.

Contributions. The contributions of this study are summarized as follows:

– Assuming the existence of shared attributes among vertically distributed tables, we found two remarkable findings. First, the shared attributes A^* enable us to estimate joint probability among distributed attributes through joining statistics as A^* as a join key. Second, the global MRF can be more accurately constructed through the ensure consistency technique using A^* despite increasing the number of parties.

– We propose MRF-JOIN, a differentially private data synthesis framework in VFL. MRF-JOIN allows an untrusted central server to construct a global MRF by integrating differentially privately protected MRF from each data party, making the global MRF less affected by domain size in terms of error.
– We conduct experiments on real-world datasets and demonstrate that MRF-JOIN can generate synthetic data with higher accuracy for large domain-size datasets than existing methods in VFL settings. Additionally, MRF-JOIN has the lowest computational cost among the marginal-based approaches and lower communication cost than the GAN-based approaches.

2   Preliminaries

2.1   Problem Definition

Let D be a set of data tuples $\{x^{(1)}, \ldots, x^{(n)}\}$. Each tuple consists of values of a set of attributes $\mathcal{A} = \{A_1, \ldots, A_d\}$. Each attribute $A_j, \forall j \in [d]$ has domain size u_j. Without loss of generality, we denote the domain of A_j as $[u_j] \triangleq \{1, \ldots, u_j\}$. With $M \subset \mathcal{A}$, $x_M^{(l)}$ denotes the values of tuple $x^{(l)}$ on an attribute set M. Let $T_{D,M}$ be the counts of occurrences of all possible value tuples of attributes M in D. That is, $T_{D,M}$ is a vector of length $\prod_{A_j \in M} u_j$ and each element is defined as

$$T_{D,M}[\mathbf{v}] = \sum_{l \in [n]} \mathbb{I}(x_M^{(l)} = \mathbf{v}), \quad \forall \mathbf{v} \in \prod_{A_j \in M} [u_j].$$

$T_{D,M}$ is referred to as the *contingency histogram* of M.

We consider a system constituted by m local parties and an untrusted central server. Each local party $\mathcal{P}_i, \forall i \in [m]$, possesses user data $D_i = \{x_{\mathcal{A}_i}^{(1)}, \ldots, x_{\mathcal{A}_i}^{(n)}\}$ with a subset of attributes $\mathcal{A}_i \subset \mathcal{A}$. User data is aligned across these m local parties by some record ID (e.g., social security number or phone number) using private set intersection [12]. If all data parties' data can be integrated, there exists a global dataset $D = (D_1 | \ldots | D_m)$ with attributes $\mathcal{A} = \cup_{i \in [m]} \mathcal{A}_i$.

Vertically Shared Attributes. We define two types of shared attribute set $\mathcal{A}^*$ in the vertical federated setting as shown in Fig. 2(a) and 2(b).

(a) Star Model. The case where the shared attributes of any combination of $\mathcal{A}_i$ and $\mathcal{A}_j$ are the same is called the Star Model. That is, $\mathcal{A}^* = \mathcal{A}_1 \cap \mathcal{A}_2 = \mathcal{A}_2 \cap \mathcal{A}_3 = \ldots = \mathcal{A}_{m-1} \cap \mathcal{A}_m$.
(b) Chain Model. We define the Chain Model as a case where shared attributes among local parties are different for each pair of local parties.

Adversary Model. The central server is assumed to be honest but curious, meaning it will execute the protocol correctly but will try to infer private information from the input dataset. However, we assume that none of the data parties is interested in colluding with the central server. External adversaries are

considered to be all third-party data analysts who aim to infer private information from the synthetic dataset and the intermediate results generated during
communication between local parties and the server.

2.2   Differential Privacy

Differential privacy (DP) is a rigorous privacy notion that quantifies the privacy
loss of algorithms by analyzing the statistical difference between the algorithm
outputs on neighboring datasets differing on only one record.

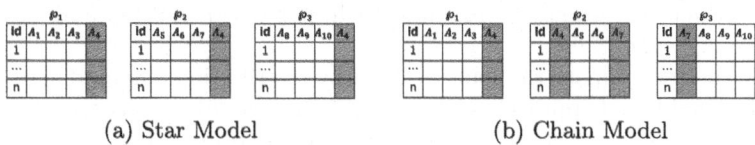

(a) Star Model   (b) Chain Model

Fig. 2. Vertically shared attributes

Definition 1 (Differential Privacy [5]). *A randomized mechanism* $\mathcal{M}$ *satisfies* (ϵ, δ)-*differential privacy if for any neighboring datasets* D, $D' \in \mathcal{D}$ *that
differ on only one record, their outputs fall in any* $R \subset Range(\mathcal{M})$ *with probability*

$$Pr[\mathcal{M}(D) \subseteq R] \leq \exp(\epsilon) Pr[\mathcal{M}(D') \subseteq R] + \delta. \qquad (1)$$

DP is a popular privacy notion because the privacy loss is composable. Basi-
cally, with any two algorithms f and g which satisfy (ϵ_1, δ_1)-DP and (ϵ_2, δ_2)-DP
respectively, the sequential use of $f \odot g$ on a dataset satisfies $(\epsilon_1 + \epsilon_2, \delta_1 + \delta_2)$-DP.

Gaussian Mechanism. To make an algorithm $\mathcal{M}$ differentially private, a typical approach is to inject noise into $\mathcal{M}$ so that its output distribution satisfies
Eq. (1). The amount of noise needed depends on the parameters ϵ and δ of DP,
the noise distribution, as well as how sensitive $\mathcal{M}$ is with respect to the addition
or omission of a tuple in its input. In this paper, we adopt the Gaussian mechanism [1], which injects Gaussian noise based on ℓ_2 sensitivity of the algorithm,
as defined in the following.

Definition 2 (ℓ_2 Sensitivity [6]). *Let* f *be a function that maps its input
data to a* h-*dimensional vector in* $\mathbb{R}^h$. *The* ℓ_2 *sensitivity of* f, *denoted as* $S(f)$,
is defined as $S(f) = \max_{D,D'} ||f(D) - f(D')||_2$.

3   Method

In the VFL setting, each party may have shared attributes such as age or gender.
Under this assumption, each party can approximate the joint probability distribution as shown in Fig. 1 to combine the data distributions of the global dataset

D. Especially to actualize better data synthesis for the distributed datasets, we introduce 1) *joint probability estimation* using the shared attributes, and 2) *consistency recovery* around the shared attributes.

3.1   Graph Merge on Shared Attributes

We here introduce our first finding that joint probability distributions can be joined by assuming the shared attributes. This finding is based on employing (privatized) MRF [2,21]. In the vertical federated setting, for example, party $\mathcal{P}_1$ has a set $\mathcal{A}_1 = \{A_1, \ldots, A_{d_1}\}$ and party $\mathcal{P}_2$ has a set $\mathcal{A}_2 = \{A_{d_1+1}, \ldots, A_{d_1+d_2}\}$, so that each party can calculate distribution $\mathbb{P}[A_1, \ldots, A_{d_1}]$ and $\mathbb{P}[A_{d_1+1}, \ldots, A_{d_1+d_2}]$.

How can we estimate the joint probability distribution among different parties? Marginal-based approaches [2,21] generate attribute graphs, such that each edge in the graph connects two highly correlated attributes. In the typical vertically federated setting, there is no edge between the attribute graphs constructed by parties $\mathcal{P}_1$ and $\mathcal{P}_2$, so there is no information connecting $\mathbb{P}[A_1, \ldots, A_{d_1}]$ and $\mathbb{P}[A_{d_1+1}, \ldots, A_{d_1+d_2}]$. Opposed to the typical setting, this paper assumes shared attributes. That is $\mathcal{A}_1 = \{A_1, \ldots, A_{d_1}, A^*\}$ and $\mathcal{A}_2 = \{A_{d_1+1}, \ldots, A_{d_1+d_2}, A^*\}$. Generating MRF$_1$ and MRF$_2$ from these attribute sets, respectively, A^* separates into $\mathcal{A}_1 \setminus A^*$ and $\mathcal{A}_2 \setminus A^*$. A locally crafted MRF contains the correlation information between the shared attribute A^* and other attributes. Therefore, by assuming the existence of A^*, the MRF constructed by each party separately can be viewed as one of the MRFs that can be constructed from the union set of attributes of each party. Consequently, two local MRFs that share the attribute A^* inherently include the correlation information between distributed attributes via A^*. Finally, we can generate a global MRF that preserves the correlation among distributed attributes without incurring additional privacy or communication costs beyond those required for local MRFs. In the overall algorithm, we introduce GraphMerge that incorporates locally crafted privatized MRFs on the central server. An example of the joined MRF is illustrated in Fig. 1.

3.2   Ensure Consistency

Next, we introduce our second finding that we can recover consistency by leveraging the duplication of shared attributes across parties.

Assuming that $M_1 = \{A_1, A_2, A^*\}$ for $\mathcal{P}_1$'s marginal and $M_2 = \{A_3, A_4, A^*\}$ for $\mathcal{P}_2$'s marginal, then $\mathbb{P}[A_1, A_2, A_3, A_4, A^*] \approx \mathbb{P}[A_1, A_2, A^*] \cdot \mathbb{P}[A_3, A_4|A^*]$ is calculated in the server. The required $\mathbb{P}[A^*]$ can be calculated from either the MRF parameters of $\mathcal{P}_1$ or $\mathcal{P}_2$. Since independent noises are added to the two parameters, the two different ways to compute distribution $\mathbb{P}[A^*]$ most likely have different results. A similar situation occurs when building a marginal table under local differential privacy [15,23]. We improve the accuracy of the synthesis data by aggregating $\mathbb{P}[A^*]$ from each party. Let T_{D_i, A^*} be the marginal A^* from the parameters of party $\mathcal{P}_i$.

Table 1. Ensure consistency and non-negativity

(a) T_{D_1,A^*} from $\mathcal{P}_1$

A^*	count
0	60
1	40

(b) T_{D_2,A^*} from $\mathcal{P}_2$

A^*	count
0	50
1	50

(c) Estimated T_{D,A^*}

A^*	count
0	55
1	45

(d) $T_{D_1,M=\{A_1,A^*\}}$ from $\mathcal{P}_1$

A_1	A^*	count
0	0	30
0	1	20
1	0	0
1	1	20
2	0	30
2	1	0

(e) Updated $T_{D_1,M=\{A_1,A^*\}}$

A_1	A^*	count
0	0	$30 + \frac{55-60}{3} \approx 28.3$
0	1	$20 + \frac{45-40}{3} \approx 21.7$
1	0	$0 + \frac{55-60}{3} \approx -1.7$
1	1	$20 + \frac{45-40}{3} \approx 21.7$
2	0	$30 + \frac{55-60}{3} \approx 28.3$
2	1	$0 + \frac{45-40}{3} \approx 1.7$

(f) Correct negatives

A_1	A^*	count
0	0	28.3
0	1	21.7
1	0	0
1	1	$21.7 - 1.7 = 20$
2	0	28.3
2	1	1.7

Recovery the Shared Attribute Against Gaussian Mechanism. The first step is to compute the best approximation for the marginal table T_{D,A^*}. We estimate T_{D,A^*} for any $a \in [u_{A^*}]$ by averaging. Since the Gaussian noise is sampled from a distribution with zero mean, the m independent noises added to T_{D_i,A^*} cancel each other out by adding them together to approximate a more accurate T_{D,A^*}. Examples are shown in Table 1(a), 1(b), and 1(c). Table 1(c) is the average of Table 1(a) and Table 1(b).

Overall Consistency. In the next step, we correct the overall marginal based on this estimated T_{D,A^*}. We update T_{D_i,M_i} for all $c \in [u_j]$ using the averaged T_{D,A^*} as follows:

$$T_{D_i,M_i}(c) \leftarrow T_{D_i,M_i}(c) + \frac{T_{D,A^*}(a) - T_{D_i,M_i}(a)}{\Pi_{A_j \in M_i \setminus A^*} u_j} \tag{2}$$

Table 1(e) that updated Table 1(d) agrees on A^* without changing the marginals of attributes not involved in the consistency, i.e., T_{D_1,A_1}.

Correcting Negative Entries. In addition, the noisy marginals may contain negative values. We employ Ripple non-negativity method in CALM [23]. This method turns negative counts into 0 while decreasing the counts for its neighbors to maintain the overall count constant. Specifically, given an k-way marginal table T_{M_i}, for any $\mathbf{v} \in \prod_{A^j \in M_i \setminus A^*} [u_j]$ with $T_{M_i}[\mathbf{v}] < 0$, we set the entry to 0 and subtract $\frac{|c|}{h}$ from each of its h neighboring cells, defined as the cells obtained by changing one of the attributes' category to other categories, and h is determined by the number of categories of each attribute in M_i. Table 1(f) is a marginal, corrected for the negative numbers in Table 1(e). Applying Ripple non-negativity step to the marginals might lead to inconsistencies. To resolve this problem, we run the consistency step after the non-negativity step several times.

3.3   Overall Algorithm

In the VFL setting, each party may have shared attributes such as age or gender. Under this assumption, Fig. 1 and Algorithm 1 visualize the workflow of MRF-JOIN consisting of LocMRF, GraphMerge and ConsistencyRecovery.

LocMRF **in Local Parties.** Each local party $\mathcal{P}_i$ directly applies the PrivMRF [2] approach to construct MRF_i. We define the total privacy budget as g. Each party has a privacy budget $\frac{g}{m}$ to construct MRF_i. This involves completing the training with only the attributes that each party possesses. When constructing MRF_i, internal results such as the attribute graph $\mathcal{G}_i$ and the refined marginal set $\mathcal{S}_i$ are generated. MRF_i, $\mathcal{G}_i$, and $\mathcal{S}_i$ are then sent to the central server.

- **Step 1: Generate Attribute Graph.** Each party generates an attribute graph $\mathcal{G}_i$ by greedily linking each attribute pair (A_j, A_k) in descending order of noisy R-score $R(A_j, A_k)$ as follows: $R(A_j, A_k) = \frac{n}{2}||\mathbb{P}[A_j, A_k] - \mathbb{P}[A_j]\mathbb{P}[A_k]||_1 + \mathcal{N}(0, \sigma_{R_i}^2)$.
- **Step 2: Choose Candidate Marginal Set.** The local party first generate a noisy version $\tilde{n}$ of the number n of tuples in D_i as $\tilde{n} = n + \mathcal{N}(0, \sigma_{\mathcal{U}_i}^2)$. Next they sample a set of candidate marginals $\mathcal{U}$ from the cliques of triangulated $\mathcal{G}$ and ensure each marginal $M \in \mathcal{U}$ satisfies θ-usefulness based on $\frac{\tilde{n}}{\Pi_{A_i \in M} u_i} \geq \theta \cdot \frac{g}{m}$. Where $\frac{g}{m}$ denotes the expected absolute value of the noise to be injected into each count of $T_{D_i, M}$.
- **Step 3: Initialize the Marginal Set.** From $\mathcal{U}$, each party selects the most highly correlated marginal for each attribute to constitute an initialized marginal set $\mathcal{S}_i$. This is used to estimate the parameters Θ of the MRF. The MRF models the distribution of arbitrary tuple x as: $\mathbb{P}[x] \approx \Pi_{M \in \mathcal{S}_i} \exp(\Theta_M[x_M])$, where Θ_M denotes the sub-vector of Θ corresponding to M, and $\Theta_M[x_M]$ is the element corresponding to x_M. Each party obtains a noisy distribution vector $\tilde{T}_{D_i, M}$ by injecting i.i.d. Gaussian noise $\mathcal{N}(0, \sigma_{M_i}^2)$ into each entry in $T_{D_i, M}$.
- **Step 4: Refine the Marginal Set.** Refine the marginal set $\mathcal{S}_i$ by inserting marginals that cannot be accurately inferred by the MRF and iteratively refine the estimation of MRF. This iteration time is defined as t. In each iteration, k marginals from the candidate marginal set $\mathcal{U}$ are randomly sampled. The sampled marginal set is $\mathcal{U}'$. For each marginal $M \in \mathcal{U}'$, Step 4 computes the ℓ_1 distance between $\mu_{i, M}$ and $T_{D_i, M}$, and injects Gaussian noise $\mathcal{N}(0, \sigma_{h_i}^2)$ into it. Each iteration in Step 4 also selects a marginal M' from $\mathcal{U}'$ and computes a noisy version $\tilde{T}_{D, M'}$ of its marginal distribution, using Gaussian noise $\mathcal{N}(0, \sigma_{M_i}^2)$.

GraphMerge **on the Central Server.** When the clique sets of $\mathcal{P}_i$ and $\mathcal{P}_j$ are $\mathcal{S}_i$ and $\mathcal{S}_j$, respectively, data synthesis is possible from the sum set of $\mathcal{S}_i$ and $\mathcal{S}_j$ as in PrivMRF. PrivMRF constructs a junction tree [19] from $\mathcal{S}$ to facilitate the approximation of $\mathbb{P}[A_1, A_2, \ldots, A_d]$. Formally, each C_i is referred to as a clique. In particular, a junction tree consists of an ordered set of marginals $\mathcal{S} = \{C_1, C_2, \ldots\}$ that satisfies the following two conditions:

Algorithm 1. MRF-JOIN

Input: The partitioned dataset $D = \{D_i, i \in [m]\}$, domain $([u_1] \times \ldots \times [u_d])$, shared attribute set $\mathcal{A}^*$, maximal clique size τ, total privacy budget g is divided as $g_0 = \frac{g}{m}$.

Output: Synthesized data $\hat{D}$.

1: Each local party $\mathcal{P}_i$:
   (a). constructs local MRF: $\{\text{MRF}_i, \mathcal{G}_i, \mathcal{S}_i\} \leftarrow \text{LocMRF}(D_i, \tau, g_0)$.
   (b). sends $\{\text{MRF}_i, \mathcal{G}_i, \mathcal{S}_i\}$ to server.

2: Central server:
   (a). generates global graph: $\mathcal{G} \leftarrow \text{GraphMerge}(\{\mathcal{G}_i, \mathcal{S}_i | i \in [m]\})$.
   (b). $\{\text{MRF}_1, \ldots, \text{MRF}_m\} \leftarrow \text{ConsistencyRecovery}(\{\text{MRF}_1, \ldots, \text{MRF}_m\}, \mathcal{A}^*)$

3: Central server:
   (a). samples $\hat{D}$ based on MRF_i and $\mathcal{G}$.

---

1. Each attribute in D is contained in at least one marginal in $\mathcal{S}$;
2. For any $i > 1$, the attribute set $S_i = C_i \bigcap(\bigcup_{j=1}^{i-1} C_j)$ appears in one of $C_1, C_2, \ldots, C_{i-1}$. For example, $C_4 = \{A_1, A_2, A_3\}, C_1 = \{A_2, A_5\}, C_2 = \{A_2, A_4\}, C_3 = \{A_3, A_4\}$, then $S_4 = \{A_2, A_3\}$.

By the property of the junction tree, we can approximate the data distribution in D as:

$$\mathbb{P}[A_1, A_2, \ldots, A_d] \approx \mathbb{P}[C_1] \cdot \Pi_{i=2}^{|\mathcal{S}|} \mathbb{P}[C_i \setminus S_i | S_i] \tag{3}$$

Since $\mathcal{S}_i$ contains at least one attribute in $\mathcal{A}_i$, $\mathcal{S} = \mathcal{S}_1 \cup \ldots \cup \mathcal{S}_m$ satisfies the first condition in our scenario. The second condition is that the attribute set $S_k = C_k \bigcap(\bigcup_{l=1}^{k-1} C_l)$ appears in one of $C_1, C_2, \ldots, C_{k-1}$ for any $k > 1$. This condition is satisfied if $A^* \in \mathcal{A}^*$ is common to both $\mathcal{P}_i$ and $\mathcal{P}_j$ since A^* is always included at least once in each clique of $\mathcal{P}_i$ and $\mathcal{P}_j$, making MRF-JOIN capable of data synthesis. A^* can be identified through the use of the private set intersection.

ConsistencyRecovery on the Central Server. Section 3.2 described the estimation of T_{D,A^*} and the consistency recovery of the marginal including A^*. MRF-JOIN enumerates the marginals containing A^* and modifies the marginals based on the estimated T_{D,A^*} for each of them. Then, we turn the negative counts into 0 while decreasing the counts for its neighbors to maintain the overall count constant.

Model for More than Two Local Parties. As shown in Fig. 2, two scenarios can be considered in MRF-JOIN. In Star Model, each MRF_i is joined conditional on A_4. In Chain Model, as an example in Fig. 2(b), $\mathbb{P}[A_1, A_2, A_3, A_4]$, $\mathbb{P}[A_4, A_5, A_6, A_7]$ and $\mathbb{P}[A_7, A_8, A_9, A_{10}]$ approximated by each party approximate the joint probability $\mathbb{P}[A_1, \ldots, A_{10}]$ conditional on A_4 or A_7.

3.4 Privacy and Communication Cost

In MRF-JOIN, the server builds a global MRF from the privacy-protected MRF_i without additional privacy budget. Therefore, we analyze the total privacy when m local parties execute LocMRF. Lemma 2 utilizes Lemma 1 from [2].

Lemma 1 ([2]). *For k queries $Q_1, Q_2, \ldots, Q_k$ with ℓ_2 sensitivity $S(Q_1), S(Q_2),$ $\ldots, S(Q_k)$, we inject independent Gaussian noise of standard deviation σ_i into the query result of Q_i. Let*

$$g = \sum_{i=1}^{k} \frac{S(Q_i)^2}{\sigma_i^2} \tag{4}$$

Then, the composition of the k queries satisfies (ϵ, δ)-DP if:

$$\Phi\left(\frac{\sqrt{g}}{2} - \frac{\epsilon}{\sqrt{g}}\right) - exp(\epsilon) \cdot \Phi\left(-\frac{\sqrt{g}}{2} - \frac{\epsilon}{\sqrt{g}}\right) \leq \delta \tag{5}$$

Lemma 2 (Privacy of MRF-JOIN). *For m local parties with the number of attributes $d_1, d_2, \ldots d_m$, each party inject Gaussian noise with variances $\sigma_{R_i}^2$, $\sigma_{\mathcal{U}_i}^2$, $\sigma_{h_i}^2$ and $\sigma_{M_i}^2$, total privacy budget g can be set as follows in MRF-JOIN.*

$$g = \sum_{i=1}^{m}\left(\frac{2d_i(d_i - 1)}{\sigma_{R_i}^2} + \frac{1}{\sigma_{\mathcal{U}_i}^2} + \frac{tk}{\sigma_{h_i}^2} + \frac{d_i + t}{\sigma_{M_i}^2}\right). \tag{6}$$

Here, $\sigma_{R_i}^2$ is the variance of Gaussian noise into each R-score, $\sigma_{\mathcal{U}_i}^2$ is the variance of Gaussian noise into the number of tuples in D_i, $\sigma_{h_i}^2$ is the variance of Gaussian noise into ℓ_1 distance between $\mu_{i,M}$ and $T_{D_i,M}$, and $\sigma_{M_i}^2$ is the variance of Gaussian noise into $T_{D_i,M}$ in each local party. Then, MRF-JOIN satisfies (ϵ, δ)-differential privacy if

$$\Phi\left(\frac{\sqrt{g}}{2} - \frac{\epsilon}{\sqrt{g}}\right) - exp(\epsilon) \cdot \Phi\left(-\frac{\sqrt{g}}{2} - \frac{\epsilon}{\sqrt{g}}\right) \leq \delta \tag{7}$$

Proof. In MRF-JOIN, each local party $\mathcal{P}_i$ access the input dataset D_i in only four places: R-scores, the number of tuples n, L_1 distances $\tilde{h}$, and marginal distributions $T_{D_i,M}$. The number of queried R-scores is equal to the number of attribute pairs, which is equal to $d_i(d_i - 1)/2$. Step 4 in LocMRF has t iterations and each iteration queries k L_1 distances. Therefore, we query $t \cdot k$ L_1 distances in total. Recall that we identify a marginal for each attribute in Step 3 and we insert t additional marginals in Step 4. Therefore, the number of queried marginal distributions is $(d_i + t)$. In addition, we utilize the L_2 sensitivity $S(R) = 2$ and $S(h) = 1$ from Lemma 4 and 5 in [2]. For other queries, the L_2 sensitivity are 1. We have

$$g = \sum_{i=1}^{m}\frac{d_i(d_i - 1)S(R)^2}{2\sigma_{R_i}^2} + \frac{1}{\sigma_{\mathcal{U}_i}^2} + \frac{tkS(h)^2}{\sigma_{h_i}^2} + \frac{d_i + t}{\sigma_{M_i}^2})$$

$$= \sum_{i=1}^{m}\left(\frac{2d_i(d_i - 1)}{\sigma_{R_i}^2} + \frac{1}{\sigma_{\mathcal{U}_i}^2} + \frac{tk}{\sigma_{h_i}^2} + \frac{d_i + t}{\sigma_{M_i}^2}\right).$$

Let g be such that Eq. (7) holds. This condition combining with Lemma 1 shows MRF-JOIN is (ϵ, δ)-differential privacy.

In our implementation of MRF-JOIN, we set its parameters as follows. First, given privacy parameters ϵ and δ, we compute the largest g that satisfies Eq. (7). We then assign a $\frac{g}{m}$ privacy budget to each local party. Then, we set σ_{R_i}, $\sigma_{\mathcal{U}_i}$, σ_{h_i}, σ_{M_i} so that $\frac{2d_i(d_i-1)}{\sigma_{R_i}^2} = 0.1 \cdot \frac{g}{m}$, $\frac{1}{\sigma_{\mathcal{U}_i}^2} = 0.01 \cdot \frac{g}{m}$, $\frac{t \cdot k}{\sigma_{h_i}^2} = 0.1 \cdot \frac{g}{m}$, $\frac{d_i+t}{\sigma_{M_i}^2} = 0.79 \cdot \frac{g}{m}$. In addition, we set $t = \lfloor 0.6 \cdot d_i \rfloor, k = 50$, and $\theta = 4$; these values are chosen based on our experiments.

Discussion on the Vertical Federated Setting. Figure 2 shows two possible settings for two or more parties. We have explained that the privacy budget of MRF-JOIN is affected by the number of attributes of each party in Lemma 2. Since the number of attributes for each party is the same in both settings (a) and (b), the Chain Model, where the shared attributes differ depending on the combination of parties, consumes the same budget as in setting (a). Furthermore, the more parties there are, the smaller the budget allocated to each party.

Communication Cost. In MRF-JOIN, one communication round is required between each party $\mathcal{P}_i$ and the central server. The communication includes local MRF information $\{\text{MRF}_i, \mathcal{S}_i, \mathcal{G}_i\}$. MRF_i is parameterized by a vector Θ with length $\sum_{M \in \mathcal{S}_i} \prod_{A^j \in M} u_j$, controlled by the maximal clique size τ for each local MRF. $\mathcal{G}_i$ is represented by a $(|\mathcal{A}_i| \times |\mathcal{A}_i|)$-dimensional adjacency matrix, with $|\mathcal{A}_i| < d$. The information in $\mathcal{S}_i$, which contains several attribute tuples, can be ignored in terms of communication costs. Considering a total of m parties, the communication cost of MRF-JOIN is $O(d^2) + O(m\tau)$.

4   Experiments

As discussed in Sect. 1, we attempt to actualize a privacy-preserving data synthesis in the vertically distributed setting having three essential properties. To confirm how much these requirements are satisfied, this section answers the following questions via extensive empirical evaluations:

Q1: How much the built privatized model can generate clear data even when noise is injected to ensure differential privacy?

Q2: How computationally efficient are the local party and the central server?

Q3: How efficient is communication between local parties and the central server?

To answer these questions, we first compare the accuracy of synthetic data generated by MRF-JOIN with baseline methods. Then, we examine the impact of the number of parties, the execution time, and the communication volume. The experiments were conducted on a 20-core Intel(R) Xeon(R) Gold 5115 CPU @ 2.40 GHz processor with 196 GB of RAM and four Nvidia GTX 1080 Ti GPUs.

Datasets. We use four benchmark datasets that are also used in previous work [2]. NLTCS consists of 21,574 records from the National Long Term Care Survey [14]. ACS contains 47,461 records of personal information obtained from IPUMS [17]. Adult consists of 45,222 records from the 1994 US Census [4]. BR2000 contains 38,000 census records collected from Brazil in 2000 [17]. The domain sizes of NLTCS, ACS, Adult, and BR2000 are approximately 6.55×10^4, 8.39×10^6, 9.06×10^{14}, and 3.23×10^9, respectively. We split the dataset so that each local party has the same number of attributes.

Metrics. We evaluate performance based on two metrics. By default, our results are validated in a two-party setting.

- **5-way TVD.** We randomly sample 300 marginals with 5 attributes each from the synthetic data. For each marginal M, we compare the total variation distance (TVD) between the noisy and original versions of M, defined as $\frac{1}{n}\|T_{\tilde{D},M} - T_{D,M}\|_1$. We report the average 5-way TVD based on five iterations.
- **Misclassification rate.** We employ synthetic data to train SVM classifiers to predict specific attributes based on all other attributes. We use 80% of the raw data to generate synthetic data and train the classifier, while the remaining data is used as the test set to report the misclassification rate. We utilize 5-fold cross-validation and report the average.

Compared Methods. We compare the following methods.

- **Centralized** refers to PrivMRF [2] in the centralized setting. We set $\delta = 10^{-5}$. Internal parameters θ, k, and t are default values.
- **VertiMRF** is considered the best-performing marginal-based method in the VFL setting. In our experiments, we use default values for VertiMRF, setting the repetition number of the DP FM sketch to $t = 2000$ and $\delta = 1/n$. The shared attributes are not assumed.
- **VFLGAN** is considered the best-performing GAN-based method in the VFL setting. The network structure of VFLGAN follows the configuration described in the original paper [20]. The shared attributes are not assumed.
- **MRF-JOIN** w/o CO implies MRF-JOIN without any treatment of consistency, i.e., $\mathbb{P}[A^*]$ is not estimated, but a one-party value is adopted as representative. We set $\delta = 10^{-5}$.
- **MRF-JOIN** w/ CO is method incorporates all of our suggestions shown in Fig. 1. That is, it includes the server-side estimation of T_{D,A^*} and ensures overall consistency.

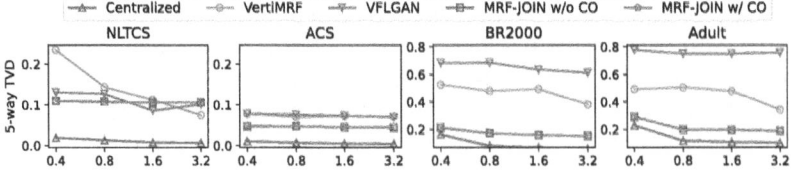

Fig. 3. 5-way TVD vs. privacy budget ϵ

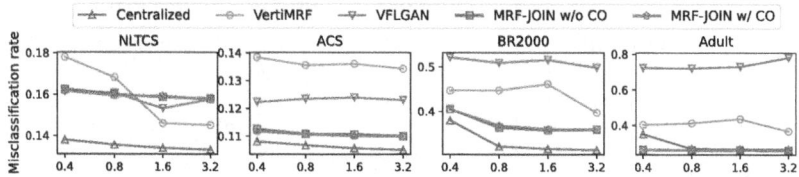

Fig. 4. SVM misclassification rate vs. privacy budget ϵ

4.1 Q1: Noise Resistance

5-way TVD. MRF-JOIN outperforms other vertical methods in the Adult, BR2000, and ACS datasets as shown in Fig. 3. MRF-JOIN performed closest to the central setting on ACS, BR2000, and Adult. Except for datasets with small domain sizes, i.e., NLTCS, the poor performance of VertiMRF is noticeable. Only 40% of each party's privacy budget is allocated for LocMRF, and 40% of the budget is allocated for FM-sketch generation. A larger domain size leads to a smaller average count in a contingency histogram, thereby deriving a more significant cardinality estimation error. For ACS, BR2000, and Adult, VFLGAN showed the lowest performance.

Misclassification Rate. Figure 4 presents the average misclassification rates of the SVM classifiers trained on the synthetic data. MRF-JOIN consistently outperforms other vertical methods. For the BR2000 dataset, MRF-JOIN showed a 10% improvement in accuracy by increasing the privacy budget from $\epsilon = 0.4$ to 3.2. These findings align with the results shown in Fig. 3, illustrating the effectiveness of MRF-JOIN and its robustness to large domain sizes.

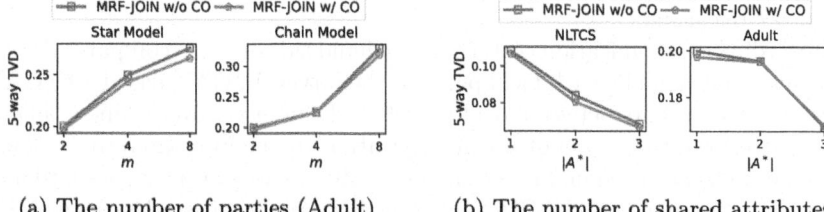

(a) The number of parties (Adult)   (b) The number of shared attributes

Fig. 5. Impact of party number and the number of shared attributes ($\epsilon = 0.8$)

Impact of the Number of Parties and Shared Attributes. Figure 5(a) demonstrates that as the number of parties m increases, the TVD results also increase. When attributes are partitioned to more parties, the ϵ allocated for LocMRF is smaller, and the global MRF is noisier. In the Star Model, the effect of the consistency recovery appears as the number of parties increases. This

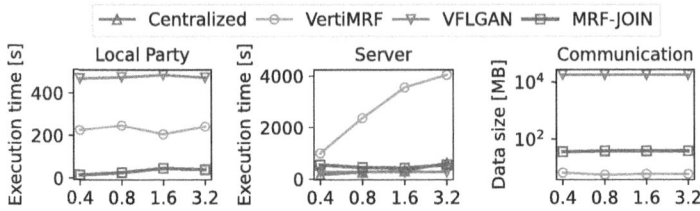

Fig. 6. Computational and communication efficiency on Adult dataset

is because the noise in T_{D,A^*} tends to cancel out as the number of parties is increased. In the Chain Model, even if m increases, the effect is difficult to see because the number of parties involved in the estimation of each A^* is small.

Figure 5(b) shows the average TVD of the 5-way marginals when the number of attributes $|\mathcal{A}^*|$ shared by each party is $\{1, 2, 3\}$. In both datasets, the TVD improved with an increase in $|\mathcal{A}^*|$. This is because the increase in $|\mathcal{A}^*|$ gives more hints for approximating the joint probability distribution $\mathbb{P}[A_1, \ldots, A_d]$ from $\mathtt{MRF}_i$ constructed by each party.

4.2   Q2: Computational and Q3: Communication Efficiency

The left and center of Fig. 6 compare the computation time and communication costs of the four methods on Adult. The execution time is measured with the GPU enabled for both local parties and the server. Generally speaking, the computation time of the marginal-based approaches (i.e., PrivMRF, VertiMRF, MRF-JOIN) tends to increase when ϵ is large. To explain, observe that a large ϵ results in a smaller amount of noise in each marginal cell, and hence, there would be a larger number of marginals satisfying θ-usefulness. The shorter server execution time in MRF-JOIN compared to VertiMRF is that there is no MRF parameter relearning.

Regarding communication cost, the right-hand side of Fig. 6 compares the size of elements uploaded by each local party to the server. In GAN-based VFLGAN, communication occurs between local parties and the server every time gradients are updated, resulting in a higher communication volume compared to marginal-based methods. For marginal-based methods, the size of $\mathtt{MRF}_i$ increases for larger ϵ. On the other hand, VertiMRF including FM-sketches costs less than MRF-JOIN. This is because VertiMRF built $\mathtt{MRF}_i$ with a smaller privacy budget than that of MRF-JOIN.

5   Discussion and Limitation

MRF-JOIN joins the graphs of each local party with respect to A^*. The shape of the graph, such as the number of edges, differs from that of a single-party graph. For example, even if the correlation between attributes A_i of $\mathcal{P}_1$ and A_j of $\mathcal{P}_2$ is strong, there are no edges between the two nodes in MRF-JOIN. Instead, the privacy budget is reduced; in VertiMRF, cardinality estimation is performed at the expense of budget to connect A_i and A_j.

6   Conclusion

In this study, we proposed MRF-JOIN, a novel approach for generating differentially private synthetic data in the VFL setting. By leveraging locally constructed MRF and shared attributes, the central server can build a global MRF representing the overall correlations without parameter relearning. We also analyzed the privacy budgets of m parties constructing MRF locally. Experimental results show that MRF-JOIN can generate synthetic data with higher accuracy for large domain-size datasets than existing methods in VFL settings. Additionally, MRF-JOIN has the lowest computational cost among the marginal-based approaches and lower communication cost than the GAN-based approaches.

References

1. Balle, B., Wang, Y.X.: Improving the gaussian mechanism for differential privacy: analytical calibration and optimal denoising. In: International Conference on Machine Learning, pp. 394–403. PMLR (2018)
2. Cai, K., Lei, X., Wei, J., Xiao, X.: Data synthesis via differentially private Markov random fields. Proc. VLDB Endow. **14**(11), 2190–2202 (2021)
3. Chen, D., Orekondy, T., Fritz, M.: GS-WGAN: a gradient-sanitized approach for learning differentially private generators. In: Advances in Neural Information Processing Systems, vol. 33, pp. 12673–12684 (2020)
4. Dua, D., Graff, C.: UCI machine learning repository (2017). http://archive.ics.uci.edu/ml
5. Dwork, C., McSherry, F., Nissim, K., Smith, A.: Calibrating noise to sensitivity in private data analysis. In: Halevi, S., Rabin, T. (eds.) TCC 2006. LNCS, vol. 3876, pp. 265–284. Springer, Heidelberg (2006). https://doi.org/10.1007/11681878_14
6. Dwork, C., Roth, A., et al.: The algorithmic foundations of differential privacy. Found. Trends® Theor. Comput. Sci. **9**(3–4), 211–407 (2014)
7. Frigerio, L., de Oliveira, A.S., Gomez, L., Duverger, P.: Differentially private generative adversarial networks for time series, continuous, and discrete open data. In: Dhillon, G., Karlsson, F., Hedström, K., Zúquete, A. (eds.) SEC 2019. IAICT, vol. 562, pp. 151–164. Springer, Cham (2019). https://doi.org/10.1007/978-3-030-22312-0_11
8. Gaboardi, M., Arias, E.J.G., Hsu, J., Roth, A., Wu, Z.S.: Dual query: practical private query release for high dimensional data. In: International Conference on Machine Learning, pp. 1170–1178. PMLR (2014)
9. Hardt, M., Ligett, K., McSherry, F.: A simple and practical algorithm for differentially private data release. In: Advances in Neural Information Processing Systems, vol. 25 (2012)
10. Jiang, X., Zhang, Y., Zhou, X., Grossklags, J.: Distributed GAN-based privacy-preserving publication of vertically-partitioned data. Proc. Priv. Enhancing Technol. (2023)

11. Jordon, J., Yoon, J., Van Der Schaar, M.: Pate-GAN: generating synthetic data with differential privacy guarantees. In: International Conference on Learning Representations (2018)
12. Kolesnikov, V., Kumaresan, R., Rosulek, M., Trieu, N.: Efficient batched oblivious PRF with applications to private set intersection. In: Proceedings of the 2016 ACM SIGSAC Conference on Computer and Communications Security, pp. 818–829 (2016)
13. Liu, Y., et al.: Federated forest. IEEE Trans. Big Data **8**(3), 843–854 (2020)
14. Manton, K.G.: National long-term care survey: 1982, 1984, 1989, 1994, 1999, and 2004. Inter-university Consortium for Political and Social Research (2010)
15. Qardaji, W., Yang, W., Li, N.: Priview: practical differentially private release of marginal contingency tables. In: Proceedings of the 2014 ACM SIGMOD International Conference on Management of Data, pp. 1435–1446 (2014)
16. Smith, A., Song, S., Guha Thakurta, A.: The flajolet-martin sketch itself preserves differential privacy: private counting with minimal space. In: Advances in Neural Information Processing Systems, vol. 33, pp. 19561–19572 (2020)
17. Steven, R., Katie, G., Ronald, G., Josiah, G., Matthew, S.: IPUMS USA: version 6.0. University of Minnesota, Minneapolis (2015)
18. Vietri, G., Tian, G., Bun, M., Steinke, T., Wu, S.: New oracle-efficient algorithms for private synthetic data release. In: International Conference on Machine Learning, pp. 9765–9774. PMLR (2020)
19. Wainwright, M.J., Jordan, M.I., et al.: Graphical models, exponential families, and variational inference. Found. Trends® Mach. Learn. **1**(1–2), 1–305 (2008)
20. Yuan, X., Yang, Y., Gope, P., Pasikhani, A., Sikdar, B.: Vflgan: vertical federated learning-based generative adversarial network for vertically partitioned data publication. arXiv preprint arXiv:2404.09722 (2024)
21. Zhang, J., Cormode, G., Procopiuc, C.M., Srivastava, D., Xiao, X.: Privbayes: private data release via Bayesian networks. ACM Trans. Database Syst. (TODS) **42**(4), 1–41 (2017)
22. Zhang, X., Ji, S., Wang, T.: Differentially private releasing via deep generative model (technical report). arXiv preprint arXiv:1801.01594 (2018)
23. Zhang, Z., Wang, T., Li, N., He, S., Chen, J.: Calm: consistent adaptive local marginal for marginal release under local differential privacy. In: Proceedings of the 2018 ACM SIGSAC Conference on Computer and Communications Security, pp. 212–229 (2018)
24. Zhang, Z., et al.: Privsyn: differentially private data synthesis. In: 30th USENIX Security Symposium (USENIX Security 2021), pp. 929–946 (2021)
25. Zhao, F., Li, Z., Ren, X., Ding, B., Yang, S., Li, Y.: Vertimrf: differentially private vertical federated data synthesis. arXiv preprint arXiv:2406.19008 (2024)

Efficient Source Selection for Federated SPARQL Queries Using Adjacent Predicate Information

Yudai Ogura[1]📷, Tadashi Masuda[2]📷, and Toshiyuki Amagasa[2](✉)📷

[1] Graduate School of Science and Technology, University of Tsukuba,
Tsukuba, Japan
`ogura@kde.cs.tsukuba.ac.jp`
[2] Center for Computational Sciences, University of Tsukuba, Tsukuba, Japan
`masuda@kde.cs.tsukuba.ac.jp, amagasa@cs.tsukuba.ac.jp`

Abstract. With the growing adoption of Linked Open Data (LOD),
many distributed knowledge bases have been developed using the RDF
format. These knowledge bases each have unique characteristics, and
querying across them can reveal richer and more comprehensive insights.
To support such cross-repository querying, federated RDF query processing is essential. However, the performance of federated query execution
largely depends on efficient source selection—the process of identifying
which data sources are relevant to a given query. In this paper, we propose a novel source selection method for federated RDF queries based
on a new summarization technique. Our approach precomputes the existence of shared elements between predicate pairs within each data source,
considering three types of join patterns: subject–subject, subject–object,
and object–object. These relationships are stored in a matrix-form summary for each data source. During query execution, our method identifies predicate pairs from triple patterns that share join keys within a
basic graph pattern (BGP) and uses the precomputed summaries to efficiently determine the relevant data sources. Experimental results on the
FedBench benchmark show that our method improves the efficiency of
source selection and significantly reduces overall query execution time.

Keywords: RDF · Knowledge base · Federated RDF query ·
Summary

1 Introduction

The rapid growth of Linked Open Data (LOD) has made hundreds of interconnected knowledge bases—accessible via SPARQL endpoints—available on the
web. While this opens up new possibilities for data integration and knowledge
discovery, executing SPARQL queries across these large-scale, distributed data
sources remains a significant challenge. One of the key difficulties lies in identifying the relevant data sources needed to answer a given query efficiently.

R. Wrembel et al. (Eds.): DEXA 2025, LNCS 16047, pp. 115–129, 2026.
https://doi.org/10.1007/978-3-032-02088-8_8

Most existing approaches to federated RDF query processing rely on triple pattern-wise source selection (TPWSS), where candidate sources are selected independently for each triple pattern in the query [2,10,13]. However, this strategy often results in the inclusion of sources that produce intermediate results irrelevant to the final answer. This not only increases network traffic but also degrades overall query performance.

To address this, HiBISCuS [9] introduced a join-aware source selection strategy that summarizes data sources based on the authority part (i.e., the domain name) of subject and object URIs associated with each predicate. At query time, HiBISCuS selects sources only if their predicates share common authorities in join positions. While effective in filtering out clearly irrelevant sources, this method cannot distinguish between multiple sources that use the same URI authority. As a result, it often selects more sources than necessary. Some extended versions of HiBISCuS attempt to improve accuracy by storing full URI strings in the summaries. However, these enhancements come at the cost of reduced compression and higher computation due to expensive string matching operations, ultimately harming performance [4].

In this paper, we propose a novel source selection method called **MAPS (Matrix-format Adjacent Predicates Summary)**, which overcomes these limitations by leveraging adjacent predicates within the data. Our method begins by extracting all predicates from a given set of data sources and generating all possible predicate pairs. For each pair, we precompute whether the pair shares common elements under different join patterns—subject–subject, subject–object, and object–object—and store the results in a matrix-form summary.

At query time, we identify predicate pairs between triple patterns that share a join key within a basic graph pattern (BGP) and use the corresponding MAPS to determine the relevant data sources. This matrix-based representation enables both high compression and fast lookups, avoiding costly string comparisons while improving the precision of source selection.

To validate the effectiveness of our approach, we conducted a comprehensive evaluation comparing MAPS with existing techniques. We assessed performance across five key metrics: source selection time, number of selected sources, number of SPARQL ASK queries issued during source selection, overall query execution time, and summary generation time/compression ratio. The results demonstrate that our method outperforms existing approaches in both accuracy and efficiency.

The remainder of this paper is structured as follows. Section 2 provides an overview of related work. Section 3 presents the necessary background knowledge for understanding our proposed method. Section 4 details the proposed approach. Section 5 describes the evaluation methodology and presents the experimental results. Finally, Sect. 6 concludes the paper and discusses future research directions.

2 Preliminary Definitions

An RDF resource is identified using a Uniform Resource Identifier (URI). Each URI consists of multiple components: scheme, authority, path, query, and fragment. In this paper, we collectively refer to the scheme and authority components as authority.

For the following discussion, we use an example where a query in Fig. 2 is executed against the data sources shown in Fig. 1.

Definition 1 (Relevant Source Set). *A data source $d \in D$ is considered relevant to a triple pattern $tp_i \in TP$ if at least one triple contained in d matches tp_i. The relevant source set $R_i \subseteq D$ for a triple pattern tp_i consists of all data sources that are relevant to that specific triple pattern. For example, in Fig. 2, the relevant source set for tp_2 is {DBpedia, MusicBrainz, LMDB}.*

However, a relevant source does not necessarily contribute to the final result set of query q. This is because the results obtained from a certain source d for a triple pattern tp_i may be filtered out when performing joins with results from other triple patterns in query q.

Definition 2 (Optimal Source Set). *The optimal source set $O_i \subseteq R_i$ for a triple pattern $tp_i \in TP$ consists of only those relevant sources $d \in R_i$ that actually contribute to computing the complete query result set. For example, in Fig. 2, the optimal source set for tp_2 is {MusicBrainz}.*

DBPedia

@prefix auth1:<http://auth1/> @prefix auth2:<http://auth2/> @prefix auth3:<http://auth3/>		
auth2:MTV	auth3:casting	auth2:Nirvana
auth2:Radiohead	auth1:create	auth2:Creep
auth1:WoodyAllen	auth1:occupation	auth3:FilmDirector

MusicBrainz

@prefix auth1:<http://auth1/> @prefix auth2:<http://auth2/>		
auth2:Bjork	auth1:create	auth2:debut
auth2:JohnLennon	auth1:create	auth2:InMyLife
auth2:Beatles	auth1:born_in	auth1:England
auth2:Bjork	auth1:born_in	auth1:Iceland

LMDB

@prefix auth1:<http://auth1/> @prefix auth2:<http://auth2/> @prefix auth3:<http://auth3/>		
auth3:DancerInTheDark	auth3:casting	auth2:Bjork
auth3:WoodyAllen	auth1:create	auth3:AnnieHall
auth3:Trainspotting	auth3:casting	auth3:EwanMcGregor

Fig. 1. Data Sources

3   Related Work

Information source selection methods for federated RDF queries can generally be classified into three categories: summary-based, non-summary-based, and hybrid approaches [8].

Summary-based methods, such as DARQ [6] and ADERIS [3], rely solely on pre-computed summaries to perform source selection. This typically enables faster execution. However, because these methods select sources based only on predicates—without considering the subjects or objects of each triple pattern—their accuracy can be limited [8].

In contrast, non-summary-based methods like FedX [13] do not rely on pre-stored summaries. Instead, they dynamically select sources by issuing SPARQL ASK queries against live data. These queries consider not just predicates but also subjects and objects, often resulting in more accurate source selection [8]. However, because they involve a large number of ASK queries, they may incur longer query execution times [8].

Hybrid methods combine elements of both summary-based and non-summary-based approaches, using both summaries and ASK queries for source selection. Prominent examples include DAW [10], TopFed [11], and SPLENDID [2]. These methods typically perform source selection based on fixed values in each triple pattern. For instance, in the query shown in Fig. 2, tp1 has the fixed predicate `auth3:casting`, tp2 has `auth1:create`, and tp3 has `auth1:born_in`. Based on these fixed predicates, the method might assign tp1 to {DBpedia, LMDB}, tp2 to {DBpedia, MusicBrainz, LMDB}, and tp3 to {MusicBrainz}. However, because these methods do not take into account joins between triple patterns during source selection, they may retrieve intermediate results from sources that do not contribute to the final answer, potentially increasing overall execution time.

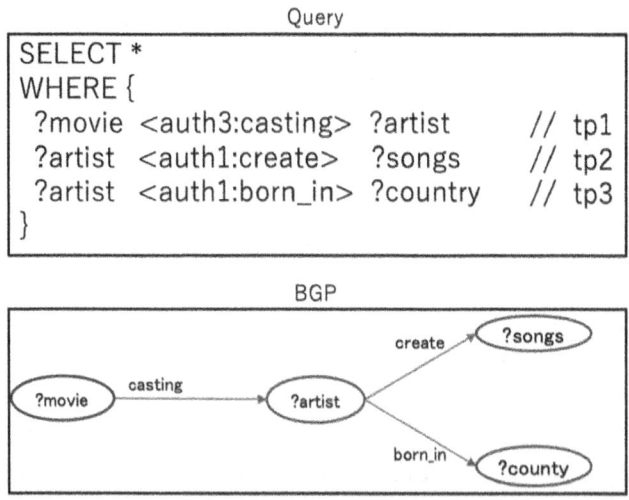

Fig. 2. SPARQL Query and Its BGP

To address this limitation, HiBISCuS [9] introduced a join-aware source selection method. It summarizes each data source by recording the domain part (referred to as the *authority*) of the subject and object URIs for each predicate. During source selection, HiBISCuS uses these summaries to identify shared authorities across join keys and exclude data sources that are unlikely to contribute to the final result.

For example, in the query shown in Fig. 2, the common authority among the object of `auth3:casting`, the subject of `auth1:create`, and the subject of `auth1:born_in` is `<auth2>`. Based on this, the selected sources are {DBPedia, LMDB} for tp1, {DBPedia, MusicBrainz} for tp2, and {MusicBrainz} for tp3. This consideration of joins eliminates LMDB as a source for tp2. However, intermediate results obtained from DBPedia by tp1 and tp2 are ultimately excluded, as the authority matches but the string following the authority in the URI does not.

Furthermore, methods such as TBSS [7] and SemCat [4] have been developed as extensions of the HiBISCuS summary format. These methods retain the string following the authority in URIs, thereby improving the accuracy of source selection. However, they suffer from trade-offs, such as lower summary compression rates and increased computational costs due to string matching, which can degrade performance. While these methods enhance source selection accuracy by extending the HiBISCuS summary format, they do not completely eliminate the selection of sources that do not contribute to the final results.

In this study, we propose a new source selection method for efficient federated RDF querying. Instead of using the HiBISCuS summary format, our approach records precomputed results in a matrix as a novel summarization technique for source selection.

4 Proposed Method

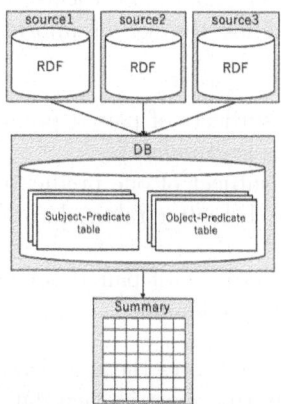

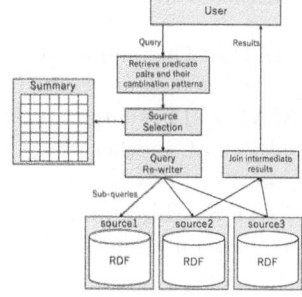

Fig. 4. Overview of the query process

Fig. 3. Overall process of summary creation

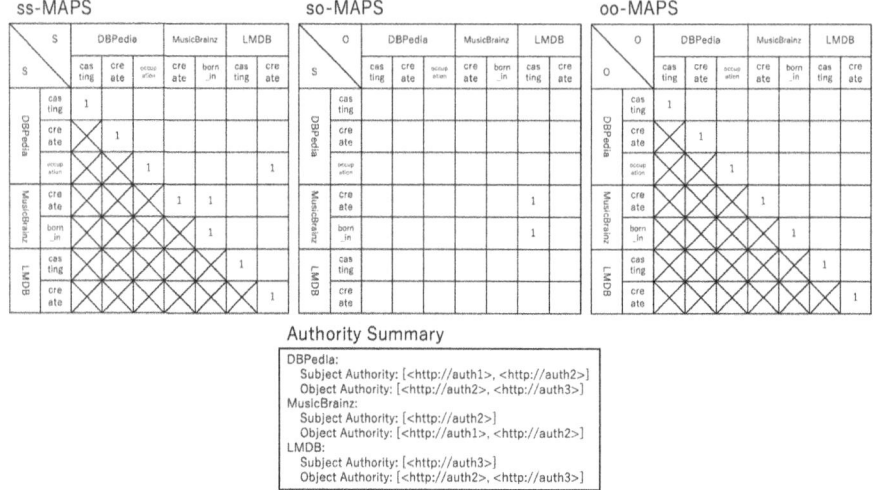

Fig. 5. Summary

We propose an efficient method for federated RDF query processing that pre-computes summaries for source selection, resolving the trade-off between accuracy and performance seen in URI string-matching approaches like HiBISCuS [9], TBSS [7], and SemCat [4].

4.1   Matrix Format Adjacent Predicates Summary (MAPS)

We propose a novel summarization of RDF data, *matrix format adjacent predicates summary* (MAPS). Specifically, we extract neighboring predicates that share the subject or object in common, generating three types of summaries: ss (subject-subject)-MAPS, so (subject-object)-MAPS, and oo (object-object)-MAPS. Figure 5 shows the summaries generated for the set of information sources depicted in Fig. 1.

- **ss-MAPS**: This summary stores whether the subjects of paired predicates share common elements (hereafter referred to as *ss-common*).
- **so-MAPS**: This summary stores whether the subject of one predicate and the object of another share common elements (hereafter referred to as *so-common*).
- **oo-MAPS**: This summary stores whether the objects of paired predicates share common elements (hereafter referred to as *oo-common*).

Additionally, we maintain *authority summary* that stores the set of authorities of subjects and the set of authorities of objects that exist in each dataset.

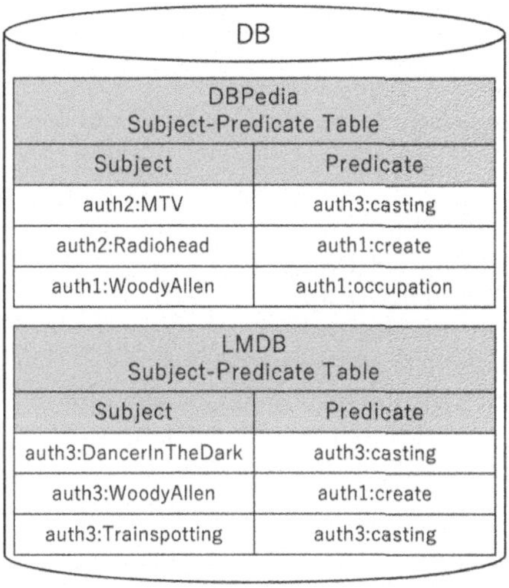

DB	
DBPedia **Subject-Predicate Table**	
Subject	Predicate
auth2:MTV	auth3:casting
auth2:Radiohead	auth1:create
auth1:WoodyAllen	auth1:occupation
LMDB **Subject-Predicate Table**	
Subject	Predicate
auth3:DancerInTheDark	auth3:casting
auth3:WoodyAllen	auth1:create
auth3:Trainspotting	auth3:casting

Fig. 6. Database for generating summaries.

Algorithm 1. Database Creation Algorithm

Require: datasources D // list of available datasources

1: **for** each $pair \in predicatesPairs$ **do**
2:     $SbjPrd = \text{fetchSbjPrd}(D)$
3:     $ObjPrd = \text{fetchObjPrd}(D)$
4:     registerTable($SbjPrd$)
5:     registerTable($ObjPrd$)
6: **end for**

4.2 Summary Generation

This section explains the method for generating summaries. The overall process of Matrix format Summary (MAPS) generation is illustrated in Fig. 3. First, the data required for summary generation is stored in a local database from the information sources. Then, MAPS is generated using the data stored in the database. Additionally, simplified code for each process is provided in Algorithm 1 and Algorithm 2.

The method for creating the database is explained in detail according to Algorithm 1. First, from each information source, it retrieves subject-predicate pairs and object-predicate pairs (Lines 2–3). Then, the retrieved data is registered in the database as Subject-Predicate Tables and Object-Predicate Tables (Lines 4–5). For example, when creating the database shown in Fig. 6 from the information sources in Fig. 1, the Subject-Predicate Tables from DBPedia and LMDB are stored. This process reduces the number of accesses to large-scale information sources. Additionally, since only the data necessary for MAPS generation is extracted, computational costs are reduced.

Algorithm 2. Summary Generation Algorithm

Require: datasources D // list of available datasources

1: $DPairs = \text{extractDPairs}(D)$
2: **for** each $pair \in DPairs$ **do**
3:    $d1 = \text{Pair}[0]$
4:    $d2 = \text{Pair}[1]$
5:    $SPd1 = \text{fetchSbjPrdFromDB}(d1)$
6:    $SPd2 = \text{fetchSbjPrdFromDB}(d2)$
7:    $OPd1 = \text{fetchObjPrdFromDB}(d1)$
8:    $OPd2 = \text{fetchObjPrdFromDB}(d2)$
9:    $ssPairs = \text{SPd1 INNER JOIN SPd2 ON sbj}$
10:   $soPairs = \text{SPd1 INNER JOIN OPd2 ON [sbj, obj]}$
11:   $osPairs = \text{OPd1 INNER JOIN SPd2 ON [obj, sbj]}$
12:   $ooPairs = \text{OPd1 INNER JOIN OPd2 ON obj}$
13: **end for**

Algorithm 3. Authority Summary Generation Algorithm

Require: datasources D // list of available datasources

1: **for** each $d_i \in D$ **do**
2:    $sbjAuthList = \text{extractSbjAuthList}(d_i)$
3:    $objAuthList = \text{extractObjAuthList}(d_i)$
4: **end for**

Next, the method for generating MAPS is explained in detail according to Algorithm 2. First, all pairs of target information sources are retrieved (Line 1). Then, for each pair of information sources, the Subject-Predicate Tables and Object-Predicate Tables are retrieved (Lines 5–8).

For ss-common predicate pairs, the Subject-Predicate Tables from both information sources are joined using the subject as the join key (Line 9). For so-common predicate pairs, the Subject-Predicate Table from one source and the Object-Predicate Table from the other are joined using the subject and object as the join keys (Lines 10–11). For oo-common predicate pairs, the Object-Predicate Tables from both information sources are joined using the object as the join key (Line 12).

For example, when generating MAPS from the database in Fig. 6, the Subject-Predicate Tables from DBPedia and LMDB are joined using the subject as the join key. As a result, the predicate pair (auth1:occupation, auth1:create) is identified as an ss-common predicate pair and stored in MAPS.

Since ss-MAPS and oo-MAPS form symmetric matrices, it is unnecessary to swap the left and right tables when retrieving join results. However, for so-MAPS, the left and right tables must be swapped to obtain the join results. This is because the relationship depends on whether the subject and object used as join keys correspond to the respective predicates in the predicate pair.

Algorithm 3 presents the overall algorithm for Authority Summary generation. This algorithm extracts all subject authorities and object authorities from the target information sources.

Finally, since the matrices representing each join pattern are expected to be sparse, only the row index and column index pairs of non-zero elements are stored when saving the summary. Similarly, we can exploit space-efficient index structure like k^2-tree [1] to maintain the summaries.

4.3 Query Processing

Algorithm 4. Source Selection Algorithm

Require: S_M//MAPS; S_A//Authority Summary; Q//query
1: $triples = $ extractTriples(Q)
2: $R = $ initializeTripleDatasetDict($triples$)
3: $joinTriplePairs = $ extractJoinTriplePairs($triples$)
4: **for** each $pair \in joinTriplePairs$ **do**
5:    $R = $ updateDict($R, pair, S_M, S_A$)
6: **end for**
7: **for** each $triple \in triples$ **do**
8:    $s, p, o = triple$
9:    $R_{tp_i} = R.get(triple)$
10:    $I = R_{tp_i}.get(1)$
11:    **for** each $d_set \in R_{tp_i}$ **do**
12:       $I = I \cap d_set$
13:       **if** $size(I) > 1$ and (bound(s) or bound(o)) **then**
14:          $label = ASK(s, p, o, I)$
15:       **end if**
16:    **end for**
17: **end for**

The overall query process is illustrated in Fig. 4. When a query is received, the system first references the summary and selects the data sources to be accessed. Then, the query is rewritten to match the federated RDF query processing. Finally, the rewritten query is executed to obtain the results.

The following sections explain the method for selecting the data sources to be accessed. Algorithm 4 presents the source selection algorithm.

The source selection process based on the summary for the given query is performed in the following steps. At the end of each step, an example explanation is provided using the query shown in Fig. 2.

1. Initialize the variable R to store relevant data sources for each triple (Line 2). E.g., $R = \{\mathtt{tp1} : \{\}, \mathtt{tp2} : \{\}, \mathtt{tp3} : \{\}\}$
2. We decompose the triple patterns into binary pairs according to the join keys (Line 3).

E.g., the join key is ?artist, and by using ?artist as the basis, we get the following pairs of triple patterns: $\{(\text{tp1}, \text{tp2}), (\text{tp1}, \text{tp3}), (\text{tp2}, \text{tp3})\}$.

3. Extract the predicate pairs and join patterns for each paired triple pattern. Query the corresponding MAPS for the extracted predicate pairs to check if results exist for the given join pattern and add the results to R (Lines 4–6). Additionally, if the subject or object is specified, use the authority summary for further refinement.

   – (tp1, tp2): The predicate pair is (auth3:casting, auth1:create), and the join pattern is *object–subject*. In this case, the order is swapped to *subject–object*, and look up so-MAPS for the following pairs:
   (DBPedia:create, DBPedia:casting),
   (DBPedia:create, LMDB:casting),
   (MusicBrainz:create, DBPedia:casting),
   (MusicBrainz:create, LMDB:casting),
   (LMDB:create, DBPedia:casting), and (LMDB:create, LMDB:casting).
   As a result, we get
   (MusicBrainz:create, LMDB:casting) and $R = \{\text{tp1} : \{\text{LMDB}\}, \text{tp2} : \{\text{MusicBrainz}\}, \text{tp3} : \{\}\}$.

   – (tp1, tp3): The predicate pair is (auth3:casting, auth1:born_in), and the join pattern is object-subject. Similarly, we change the order to subject-object, and look up so-MAPS for the following pairs:
   (MusicBrainz:born_in, DBPedia:casting)   and
   (MusicBrainz:born_in, LMDB:casting).   As   a
   result, we get (MusicBrainz:born_in, LMDB:casting) and $R = \{\text{tp1} : \{\text{LMDB}, \text{LMDB}\}, \text{tp2} : \{\text{MusicBrainz}\}, \text{tp3} : \{\text{MusicBrainz}\}\}$.

   – (tp2, tp3): The predicate pair is (auth1:create, auth1:born_in) and the join pattern is subject-subject. Then, it looks up ss-MAPS for the following pairs: (DBPedia:create, MusicBrainz:born_in),
   (MusicBrainz:create, MusicBrainz:born_in), and
   (LMDB:create, MusicBrainz:born_in). As a result, we get
   (MusicBrainz:create, MusicBrainz:born_in) and
   $R = \{\text{tp1} : \{\text{LMDB}, \text{LMDB}\}, \text{tp2} : \{\text{MusicBrainz}, \text{MusicBrainz}\}, \text{tp3} : \{\text{MusicBrainz}, \text{MusicBrainz}\}\}$.

4. Obtain the common data sources for each triple pattern. Furthermore, if the subject or object is specified and additional refinement is possible, use SPARQL ASK queries to further filter the data sources. The final set of selected data sources is determined based on these results.
   ex)
   The following shows the collection of candidate sets for each triple pattern and the computation of their common data sources:
   $I_{tp1} = \{\text{LMDB}\}$, $I_{tp2} = \{\text{MusicBrainz}\}$, $I_{tp3} = \{\text{MusicBrainz}\}$

The above operations do not affect recall because join processing works like a logical "AND": data retrieved for one part of the query must also successfully join with data from other parts to be included in the final result. If it cannot be joined, it is automatically excluded and does not impact the completeness of the results.

5   Experiment

In this section, we present the results of our performance evaluation of the proposed method by comparing it with a baseline approach using nine datasets and multiple evaluation metrics. As a baseline method, we use HiBISCuS [9]. Additionally, we conducted experiments using NumPy's ndarray and k^2-tree as data structures for storing MAPS. Furthermore, the memory usage of each data structure is reported in Table 3. k^2-tree [1] is a compression technique for storing large sparse matrices by recursively partitioning the target matrix and recording only the regions containing non-zero elements using a tree structure. This approach significantly reduces storage requirements while maintaining efficient access to the data.

For evaluation, we utilized FedBench [12], which is widely used for evaluating federated RDF queries [2,5,7–10]. The nine information sources from FedBench were uploaded to a Fuseki server for testing.

The proposed method is evaluated based on the following five performance metrics:

- Number of selected data sources
- Total number of SPARQL ASK queries
- Execution time of source selection
- Query execution time
- Summary generation time/compression rate

The results are presented in Tables 1 and 2. The reported values for source selection execution time and query execution time are the average of five measurements.

5.1   Experimental Environment

All experiments were conducted using a PC with Intel Core i7-13700 CPU and 32 GB RAM running Ubuntu 24.04 LTS. The proposed method was implemented using Python 3.8.19. For consistency, the baseline method was also re-implemented in Python 3.8.19.

5.2   Experimental Results

The experimental results are summarized in Table 1. Bold values indicate cases where the proposed method outperforms HiBISCuS.Underlined values indicate cases where the performance difference between storing MAPS in NumPy's ndarray and k^2-Tree is notable.

Number of Selected Data Sources. The number of selected data sources refers to the total number of datasets selected for querying for each triple pattern. The results are shown in the SSN column of Table 1. In CD1, CD2, CD3, CD4, CD5, and LS4, no improvement was observed, and the selection was the same as HiBISCuS. However, in LS2, LS5, and LS7, the number of selected sources was reduced, demonstrating the effectiveness of the proposed method.

Total Number of SPARQL ASK Queries. The total number of SPARQL ASK queries refers to the total number of queries sent during the source selection process. The results are shown in the AN column of Table 1.

For CD3, CD4, CD5, LS4, LS5, and LS7, HiBISCuS already did not send any SPARQL ASK queries, so the results remained the same. However, in CD1, CD2, and LS2, where HiBISCuS did send SPARQL ASK queries, the proposed method reduced the number of SPARQL ASK queries by 11, 9, and 16 in CD1, CD2, and LS2 respectively, thereby enhancing efficiency.

Source Selection Execution Time. The execution time for source selection is measured as the time taken from receiving a query to selecting the relevant data sources using the summaries. The results are shown in the SST column of Table 1. For queries where the number of SPARQL ASK queries was 0, using NumPy-based MAPS, the proposed method achieved an average speed-up of 270x and a maximum speed-up of 500x, demonstrating significant improvements. For queries where SPARQL ASK queries were still required, using k^2-Tree-based MAPS, LS2 saw a 53x speed-up, and CD1 was 2.7x faster.

Query Execution Time. Query execution time is measured as the time taken from sending the query to the endpoint until the result set is received. The results are shown in the QET column of Table 1. For queries where no improvement in source selection was observed (CD1, CD2, CD3, CD4, CD5, LS4), the difference in query execution time was within the margin of error, and no significant improvement was observed. For queries where source selection was improved (LS2, LS5, LS7), execution time was also significantly improved. In LS5, HiBISCuS exceeded the execution time limit and failed to return a result, whereas the proposed method successfully returned a result in 6148 ms. In LS7, query execution time was reduced from 11738 ms to 2759 ms, demonstrating substantial improvement.

Summary Generation Time and Compression Rate. The summary generation time and compression rate are shown in Table 2. The summary generation time was reduced by approximately 20 min compared to HiBISCuS. The compression rate was 0.2% lower than HiBISCuS but remained highly efficient overall.

5.3   Discussion

This section discusses the results obtained in Sect. 5.2.

Number of Selected Data Sources. In the Number of Selected Data Sources, the proposed method demonstrated performance equivalent to or better than HiBISCuS. Since the proposed method performs precomputations using the

Table 1. Experimental Results

Query	HiBISCuS				Proposed Method(NumPy)				Proposed Method(k^2-Tree)			
	SSN	AN	SST	QET	SSN	AN	SST	QET	SSN	AN	SST	QET
CD1	4	18	103.66	115	4	**7**	**60.38**	87	4	7	**37.89**	83
CD2	3	9	34.87	12	3	0	**0.16**	12	3	0	**0.29**	11
CD3	5	0	51.76	29	5	0	**0.30**	20	5	0	**0.70**	20
CD4	5	0	119.23	30	5	0	**0.23**	19	5	0	**0.56**	21
CD5	4	0	75.49	53	4	0	**0.14**	36	4	0	**0.32**	38
LS2	7	18	237.49	194	**3**	**2**	**13.29**	30	4	**1**	**4.48**	**93**
LS4	7	0	14.41	29	7	0	**0.27**	18	7	0	**0.74**	21
LS5	8	0	52.07	-	**6**	0	**0.24**	**6148**	6	0	**0.81**	**6205**
LS7	5	0	30.76	11738	**4**	0	**0.16**	**2759**	**4**	0	**0.26**	**3131**

SSN: Number of selected data sources, AN: Number of SPARQL ASK queries, SST:
Source selection time (ms), QET: Query execution time (ms).

Table 2. Summary Generation Time and Compression Rate

	HiBISCuS	Proposed Method
Summary Generation Time (min)	38	19
Compression Rate (%)	99.997	99.78

entire URI string, information sources that generate intermediate results discarded during the join process are not selected. This means that when the number of selected sources in the proposed method is the same as in HiBISCuS, the optimal set of sources has already been selected.

In the Total Number of SPARQL ASK Queries, the proposed method also demonstrated performance equivalent to or better than HiBISCuS. This improvement is likely due to the higher accuracy of source selection using the proposed method's summary, which reduced the need to send SPARQL ASK queries compared to HiBISCuS.

In the Source Selection Execution Time, the proposed method outperformed HiBISCuS in all queries. Generally, a large number of SPARQL ASK queries increases network load, leading to longer overall query execution times. Therefore, to clarify the evaluation, we separately analyzed queries that required

Table 3. Memory Usage

	NumPy	k^2-Tree
Memory Usage (MB)	49.07	1.69

SPARQL ASK queries and those that did not. For queries where no SPARQL ASK queries were issued, the significant improvement is attributed to the difference in search mechanisms: HiBISCuS searches for common elements using string matching. The proposed method searches using matrix indices. This highlights the effectiveness of the proposed method's summary format. In comparing NumPy ndarray and k^2-Tree as data structures for storing summaries, NumPy ndarray generally performed better. A likely reason is that the matrix size was not large enough for k^2-Tree to demonstrate its advantages effectively.

In the Query Execution Time, the proposed method also demonstrated performance equivalent to or better than HiBISCuS. For queries with similar results to HiBISCuS, the number of selected sources was the same. For queries where the proposed method outperformed HiBISCuS, the number of selected sources was also improved, indicating reasonable and expected results.

In the Summary Generation Time and Compression Rate, the proposed method again demonstrated performance equivalent to or better than HiBIS-CuS. In real-world data sources, fast summary generation is essential to reflect up-to-date data in source selection. The high-speed summary generation of the proposed method is therefore highly beneficial. A high compression rate is crucial for efficient summary lookups during source selection. Although the compression rate of the proposed method was slightly lower than HiBISCuS, the proposed summary format retains richer information, enabling more accurate source selection than HiBISCuS.

This suggests that the summary format of the proposed method is beneficial for source selection. Regarding the data structure for MAPS, the NumPy-based implementation generally outperformed the alternative. However, k^2-Tree-based MAPS exhibited higher memory efficiency (Table 3), it is expected to be effective when handling a larger number of data sources.

6   Conclusion

In this study, we proposed a source selection method using MAPS, a novel summary format, to enable efficient querying across multiple information sources with large-scale RDF data. The evaluation experiments demonstrated that the proposed method achieved performance equivalent to or superior to existing methods across all evaluation metrics. These results indicate that the proposed method successfully resolves the trade-off between source selection accuracy, execution time, and compression rate, enabling more efficient source selection. For future work, we plan to conduct experiments using a larger number of information sources and explore strategies to manage the increase in summary generation time.

Acknowledgments. This paper is based on results obtained from the project, "Research and Development Project of the Enhanced infrastructures for Post-5G Information and Communication Systems" (JPNP20017), commissioned by the New Energy and Industrial Technology Development Organization (NEDO), JST CREST Grant Number JPMJCR22M2, and JSPS KAKENHI Grant Number JP23K24949.

References

1. Brisaboa, N.R., Ladra, S., Navarro, G.: k^2-trees for compact web graph representation. In: Karlgren, J., Tarhio, J., Hyyrö, H. (eds.) SPIRE 2009. LNCS, vol. 5721, pp. 18–30. Springer, Heidelberg (2009). https://doi.org/10.1007/978-3-642-03784-9_3
2. Gorlitz, O., Staab, S.: SPLENDID: SPARQL endpoint federation exploiting VoID descriptions. In: ISWC (2011)
3. Lynden, S., Kojima, I., Matono, A., Tanimura, Y.: ADERIS: an adaptive query processor for joining federated SPARQL endpoints. In: Meersman, R., et al. (eds.) OTM 2011. LNCS, vol. 7045, pp. 808–817. Springer, Heidelberg (2011). https://doi.org/10.1007/978-3-642-25106-1_28
4. Molli, P., Skaf-Molli, H., Grall, A.: SemCat: Source Selection Services for Linked Data. hal, hal-02931367 (2020). https://hal.science/hal-02931367
5. Ozkan, E.C., Saleem, M., Dogdu, E., Ngonga Ngomo, A.C.: UPSP: unique predicate-based source selection for SPARQL endpoint federation. In: European Semantic Web Conference, pp. 176–191 (2016)
6. Quilitz, B., Leser, U.: Querying distributed RDF data sources with SPARQL. In: Bechhofer, S., Hauswirth, M., Hoffmann, J., Koubarakis, M. (eds.) ESWC 2008. LNCS, vol. 5021, pp. 524–538. Springer, Heidelberg (2008). https://doi.org/10.1007/978-3-540-68234-9_39
7. Saleem, M.: Efficient source selection and benchmarking for SPARQL endpoint query federation. In: Studies on the Semantic Web, vol. 31, p. 91. IOS Press (2017)
8. Saleem, M., Khan, Y., Hasnain, A., Ermilov, I., Ngonga Ngomo, A.C.: A fine-grained evaluation of SPARQL endpoint federation systems. Semantic Web J. (2014)
9. Saleem, M., Ngonga Ngomo, A.C.: HiBISCuS: hypergraph-based source selection for SPARQL endpoint federation. In: European Semantic Web Conference, pp. 176–191 (2014)
10. Saleem, M., Ngonga Ngomo, A.-C., Xavier Parreira, J., Deus, H.F., Hauswirth, M.: DAW: duplicate-AWare federated query processing over the web of data. In: Alani, H., et al. (eds.) ISWC 2013. LNCS, vol. 8218, pp. 574–590. Springer, Heidelberg (2013). https://doi.org/10.1007/978-3-642-41335-3_36
11. Saleem, M., et al.: TopFed: TCGA tailored federated query processing and linking to LOD. J. Biomed. Semant. (2014)
12. Schmidt, M., Görlitz, O., Haase, P., Ladwig, G., Schwarte, A., Tran, T.: FedBench: a benchmark suite for federated semantic data query processing. In: Aroyo, L., et al. (eds.) ISWC 2011. LNCS, vol. 7031, pp. 585–600. Springer, Heidelberg (2011). https://doi.org/10.1007/978-3-642-25073-6_37
13. Schwarte, A., Haase, P., Hose, K., Schenkel, R., Schmidt, M.: FedX: optimization techniques for federated query processing on linked data. In: Aroyo, L., et al. (eds.) ISWC 2011. LNCS, vol. 7031, pp. 601–616. Springer, Heidelberg (2011). https://doi.org/10.1007/978-3-642-25073-6_38

Empathetic Response Generation in Emotional Support Conversation via Multi-stage Cascading Information Fusion

Jianwei Zhang[1]([envelope]) [iD], Shota Sato[1], Yuta Sasaki[2], and Yuhki Shiraishi[3] [iD]

[1] Iwate University, Morioka, Japan
zhang@iwate-u.ac.jp
[2] Institute of Science Tokyo, Tokyo, Japan
yubo1336@lr.pi.titech.ac.jp
[3] Tsukuba University of Technology, Tsukuba, Japan
yuhkis@a.tsukuba-tech.ac.jp

Abstract. Emotional Support Conversation (ESC) aims to ease a help-seeker' psychological distress through dialogue with a supporter. We propose CasDecNet (**Cas**cading Fusion-Guided **Dec**oding **Net**work), a novel response generation model that enhances empathy in three stages: understanding context, acquiring cognitive and affective empathy, and incorporating strategic support. Our model uses multi-dimensional external knowledge to better infer the user's situation and implement empathy effectively. Experiments on the ESConv dataset show that CasDec-Net outperforms baselines in both automatic and LLM-based evaluations, especially in terms of response diversity and empathy, indicating its promise for human-like emotional support.

Keywords: Emotional Support Conversation · Empathetic Response · Response Strategy · Response Generation · Dialogue System

1 Introduction

Emotional Support Conversation (ESC) is a task aimed at gradually alleviating mental distress through interactive dialogue between a help-seeker and a supporter. Empathy is an essential element in ESC. It refers to the psychological process of cognitively and affectively understanding another's emotions and circumstances and is recognized for its positive therapeutic impact in counseling. Previous work such as MISC [5] has reported attempts to leverage external knowledge to infer the user's mental state and generate empathetic responses. However, MISC processes empathetic elements and supportive strategies in parallel, which misaligns with the typical human empathy formation process [2]. As a result, empathy expression may be insufficient, and its integration with supportive strategies can remain incomplete.

R. Wrembel et al. (Eds.): DEXA 2025, LNCS 16047, pp. 130–135, 2026.
https://doi.org/10.1007/978-3-032-02088-8_9

Drawing inspiration from established psychological models of empathy, such as Davis's organizational model [1], we propose an empathetic response generation model called CasDecNet (**Cas**cading Fusion-Guided **Dec**oding **Net**). Specifically, during the response generation stage, this model projects the process of empathy formation onto three sequential phases: (1) Establishing an empathetic foundation by analyzing the dialogue context, (2) Cognitively and affectively recognizing the user's situation and emotions, (3) Expressing concrete supportive actions based on the acquired perspective. Through these cascading phases, CasDecNet first grasps the user's emotional state and subsequently organizes the process of building concrete supportive actions. This multi-stage approach encourages the acquisition of empathetic elements that might otherwise be overshadowed by purely parallel strategies. Specifically, by drawing on previous work such as CEM [4], our model leverages multi-dimensional external knowledge to more accurately infer and supplement the user's mental state.

We conducted experiments using the ESConv dataset to validate the effectiveness of our proposed method. Specifically, we employed MISC as the baseline and compared the generated responses using both automatic metrics and LLM-based evaluation. Automatic evaluation results indicate that our method outperforms the baseline in overall response quality, with notable improvements in reference text matching and token diversity. Furthermore, the LLM-based evaluation suggests that our method also improves empathy and supportiveness in the generated responses.

2   Related Work

MISC [5] aims to generate responses that integrate both empathetic and strategic factors. It aligns multiple types of commonsense knowledge derived from the generative model COME with contextual information, thereby enhancing the understanding of the user's mental state. Additionally, a strategy label predictor generates a probability distribution of possible support strategies, allowing the system to more effectively address the user's emotional needs. A key limitation of MISC is that it processes empathetic and strategic aspects in parallel during response generation. One potential drawback of parallel processing in response generation is that overly focusing on strategic aspects may weaken attention to empathetic information. Therefore, our approach adopts a multistage integration of various elements to enhance empathetic expression, even in strategy-oriented responses.

In recent years, considerable efforts have been made to integrate empathy into dialogue systems for generating empathic responses. Some primarily focus on emotional detection while paying less attention to the cognitive dimension. To address both facets, CEM [4] leverages multiple types of external knowledge by partitioning them into affective and cognitive domains. Instead of incorporating a user emotion identification task, as in many empathic dialogue models, our approach focuses on deepening the understanding of the user's implicit situation and integrating empathy in a way that more closely resembles human empathy

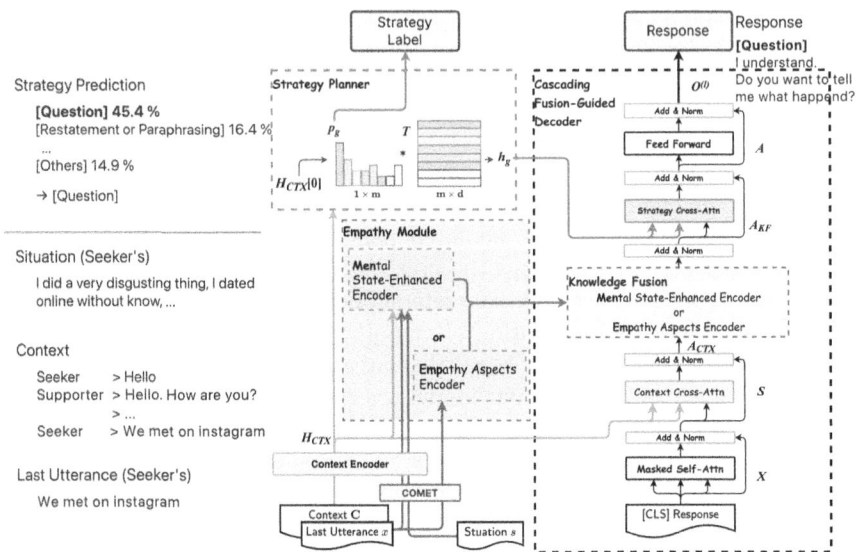

Fig. 1. Overview of the proposed CasDecNet architecture. This module is composed of Context Encoder, Strategy Planner, Empathy Module (*Mental State-Enhanced Encoder* or *Empathy Aspects Encoder*), and Cascading Fusion-Guided Decoder.

formation. By projecting a multi-stage empathy process onto the model, we seek to enable more effective support aligned with the goals of ESC.

3 Proposed Method

Figure 1 illustrates the architecture of our proposed model, CasDecNet, which is a Transformer-based encoder-decoder model consisting of four main modules: (1) Context Encoder, (2) Strategy Planner, (3) Empathy Module (*Mental State-Enhanced Encoder* or *Empathy Aspects Encoder*), and (4) Cascading Fusion-Guided Decoder. The encoding parts of (1) Context Encoder, (2) Strategy Planner, and (3) Empathy Module (*Mental State-Enhanced Encoder*) adopt the architecture proposed in MISC [5], whereas (3) Empathy Module (*Empathy Aspects Encoder*) builds upon the CEM [4] framework to incorporate both the cognitive and affective facets of empathy. Notably, (4) Cascading Fusion-Guided Decoder follows three sequential phases, which are entirely distinct from those in MISC and CEM.

To train the proposed model, we use data from the dataset employed in our experiments. Specifically, the model takes as input: (i) a dialogue context (Context C) comprising multiple utterances exchanged between the seeker (user) and the supporter, (ii) the seeker's final utterance in the given context (Last Utterance x), and (iii) a free-form textual description of the problem faced by the seeker (Situation s). We minimize the sum of two loss functions: one for the generated response and one for the predicted strategy labels.

Table 1. Statistics of the pre-processed ESConv dataset

Category	Train	Val	Test
Number of dialogues	14,117	1,764	1,764
Avg. words per dialogue	148.46	146.66	145.17
Avg. turns per dialogue	7.61	7.58	7.49
Avg. words per turn	17.25	17.09	17.11

4   Experimental Setup

For model training and evaluation, we used the pre-processed version of the Emotional Support Conversation (ESConv) dataset [3], provided by morecry/MISC[1]. Table 1 presents the statistical details of the dataset.

We employed both automatic metrics and LLM-based evaluation to assess model performance.

Automatic Evaluation. To measure the quality of generated responses, we used the following five automatic evaluation metrics: BLEU-2 (**B-2**), BLEU-4 (**B-4**), ROUGE-L (**R-L**), Distinct-1 (**D-1**), and Distinct-2 (**D-2**). BLEU-n and ROUGE-L are reference-based metrics, and Distinct-n measures response diversity. Additionally, we evaluated Top-1 strategy label classification performance using Accuracy (**Acc.**), Precision (**Pre.**), Recall (**Rec.**), and F1-score (**F1**).

LLM-Based Evaluation. To supplement automatic metrics, we conducted an LLM-based evaluation as a proxy for human judgment. The LLM-based evaluation considered five key criteria: **Naturalness (Nat.)**, **Coherence (Coh.)**, **Knowledge (Know.)**, **Supportiveness (Sup.)**, **Empathy (Emp.)**. For the LLM-based evaluation, we randomly sampled 200 generated responses from the test dataset. The evaluation was conducted using Llama-3.1-8B-Instruct, and the assigned scores was 1, 2, or 3.

For comparison, we used MISC as the baseline model. Additionally, we implemented two variants of our proposed approach.

- **Baseline MISC [5]:** An ESC system that generates responses by mixing strategic and empathetic aspects in parallel.
- **Proposed Model (A) CasDecNet+Men:** Unlike MISC's parallel fusion approach, this model applies a sequential fusion strategy in the decoder. Specifically, it incrementally fuses three types of information in the decoder. Apart from this multi-stage decoding mechanism, it inherits the basic architecture and modules from MISC.
- **Proposed Model (B) CasDecNet+Emp:** Like Model (A), this model also adopts the sequential fusion approach for decoding. However, it replaces the empathy module used in Model (A) with our newly proposed Empathy Enhancement Encoder.

---

[1] https://github.com/morecry/MISC.

Table 2. Results of Automatic Evaluation

Model	Text Generation					Strategy Label Pred.			
	B-2	B-4	R-L	D-1	D-2	Acc.	Pre.	Rec.	F1
MISC [5]	6.51	1.95	17.50	4.32	18.36	33.84	27.44	25.43	21.38
(A) **CasDecNet+Men**	7.15	**2.34**	**17.87**	3.97	16.86	**34.09**	**29.64**	**27.54**	26.39
(B) **CasDecNet+Emp**	**7.36**	2.22	17.84	**4.50**	**19.53**	33.64	28.26	27.41	**26.76**

Table 3. Results of LLM-Based Evaluation

Model	Nat.	Coh.	Know.	Sup.	Emp.
MISC [5]	2.118	2.161	2.149	2.137	2.153
(A) CasDecNet+Men	2.113	2.151	**2.158**	2.159	2.190
(B) CasDecNet+Emp	**2.121**	**2.171**	2.137	**2.161**	**2.195**
Reference	2.219	2.277	2.249	2.277	2.310

5   Experimental Results

5.1   Automatic Evaluation

Table 2 presents the results of the automatic evaluation. For each metric, we underline scores that exceed the baseline MISC and highlight the highest score in bold.

Text Generation. First, we evaluate the overall performance of response generation. As shown in Table 2, both proposed models (A) and (B) outperform the baseline MISC on multiple metrics. Notably, for reference-based metrics (**B-4, B-2, R-L**), all proposed models demonstrate improvements. These findings indicate that our multi-stage decoder approach better preserves and leverages information that might otherwise be overshadowed by a parallel-based strategy. Furthermore, the non-reference **D-n** metrics confirm that incorporating an empathy module enhances response diversity. Model (B) achieves the highest **D-1** score (4.50), indicating that leveraging multi-dimensional external knowledge, which focuses on both cognitive and affective empathy, promotes more varied word choices aligned with the user's situation and emotional state.

Strategy Label Prediction. In our ESC task, the model first predicts a strategy label to provide specific support to the user and then generates a response based on that label. Regarding accuracy (**Acc.**), Model (A) CasDecNet+Men attains the highest score (34.09%), slightly surpassing MISC (33.84%) and Model (B) CasDecNet+Emp (33.64%). Although the differences among the models are small, Model (A) appears slightly more consistent in selecting the correct label.

5.2   LLM-Based Evaluation

As shown in Table 3, our proposed Model (B) outperforms MISC on all metrics except **Knowledge (Know.)**, with particularly noticeable improvements in

Supportiveness (Sup.) and **Empathy (Emp.)**. For **Sup.**, Model (B) achieves a score of 2.161, which is 0.024 points higher than MISC, suggesting that the proposed model generates more supportive responses. Although the difference from Model (A) is small, Model (B) retains a slight advantage, suggesting that the empathy module more accurately generates supportive phrases aligned with the seeker's fundamental concerns. For **Emp.**, Model (B) achieves the highest score of 2.195, exceeding MISC by 0.042 points. This improvement is likely due to the multi-dimensional external knowledge approach, which enhances the model's emotional connection to the seeker.

6   Conclusion and Future Work

In this paper, we proposed an empathetic response generation model, CasDec-Net (**Cas**cading Fusion-Guided **Dec**oding **Net**work), to address the challenges of Emotional Support Conversation tasks. Inspired by the human empathy formation process, CasDecNet aims to accurately capture the seeker's emotional state and generate responses that facilitate concrete supportive actions. Specifically, we introduced (1) a Cascading Fusion-Guided Decoder that sequentially integrates empathetic and strategic elements, as opposed to a parallel fusion approach, to better emulate human empathy, and (2) a multi-dimensional empathy module that leverages external knowledge encompassing both cognitive and affective empathy. Comparative analyses with prior studies indicate that CasDec-Net not only enhances overall response quality but also effectively strengthens empathetic expressions and supportive content.

However, challenges remain in selecting and incorporating external knowledge that is contextually appropriate while maintaining proper empathetic expressions. In future work, we plan to further refine our approach by exploring the process of empathy formation in more extended conversational scenarios and integrating additional insights from psychological models of empathy.

Acknowledgments. This work was supported by JSPS KAKENHI Grant Numbers 22K12271.

References

1. Davis, M.H.: Empathy: A Social Psychological Approach. Westview Press (1996)
2. Hayama, D., Uemura, M., Hagihara, T.: Development of an empathetic process scale. Tsukuba Psychol. Res. **36**, 39–48 (2008)
3. Liu, S., et al.: Towards emotional support dialog systems. In: ACL-IJCNLP, pp. 3469–3483 (2021)
4. Sabour, S., Zheng, C., Huang, M.: CEM: commonsense-aware empathetic response generation. In: AAAI, vol. 36, pp. 11229–11237 (2021)
5. Tu, Q., Li, Y., Cui, J., Xian Bin, W., Wen, J.R., Yan, R.: MISC: a mixed strategy-aware model integrating COMET for emotional support conversation. In: ACL, pp. 308–319 (2022)

Unified Schema-Driven Graph Polystore: Achieving Transparency in Multi-model Integration and Migration

Fumihiro Yamashita[ID], Qiong Chang[ID], and Jun Miyazaki[(✉)][ID]

Department of Computer Science, School of Computing,
Institute of Science Tokyo, Tokyo, Japan
yamashita@lsc.c.titech.ac.jp, {q.chang, miyazaki}@comp.isct.ac.jp

Abstract. The rise of big data and social networks necessitates using multiple data models, including NoSQL, leading to complex database management due to their schemaless nature and ambiguous structures. Graph data models offer unique flexibility to facilitate. This paper proposes a graph-centric multi-model DBMS leveraging a unified schema (U-Schema) to integrate heterogeneous data into a cohesive graph. This polystore architecture provides a transparent interface for querying as a unified graph, seamlessly supporting data migration and dynamic data placement. Qualitative and quantitative evaluations show transparent query processing and significant impact of data migration on execution speed, indicating optimization potential.

Keywords: polystore system · data migration · query rewriting · property graph

1 Introduction

The rapid evolution of application requirements since the early 2000s, driven by the Internet's spread, revealed the inadequacy of relational databases (RDBs) alone [12]. This led to the emergence of diverse NoSQL databases (e.g., graph, document, column-oriented, key-value stores) [6]. NoSQL databases offer greater data storage freedom and simplify complex structures, expressing intricate relationships more naturally than RDBs' multi-table join operations.

Modern application development often uses multiple databases concurrently, optimizing them for specific purposes. However, managing these independent systems with distinct interfaces and query languages complicates usability. *Polystore systems* [2,9,14,16] address this by providing a common interface to heterogeneous data models. Yet, many existing solutions lack location and migration transparency, requiring users to explicitly manage distributed and schemaless NoSQL data.

To overcome these issues, we propose a novel graph-centric multi-model DBMS. Our system, leveraging a unified schema model (U-Schema), seamlessly

R. Wrembel et al. (Eds.): DEXA 2025, LNCS 16047, pp. 136–143, 2026.
https://doi.org/10.1007/978-3-032-02088-8_10

```
1 // Query for RDB
2 T1(x int, y int)@rdb = (SELECT x, y FROM A)
3 // Query for Document DB
4 T2(x int, z array)@mongo = {*
5 db.B.find({$lt: {x, 10}}, {x:1, z:1})
6 *}
7 SELECT T1.x, T2.z
8 FROM T1, T2
9 WHERE T1.x = T2.x AND T1.y <= 3
```

Fig. 1. Example query for CloudMdsQL

integrates heterogeneous data into a single, cohesive graph. This architecture ensures true location and high migration transparency, allowing users to query the entire data landscape as a unified graph without being affected by dynamic data migration. We rigorously evaluate our system qualitatively and quantitatively, demonstrating improvements in transparency, flexibility, and performance.

2   Related Work

2.1   U-Schema

U-Schema [3] is a unified schema model for RDBs and four types of NoSQL databases: graph, document, key-value, and column-oriented. Although NoSQL databases are often described as schemaless, they possess implicit schemas that are dependent on stored data, termed *schema-on-read* [7]. U-Schema aims to extract these implicit schemas into a unified model.

NoSQL databases often contain data objects with varying structures within a single database. This *structural variation* is reflected in the schema, distinguishing data objects with the same label or entity based on differences in their constituent elements. For example, Person nodes in a graph database may have different property schemas, identifying them as distinct variations. U-Schema leverages this to capture the flexibility of schemaless NoSQL databases.

2.2   Polystore Systems

Few polystores support graph databases as comprehensively as our proposed method. CloudMdsQL [8] is a representative example that supports RDBs, document databases, and graph databases using a SQL-like common query language. See Fig. 1 for query examples. It processes queries as subqueries for each database, converts results into relational tables, and then integrates them with a main query. However, users are required to explicitly specify data locations, leading to low query transparency and no support for data migration.

2.3   Property Graph Model

A property graph is a graph where nodes and edges have labels and properties. Labels indicate type and group nodes/edges. Properties are attributes associated with each node or edge, representing the object. Neo4j [1] is a prominent property graph database that uses Cypher [5], a query language similar to SQL but with unique pattern-matching syntax for flexible network queries.

3   Proposed Approach

Unlike conventional methods, our system integrates data without intermediary tables. It leverages U-Schema to capture multi-model data placement and rewrites input queries into native queries for each database.

3.1   Data Structure

Our polystore system models the entire data as a graph, representing entities as nodes and relationships as edges for consistent operations. Inspired by Yasuda et al. [15], who managed data dependencies in a graph database and stored data in a key-value store for large tabular data, our approach uniquely identifies each graph node using the entity name, node ID, and field name, linking it to corresponding data. An entity denotes a set of labels, e.g., a `Person` node represents the Person entity. If a Person node has an `Age` field in a key-value store, it is linked via a key such as `Person:1:Age`.

To enable reverse lookups (filtering by stored values) in key-value stores, which typically only support value retrieval by key, we create a reverse key. For a key of the form ⟨EntityName⟩ : ⟨ID⟩ : ⟨PropertyName⟩, a corresponding key `index:EntityName:PropertyName:Value` is generated to store a set of IDs.

3.2   Schema Management

An integrated schema is defined by surveying individual database schemas based on U-Schema to integrate heterogeneous data. Both integrated and individual database schemas are prepared to determine property locations. A mapping dictionary (Fig. 2) is constructed from these schemas, holding metadata for entity properties and relationships. It is modeled as $\mathcal{M} = \{\ell_1 \mapsto E_1, \ell_2 \mapsto E_2, \dots\}$, where ℓ_i is a label and E_i contains entity information. Each E_i is a tuple (ent, P, R), where ent is the entity name, P the property set, and R the relationship set. P is defined as $\{\texttt{prop} \mapsto (\text{type}, \text{database}), \dots\}$, with `prop` being the property name, `type` its data type, and `database` a set of storage locations (e.g., {graph, kvs}). R stores target entities for edge traversal and rewriting.

```
 1  "entities": {
 2   "People": {
 3    "properties": {
 4     "Age": {
 5      "type": "integer",
 6      "database": "kvs"
 7     }
 8    },
 9    ...
10  }
```

Fig. 2. Mapping dictionary example

3.3   Query Rewriting

Our system preserves the node and edge structures while rewriting queries based on property storage locations, with reference to a mapping dictionary. For the WHERE clause, if a property is in "graph," the expression is retained. If it is in "kvs," an index key constructs a node ID list (e.g., (p.age = 30), the ID set is obtained by querying index:People:age:30, yielding [1, 2, 5]) and rewriting the Cypher query to WHERE n.ID IN [1, 2, 5]. For the RETURN clause, properties in "kvs" only return the node ID (e.g., m.ID), allowing value retrieval from the key-value store using the entity, property, and ID. Properties in "graph" require no rewriting. The rewritten Cypher query is executed on the graph database, returning expected values or IDs, which are then used to query the key-value store for corresponding values if IDs were returned (e.g., Movie:1:name and Movie:2:name).

3.4   Data Migration

Dynamic data migration is supported per entity or property, optimizing access or balancing performance and storage efficiency by moving data between key-value and graph databases. The process involves: 1) referencing the mapping dictionary to determine source and target databases; 2) transferring data (e.g., graph to KVS: write (People:IDValue:name) to KVS and register ID in index:People:name:value; KVS to graph: read ID/value from KVS, write to graph node using Cypher: MATCH (n:People) WHERE n.ID = IDValue SET n.name = value); 3) deleting old data from the source to prevent duplicates; and 4) updating the mapping dictionary's Databases field to reflect new placement (e.g., graph to KVS: remove graph, add kvs). This ensures that the data and mapping dictionaries are synchronized, enabling consistent data capture and potential optimization via automatic placement of frequently/infrequently accessed properties.

4   Evaluation

The proposed approach was implemented in Go 1.23, using antlr 4.13 [11] for graph query parsing, Neo4j 5.12 as the graph database, and BadgerDB 1.6 [4] as the key-value store. Figure 3 shows the system architecture. A qualitative assessment was performed using the polystore system evaluation framework by Ran et al. [13].

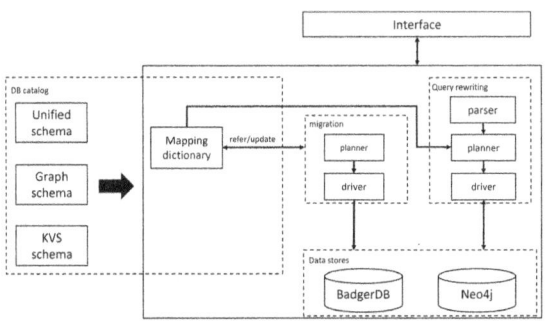

Fig. 3. System architecture

4.1   Evaluation Using Framework

The framework evaluates heterogeneity, autonomy, flexibility, transparency, and optimality, each subdivided into two or three sub-aspects. Table 1 summarizes the results, rating each sub-aspect on a three-level scale based on evaluations from Ran et al. and Wong et al.

We evaluated the effect of migration on query execution time using the LUBM dataset, which simulates a university's organizational structure and activities. The RDF data from this dataset was converted into a property graph via the Neosemantics plugin [10]. Nine LUBM queries were used, with random WHERE clause values to reduce caching effects. Four migration states were defined: (i) all data in BadgerDB, (ii) WHERE clause properties in BadgerDB and RETURN clause in Neo4j, (iii) WHERE clause in Neo4j and RETURN clause in BadgerDB, and (iv) both in Neo4j. $prop_{where}$, $prop_{return}$, and $prop_{otherwise}$ denote properties accessed in WHERE, RETURN, and other clauses, respectively; queries were executed randomly per state, and the average execution time and its breakdown (Neo4j, BadgerDB WHERE, BadgerDB RETURN, overhead) were measured. Baseline (all data in Neo4j) averaged 46.44 ms. As shown by the results in Fig. 4, state (i) had the shortest time and (iv) the longest, with all states outperforming the baseline.

Table 1. Comparison of each approach based on the evaluation framework

Major Aspect	Sub-Aspect	CloudMdsQL	Wong et al.'s Proposal	Proposed Approach
Heterogeneity	Data Store	Excellent	Average	Average
	Processing Engine	Excellent	Excellent	Excellent
	Interface	Excellent	Average	Average
Autonomy	Cooperation	Average	Excellent	Poor
	Execution	Excellent	Excellent	Excellent
	Evolution	Average	Excellent	Excellent
Flexibility	Schema	Poor	Average	Average
	Interface	Average	Poor	Average
	Architecture	Average	Average	Average
Transparency	Location	Average	Poor	Excellent
	Migration	Average	Poor	Excellent
Optimality	Federation Plan	Average	Poor	Poor
	Data Placement	Average	Poor	Excellent

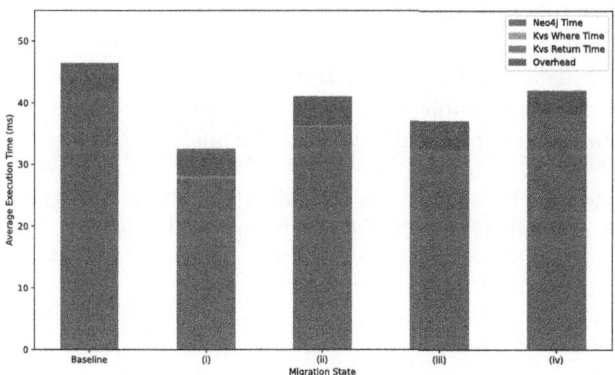

Fig. 4. Changes in query execution time by migration state

5 Discussion

Query execution times varied significantly with migration state. The shortest times occurred when all properties resided in BadgerDB (state i), while the longest was when all accessed properties were in Neo4j (state iv). Figure 4 shows that the Neo4j processing time dramatically increased in state (iv), while BadgerDB's was near 0 ms. Contrary to initial expectations that placing all accessed properties in Neo4j would be fastest, experimental results showed the opposite. This is because rewriting (e.g., `WHERE p.ns0_name = '...'` to `WHERE p.ID IN [...]`) allows BadgerDB to leverage its index more efficiently than Neo4j's filtering.

While placing all property data in BadgerDB yielded the highest in this experiment, it may not be universally optimal. Optimal data placement depends on factors such as database resource availability and network conditions, and future investigation is required. Nevertheless, our experiments show that changing data placement significantly impacts query execution time, highlighting the potential for optimization through the proposed data migration support.

6   Conclusion

This paper proposed a U-Schema-driven graph polystore system for transparent multi-model integration and migration. Our system uses a mapping dictionary to transparently rewrite graph queries, enabling users to interact with heterogeneous data as a unified graph without awareness of distributed placement. Unlike conventional methods using SQL-like subqueries, our approach accepts graph queries directly, offering more transparent processing and flexible querying capabilities. Data migration is supported on a per-property basis, hidden from users by the mapping dictionary, which addresses the limitation of conventional polystores. Experiments demonstrated that data placement significantly impacts query execution time, highlighting our system's potential for optimization and improved performance over existing solutions.

Future work includes dynamic migration optimization based on query metrics, robust transaction management for migrations, and expanding support to additional data models compatible with U-Schema.

Acknowledgments. This paper was supported by JST, CREST Grant Number JPMJCR22M2, and JSPS KAKENHI Grant Numbers JP23K28091 and JP23K28383, Japan.

References

1. Neo4j Homepage (2012). http://neo4j.org/
2. Azevedo, L.G., et al.: HKPoly: a polystore architecture to support data linkage and queries on distributed and heterogeneous data. In: Proceedings of the 20th Brazilian Symposium on Information Systems, SBSI 2024. Association for Computing Machinery, New York (2024). https://doi.org/10.1145/3658271.3658322
3. Candel, C.J.F., Sevilla Ruiz, D., García-Molina, J.J.: A unified metamodel for NoSQL and relational databases. Inf. Syst. **104**, 101898 (2022)
4. Dgraph Labs: BadgerDB: A Fast Key-Value DB in Go (2017). https://github.com/dgraph-io/badger. Accessed 22 Mar 2025
5. Francis, N., et al.: Cypher: an evolving query language for property graphs. In: Proceedings of the 2018 International Conference on Management of Data, SIGMOD 2018, pp. 1433–1445. Association for Computing Machinery, New York (2018). https://doi.org/10.1145/3183713.3190657
6. Han, J., Haihong, E., Le, G., Du, J.: Survey on NoSQL database. In: 2011 6th International Conference on Pervasive Computing and Applications, pp. 363–366 (2011). https://doi.org/10.1109/ICPCA.2011.6106531
7. Klettke, M., Störl, U., Scherzinger, S.: Schema extraction and structural outlier detection for JSON-based NoSQL data stores (2015)

8. Kolev, B., Valduriez, P., Bondiombouy, C., Jiménez-Peris, R., Pau, R., Pereira, J.: CloudMdsQL: querying heterogeneous cloud data stores with a common language. Distrib. Parallel Databases **34**, 463–503 (2016)

9. Leclercq, E., Savonnet, M.: A tensor based data model for polystore: an application to social networks data. In: Proceedings of the 22nd International Database Engineering & Applications Symposium, IDEAS 2018, pp. 110–118. Association for Computing Machinery, New York (2018). https://doi.org/10.1145/3216122.3216152

10. Moreira, E.J.V.F., Ramalho, J.C.: SPARQLing Neo4J. In: Simões, A., Henriques, P.R., Queirós, R. (eds.) 9th Symposium on Languages, Applications and Technologies (SLATE 2020). Open Access Series in Informatics (OASIcs), vol. 83, pp. 17:1–17:10. Schloss Dagstuhl – Leibniz-Zentrum für Informatik, Dagstuhl (2020). https://doi.org/10.4230/OASIcs.SLATE.2020.17. https://drops.dagstuhl.de/entities/document/10.4230/OASIcs.SLATE.2020.17

11. Parr, T., et al.: What's ANTLR (2004)

12. Stonebraker, M., Madden, S., Abadi, D.J., Harizopoulos, S., Hachem, N., Helland, P.: The end of an architectural era: it's time for a complete rewrite. In: Making Databases Work: The Pragmatic Wisdom of Michael Stonebraker, pp. 463–489 (2018)

13. Tan, R., Chirkova, R., Gadepally, V., Mattson, T.G.: Enabling query processing across heterogeneous data models: a survey. In: 2017 IEEE International Conference on Big Data (Big Data), pp. 3211–3220 (2017). https://doi.org/10.1109/BigData.2017.8258302

14. Vogt, M., et al.: Polypheny-DB: towards bridging the gap between polystores and HTAP systems. In: Gadepally, V., et al. (eds.) DMAH/Poly 2020. LNCS, vol. 12633, pp. 25–36. Springer, Cham (2021). https://doi.org/10.1007/978-3-030-71055-2_2

15. Wong, W.J., Yasuda, K., Chang, Q., Miyazaki, J.: A data model of a data lineage management system for database repair and simulation. In: Delir Haghighi, P., Greguš, M., Kotsis, G., Khalil, I. (eds.) IiWAS 2024, Part II, vol. 15343, pp. 243–248. Springer, Heidelberg (2024). https://doi.org/10.1007/978-3-031-78093-6_22

16. Yamada, H., Suzuki, T., Ito, Y., Nemoto, J.: ScalarDB: universal transaction manager for polystores. Proc. VLDB Endow. **16**(12), 3768–3780 (2023). https://doi.org/10.14778/3611540.3611563

Optimisation Methods

Group Trip Planning Query Problem
with Multimodal Journey

Dildar Ali, Suman Banerjee$^{(\boxtimes)}$, and Yamuna Prasad

Indian Institute of Technology Jammu, Jammu 181221, J & K, India
{2021rcs2009,suman.banerjee,yamuna.prasad}@iitjammu.ac.in

Abstract. In *Group Trip Planning (GTP) Query Problem*, we are given
a city road network where a number of *Point of Interest (PoI)* have been
marked with their respective categories (e.g., Cafeteria, Park, Movie The-
ater, etc.). A group of agents wants to visit one PoI from every category
from their respective starting location, and once finished, they want to
reach their respective destinations. This problem asks which PoI from
every category should be chosen so that the aggregated travel cost of the
group is minimized. This problem has been studied extensively in the last
decade, and several solution approaches have been proposed. However, to
the best of our knowledge, none of the existing studies have considered the
different modalities of the journey, which makes the problem more practical. To bridge this gap, we introduce and study the GTP Query Problem with Multimodal Journey in this paper. Along with the other inputs
of the GTP Query Problem, we are also given the different modalities of
the journey that are available and their respective cost. Now, the problem
is not only to select the PoIs from respective categories but also to select
the modality of the journey. For this problem, we have proposed an efficient solution approach, which has been analyzed to understand its time
and space requirements. A large number of experiments have been conducted using real-life datasets, and the results have been reported. From
the results, we observe that the PoIs and modality of journey recommended
by the proposed solution approach lead to much less time and cost than the
baseline methods.

Keywords: Group Trip Planning Query · Point of Interest · Dynamic
Programming · Optimization

1 Introduction

In recent times, due to the advancement of wireless internet and GPS-enabled
handheld mobile devices, capturing the location of a moving object has become
easier. This leads to the generation of a large number of datasets, and such
datasets lead to a different domain called *Spatial Databases and Spatial Data
Mining* [10,13]. One well-studied problem in the domain of spatial databases is
the *Group Trip Planning Query Problem* [1,8,10,15]. In this problem, we are
given a road network of a city where the vertex set is constituted by the set of

© The Author(s), under exclusive license to Springer Nature Switzerland AG 2026
R. Wrembel et al. (Eds.): DEXA 2025, LNCS 16047, pp. 147–162, 2026.
https://doi.org/10.1007/978-3-032-02088-8_11

PoIs, and the edge set is the road fragments connecting the PoIs. Also, the PoIs are classified into different categories, such as cafeterias, parks, movie theaters, restaurants, etc. A group of friends (referred to as agents in this paper) wants to travel from their respective starting locations to a destination location in the city, and during their journey, they want to visit one PoI from every category so that the aggregated travel cost by the group is minimized. This problem has been studied extensively in the last decade, and these can be broadly categorized into passenger-focused and transport manager-focused studies. The passenger-focused studies aim to optimize travel efficiency and experience by minimizing factors like travel time, cost, and number of transfers while considering comfort and convenience. Potthoff and Sauer [11,12] introduced the McTB approach, which optimizes three key criteria: arrival time, the number of public transit trips, and unrestricted transfer modes. Building on this, their HydRA algorithm [11] enhances query execution efficiency, supporting faster computations in multimodal networks. Delling et al. [4,6] developed efficient algorithms for multimodal routing and delays, incorporating transfer penalties and mode preferences. Further, RAPTOR, proposed by Delling et al. [5], focuses on minimizing transfer times and ensuring scalability in large public transportation networks. The transport manager-focused studies address system-wide efficiency, emphasizing operational optimization. Ceder [3] outlined methods for public transit scheduling and operations essential for integrating multiple transport modes.

Li et al. [10] was the first to study the trip planning query problem on metric graphs and proposed several approximate solutions. Subsequently, this problem has been extended to include a group of travelers instead of a single traveler by Hashem et al. [8]. Later, Ahmadi et al. [1] proposed a mixed search (both breadth and depth) strategy and used the progressive group neighbor exploration technique. Lee and Park [9] studied the trip planning problem to identify a typical meeting point such that the ride-sharing mechanism becomes effective. Additionally, several variants of the GTP Query Problem have been studied, such as incorporating PoI utility values [2] and fairness criteria [14].

To the best of our knowledge, the GTP Query Problem has not been studied considering the presence of multiple transport mediums. This paper bridges this gap by studying the GTP Query Problem with multiple transport mediums. In particular, we make the following contributions in this paper:

- We introduce and study a practical variant of the GTP Query Problem, considering the presence of multiple transport mediums. To the best of our knowledge, this is the first study in this direction.
- To address this problem, we introduce an optimal journey planning algorithm and analyze it to understand its time and space requirements.
- A large number of experiments have been carried out to establish the effectiveness and efficiency of the proposed solution approach.

The rest of this paper has been organized as follows. The background concepts and formal problem description have been described in Sect. 2. The proposed solution approach has been stated in Sect. 3. Section 4 contains the experiment evaluation of the proposed solution approach. Finally, Sect. 5 concludes our study and gives future research directions.

2 Background and Problem Definition

2.1 Road Network

In this study, we model a city road network using an undirected, weighted graph $\mathcal{G}(\mathcal{V}, \mathcal{E}, W)$, where the vertex set $\mathcal{V} = \{v_1, v_2, \ldots, v_n\}$ contains the set of PoIs in the city. The edge set $\mathcal{E}$ of $\mathcal{G}$ is constituted by the road segments joining the PoIs. Here, W is the edge weight function that maps each edge to the corresponding travel cost between the two PoIs joined by the corresponding road segment, i.e., $W : \mathcal{E} \longrightarrow \mathbb{R}^+$. For any edge $(v_i v_j)$, its weight is denoted by $W(v_i v_j)$. As per our problem context, 'vertex' and 'PoI' have been used interchangeably. Similarly, 'road segment' and 'edge' have been used interchangeably. For any vertex $v_i \in \mathcal{V}$, the neighbor of v_i is denoted by $N(v_i)$ and defined as the other PoIs that are directly linked with v_i, i.e., $N(v_i) = \{v_j : (v_i v_j) \in \mathcal{E}\}$. The cardinality of $N(v_i)$ is called the degree of v_i. A sequence of vertices $P = < v_i, v_{i+1}, \ldots, v_j >$ is said to be a path in $\mathcal{G}$ if $(v_k v_{k+1}) \in \mathcal{E}$ for all $i \leq k \leq j - 1$. For any path P in $\mathcal{G}$, let $\mathcal{V}(P)$ and $\mathcal{E}(P)$ denote the set of vertices and edges that constitute the path. The weight of the path P is defined as the sum of the edge weights of the edges that constitute the path, i.e., $W(P) = \sum\limits_{(v_p v_q) \in \mathcal{E}(P)} W(v_p v_q)$. For any two vertices v_i and v_j, the shortest path distance from v_i to v_j is denoted by $dist(v_i v_j)$. As $\mathcal{G}$ is undirected, $dist(v_i v_j) = dist(v_j v_i)$.

2.2 Group Trip Planning Query Problem

For any positive integer k, $[k]$ denotes the set $\{1, 2, \ldots, k\}$. In the GTP Query Problem, a set of ℓ many agents $\mathcal{U} = \{u_1, u_2, \ldots, u_\ell\}$ wants to travel from their respective source to the destination location. The source and destination location of the i-th agent is denoted by v_i^s and v_i^d, respectively. All the PoIs under consideration can be classified into one of k distinct categories. This can be formalized by the function $\mathcal{C} : \mathcal{V} \longrightarrow [k]$. For any $i \in [k]$, let $\mathcal{V}_i$ denotes the set of PoIs belongs to i-th category. Also, we assume that there does not exist any i such that $\mathcal{V}_i = \emptyset$. For simplicity, we assume that the source and destination locations of the agents are also part of the city road network, and each one is a PoI in $\mathcal{G}$. Hence, $\mathcal{V} = \{v_1^s, v_2^s, \ldots, v_\ell^s\} \cup \{v_1^d, v_2^d, \ldots, v_\ell^d\} \cup \mathcal{V}_1 \cup \mathcal{V}_2 \cup \ldots \cup \mathcal{V}_k$. Now, in our problem context, we define *Valid Path* in Definition 1.

Definition 1 (Valid Path). *For the agent u_i a path $< v_i^s, v_1, v_2, \ldots, v_k, v_i^d >$ in $\mathcal{G}$ is said to be a valid path if v_i^s and v_i^d are the source and destination location of u_i, and for all $j \in [k]$, $v_j \in \mathcal{V}_j$.*

We observed that in a valid path, every agent visits one PoI from every category in the predetermined sequence, and from the first to the k-th category, the path will be the same for all the agents. Consider for any $i \in [k]$, $|\mathcal{V}_i| = n_i$. Now, the number of valid paths will be $\prod\limits_{i \in [k]} n_i$. Let $\mathcal{P}$ contain the common portions of all valid paths. Now, for any valid path $p \in \mathcal{P}$, let $\mathcal{D}(p)$ denote the aggregated distance for the path. If the common path is $p =< v_1, v_2, \ldots, v_k >$ then the aggregated distance $\mathcal{D}(p)$ can be computed using Eq. 1.

$$\mathcal{D}(p) = \sum_{i \in [\ell]} dist(v_i^s v_1) + \sum_{i \in [k-1]} dist(v_i v_{i+1}) + \sum_{i \in [\ell]} dist(v_k v_i^d) \qquad (1)$$

Next, we formally define the GTP Query Problem, which has been formally stated in Definition 2.

Definition 2 (GTP Query Problem). *Given a city road network $\mathcal{G}(\mathcal{V}, \mathcal{E}, W)$ which contains all the PoIs of interest and classified into different categories, the GTP Query Problem asks to select one PoI from each category so that the aggregated travel cost of the group is minimized. This can be expressed using Eq. 2.*

$$p^* \longleftarrow \underset{p \in \mathcal{P}}{argmin}\ \mathcal{C}(p) \qquad (2)$$

Here, $p^* =< v_1^*, v_2^*, \ldots, v_k^* >$ denotes the optimal path.

2.3   Multi Modal Journey

Consider the city of our consideration, there exist p different modes of transport. In this paper, we use the terminology *'vehicle'* to refer to any conveyance used in any mode of transport. For any $i \in [p]$, let V_i denote the set of vehicles of i-th transport medium. Next, we state the notion of *route* in Definition 3.

Definition 3 (Route). *Given a city road network $\mathcal{G}(\mathcal{V}, \mathcal{E}, W)$, the route is defined as a path in the network such that there must exist at least one vehicle that covers that path in its journey.*

For any route $r =< v_i, v_{i+1}, \ldots, v_j >$, let $PoI(r)$ denote the set of PoIs that route r covers. $PoI_s(r)$ and $PoI_e(r)$ denote the starting and ending PoI of the route r. For any vehicle x that covers the route r, $t_s^x(r)$ and $t_e^s(r)$ denote the start and end time of the vehicle x for the route r, respectively.

Definition 4 (Journey Planning). *Given the city road network $\mathcal{G}(\mathcal{V}, \mathcal{E}, W)$ with the transport details and cost, two PoIs of $\mathcal{G}$ v_a and v_b, and a start time t the journey planning is defined as a sequence of routes along with the corresponding vehicles such that the following criteria got satisfied:*

- *The starting PoI of the first route and the ending PoI of the last route in the journey plan must be the start and destination of the journey.*
- *For any two consecutive routes, the starting time of the chosen vehicle of the second route should be more than the ending time of the chosen vehicle of the first route.*

In this study, we consider that for a specific modality (say i) for any two vehicles $x, y \in V_i$, such that both cover the route r, the cost of commuting between PoIs v_a and v_b by x or y will be the same. Now, we define the cost of a journey planning in Definition 5.

Definition 5 (Cost of a Journey Planning). *Given the transport details, the cost, and journey planning (i.e., the routes and the corresponding vehicles as stated in Definition 4), the cost of journey planning is defined as the sum of the costs of the routes that constitute the whole journey. For the journey planning* $\mathcal{R} = <r_1, r_2, \ldots, r_y>$, *and* $V = \{i, j, \ldots, k\}$ *the cost is denoted by* $\mathcal{C}(\mathcal{R})$ *and can be mathematically expressed as:*

$$\mathcal{C}(\mathcal{R}) = \sum_{r \in \mathcal{R}} \mathbb{C}(a, b, c) \tag{3}$$

where v_a *and* v_b *are the start and end PoI of the route* r, *and in the journey planning, this route will be traveled using a vehicle of c-th modality.*

Definition 6 (Minimum Cost Journey Planning). *Given a city road network* $\mathcal{G}(\mathcal{V}, \mathcal{E}, W)$, *the transport details (i.e., different modalities of journey with their respective routes and their associated cost) the minimum cost journey planning problem asks to choose a least cost journey planning. Mathematically, this problem can be posed as an optimization problem as follows:*

$$\mathcal{R}^* \longleftarrow \underset{\mathcal{R} \in \mathbb{R}}{argmin} \ \mathcal{C}(\mathcal{R}) \tag{4}$$

Here, $\mathbb{R}$ *denotes the set of all possible journey planning.*

Subsequently, we define the GTP Query Problem under the realm of multimodal journey and formally define our problem.

2.4   GTP Query Problem with Multi-modal Journey

In this paper, we introduce a variant of the GTP Query Problem where multiple modes of transportation exist in the city under consideration. Now, it can be observed that for every agent, the individual cost may be different, and this is due to the following reasons. Let, $<v_1^*, v_2^*, \ldots, v_k^*>$ the common path, however, for any i-th agent the path that it covers is $<v_i^s, v_1^*, v_2^*, \ldots, v_k^*, v_i^d>$. So, to commute from their respective source locations to v_1^* and from v_k^* to their respective destination location incur different costs. As the common portion of the trip, the modality of the journey has to be decided collectively. Now, we define the cost of commuting for the whole group in Definition 7.

Definition 7 (The Cost of the Group). *Given a GTP Query Problem instance, any of the transport details of the city along with the cost, the cost of commuting of the group can be defined as the sum of the following three costs:*

- *Sum of the individual costs for reaching the recommended PoI of the first category.*
- *Sum of the total cost incurred by the group from the first category of PoI to the k-th category of PoI.*
- *Sum of the individual costs for reaching the recommended destinations PoI from the k-th category of PoI.*

Hence, mathematically, this can be posed as follows

$$\mathcal{C}_R(\mathcal{U}) = \sum_{i \in [\ell]} \mathbb{C}(v_i^s, v_1^*, d_i) + \ell \cdot \sum_{j \in [k-1]} \mathbb{C}(v_j^*, v_{j+1}^*, d_j) + \sum_{s \in [\ell]} \mathbb{C}(v_k^*, v_i^d, d_s) \tag{5}$$

It can be observed that in Eq. 5, a subscript R has been used. This signifies this cost has been defined for a specific journey planning R (for all the agents collectively). Now, based on the group cost as defined in Eq. 5, we formally define the GTP Query with Multi-Modal Journey Problem in Definition 8.

Definition 8 (GTP Query with Multi Modal Journey Problem). *Given a GTP Query Problem instance along with the transport details of the city (i.e., details of different transport medium, routes, their corresponding costs, etc.), the GTP Query with Multi-Modal Journey Problem asks to recommend one PoI from every category, and the journey plan for all the agents such that the total travel cost as defined in Eq. 5 gets minimized. Mathematically, this problem can be posed as a discrete optimization problem as follows.*

$$\mathcal{R}^* \longleftarrow \underset{R \in \mathbb{R}}{argmin} \ \mathcal{C}_R(\mathcal{U}) \tag{6}$$

3 Proposed Approach

Given the city road network $\mathcal{G}(\mathcal{V}, \mathcal{E}, W)$, the transport details, and the set of PoIs, we propose a dynamic programming (DP) based solution approach that will return the minimum cost journey plan. The working of Algorithm 1 is as follows. In Line No. 1 to 5, we construct the multi-graph using the PoIs and initialize the categories of PoIs, source, destinations, and the DP dictionary. Next, in Line No. 6 to 19, the transition between source PoIs to the PoIs of the first intermediate category is performed. The transition between intermediate categories is performed, and a common path is selected in Line No. 20 to 33. Next, in Line No. 34 to 45, the transitions between the last intermediate category and the destinations are performed. Finally, in Line No. 46 to 47, the total minimum cost and the shortest paths [7] of each individual agent are extracted. In the intermediate paths, all the agents travel together, and the intermediate travel cost is multiplied by the number of agents (ℓ). The representation of PoIs in Algorithm 1 is as follows. Consider a set of ℓ many agents $\{v_1^s, v_2^s, \ldots v_\ell^s\}$ are the source vertices and $\{v_1^d, v_2^d, \ldots v_\ell^d\}$ are the destination vertices in the network $\mathcal{G}$ and the PoIs are categories into k types and they contains $p_1, p_2, \ldots, p_k$ many PoIs, respectively. Now, from source vertices to the first category of PoIs, each PoI contains $g \times \ell$ many variables to store the transport medium and costs, where g is the number of transport mediums. So, the first category contains total $p_1 \times g \times \ell$ many variables. It can be represented as a matrix below.

$$\mathcal{V}_{11} = \begin{bmatrix} \mathcal{V}_{11}^{(1,\mathcal{V}_1^s)} & \mathcal{V}_{11}^{(2,\mathcal{V}_1^s)} & \cdots & \mathcal{V}_{11}^{(n,\mathcal{V}_1^s)} \\ \mathcal{V}_{11}^{(1,\mathcal{V}_2^s)} & \mathcal{V}_{11}^{(2,\mathcal{V}_2^s)} & \cdots & \mathcal{V}_{11}^{(n,\mathcal{V}_2^s)} \\ \vdots & \vdots & \vdots & \vdots \\ \mathcal{V}_{11}^{(1,\mathcal{V}_\ell^s)} & \mathcal{V}_{11}^{(2,\mathcal{V}_\ell^s)} & \cdots & \mathcal{V}_{11}^{(n,\mathcal{V}_\ell^s)} \end{bmatrix} \quad \mathcal{V}_{21} = \begin{bmatrix} \mathcal{V}_{21}^{(1,\mathcal{V}_{11})} & \mathcal{V}_{21}^{(2,\mathcal{V}_{11})} & \cdots & \mathcal{V}_{21}^{(n,\mathcal{V}_{11})} \\ \mathcal{V}_{21}^{(1,\mathcal{V}_{12})} & \mathcal{V}_{21}^{(2,\mathcal{V}_{12})} & \cdots & \mathcal{V}_{21}^{(n,\mathcal{V}_{12})} \\ \vdots & \vdots & \vdots & \vdots \\ \mathcal{V}_{21}^{(1,\mathcal{V}_{1p_1})} & \mathcal{V}_{21}^{(2,\mathcal{V}_{1p_1})} & \cdots & \mathcal{V}_{21}^{(n,\mathcal{V}_{1p_1})} \end{bmatrix}$$

Here, $\mathcal{V}_{11}^{(1,\mathcal{V}_1^s)}$ contains the transport cost from vertex $\mathcal{V}_1^s$ to vertex $\mathcal{V}_{11}$ in the first category of PoIs via transport medium 1. Next, from the first category to the second category of PoIs, each PoI in the second category obtains g many variables. The second category contains a total of $p_2 \times g$ variables, and the cost of each transport medium will be multiplied by ℓ as a group travels with all the agents. Here, $\mathcal{V}_{21}$ is a PoI in the second category. Similarly, for all intermediate categories till the kth category, all PoIs contain g many variables. Finally, from the last category to the destination vertices, each obtains g many variables. In this case, the destination vertex cost will not be multiplied by the number of agents, as their destinations differ.

Complexity Analysis of Algorithm 1. Now, we analyze the time and space requirements of Algorithm 1. In-Line No. 1, constructing the multi-graph and ensuring connectivity by adding edges between disconnected components takes $\mathcal{O}(n + m)$. Next, in Line No. 2 to 3 scanning node attributes to classify PoIs into k categories takes $\mathcal{O}(n)$ and in Line No. 4 randomly assigning source v_i^s and destination v_i^d will take $\mathcal{O}(1)$ time. Initializing the DP dictionary in Line No. 5 will take $\mathcal{O}(n)$, and in Line No. 6, initialization of variables will take $\mathcal{O}(1)$ time. So, Line No. 1 to 6 will take $\mathcal{O}(n + m)$ time. Next, in Line No. 7 to 19 for each PoI $j \in \mathcal{P}[C_1]$ and each source $i \in v_i^s$, computing shortest path (using Dijkstra) will take $\mathcal{O}(n \log n + m \log n)$ and the total time requirement will be $\mathcal{O}(\ell \cdot p_1 \cdot n \log n + \ell \cdot p_1 \cdot m \log n)$, where ℓ and p_1 are the number of agents and number of PoIs in $\mathcal{P}[C1]$, respectively. In Line No. 20 to 33, for each PoI $j \in \mathcal{P}[C_c]$ and each PoI $i \in \mathcal{P}[C_{c-1}]$ computing shortest path will take $\mathcal{O}(n \log n + m \log n)$. Considering p_c is the number of PoIs in $\mathcal{P}[C_c]$ and p_{c-1} is the number of PoIs in $\mathcal{P}[C_{c-1}]$, the time requirements will be $\mathcal{O}(p_c \cdot p_{c-1} \cdot (n \log n + m \log n))$. Now, summing all the intermediate category $(k - 2)$, the total time requirements will be $\mathcal{O}((k-2) \cdot p \cdot p \cdot (n+m) \log n)$ i.e., $\mathcal{O}(k \cdot p^2 \cdot (n+m) \log n)$. Next, in Line No. 34 to 47 time requirements will be $\mathcal{O}(\ell \cdot p_k \cdot (n + m) \log n)$, where p_k is the number of PoIs in the last category. So, the overall time requirement for Algorithm 1 will be $\mathcal{O}(k \cdot p^2 \cdot (n + m) \log n)$.

Now, the graph $\mathcal{G}$ uses $\mathcal{O}(n + m)$ space, and the DP table stores costs for all PoIs in all categories, which will be $\mathcal{O}(k \cdot)$. Next, each path required $\mathcal{O}(p \cdot s)$ space, where s is the average length of the paths. Hence, the overall space requirements of Algorithm 1 will be $\mathcal{O}(n + m + k \cdot + p \cdot s)$, i.e., $\mathcal{O}(n + m + k \cdot)$.

An Illustrative Example. We consider the following scenario to demonstrate the working of the Algorithm 1. Let, we have a road network $\mathcal{G}(\mathcal{V}, \mathcal{E}, W)$ consists of 10 Point of Interest (PoIs) $v_1, v_2, \ldots, v_{10}$. The PoIs are categorized into five categories $\mathcal{C} = \{C_1, C_2, C_3, C_4, C_5\}$, where $C_1 = \{v_1, v_2\}$ (sources), $C_2 = \{v_3, v_4\}$ (first category), $C_3 = \{v_5, v_6\}$ (second category), $C_4 = \{v_7, v_8\}$ (third category), $C_5 = \{v_9, v_{10}\}$ (destinations). The edge weights W represent the travel costs, and each edge is labeled with multiple transport mediums like Bus (B), Car (C), Train (T), and Ferry (F). Figure 1 shows the road network and categories. Next, we initialize $DP[C_1][v_1] = 0$, $DP[C_1][v_2] = 0$ and set all other $DP[c][v_i] = \infty$. In the transition for category C_2, for each PoI in C_2, compute the minimum Cost from PoIs in C_1, considering all transport mediums. For example, for v_3, the Cost

---

Algorithm 1: An Optimal Journey Planning Algorithm

Data: The Road Network $\mathcal{G}(\mathcal{V}, \mathcal{E}, W)$, ℓ number of agents, transport details, the set of Source and Destination Points of Interest (PoIs) v_i^s and v_i^d, the set of PoI categories $\mathcal{C}$, and the set of transport mediums M.

Result: A journey (routes and corresponding vehicles) connecting the PoIs v_i^s and v_i^d with minimum cost.

```
/* Step 1: Graph Construction and Initialization                              */
```
1 Construct a multi-graph $\mathcal{G}$ using transport details, with nodes $\mathcal{V}$ and edges $\mathcal{E}$;
2 Identify the set of categories $\mathcal{C} = \{C_1, C_2, \ldots, C_k\}$ from node attributes;
3 Categorize PoIs into $\mathcal{P}[c]$ for each category $c \in \mathcal{C}$, ensuring intermediate categories have a fixed number of PoIs;
4 Randomly assign sources $v_i^s \in C_1$ and destinations $v_i^d \in C_k$;
5 Initialize a DP dictionary $DP[c][i] \leftarrow \infty$ for minimum costs to PoIs in category c from source v_i^s;

```
/* Step 2: Base Case (First Category)                                         */
```
6 $First\_Category\_Cost = \infty$, $Chosen\_PoI = \emptyset$, $Best\_Path = [\ ]$;
7 **for** *each PoI* $j \in \mathcal{P}[C_1]$ **do**
8 | $total\_cost = 0$; $path = [\ ]$;
9 | **for** *each source* i *in* v_i^s **do**
10 | | **if** *a path exists from* i *to* j **then**
11 | | | $path \leftarrow \text{ShortestPath}(\mathcal{G}, i, j, W)$;
12 | | | $path\_cost \leftarrow \text{ShortestPathCost}(\mathcal{G}, i, j, W)$;
13 | | | $total\_cost \leftarrow total\_cost + path\_cost$;
14 | | | *update path*;
15 | **if** $total\_cost < First\_Category\_Cost$ **then**
16 | | $First\_Category\_Cost \leftarrow total\_cost$;
17 | | $Chosen\_PoI \leftarrow j$;
18 | | $Best\_Path \leftarrow path$
19 $DP[C1][j] \longleftarrow \{PoI \leftarrow Chosen\_PoI, cost \leftarrow First\_Category\_Cost, paths \leftarrow Best\_Path\}$

```
/* Step 3: Transition for Intermediate Categories                            */
```
20 **for** *each category* $c \in \{2, \ldots, k-1\}$ **do**
21 | $min\_cost = \infty$, $Chosen\_PoI = \emptyset$, $Best\_Path = [\ ]$;
22 | **for** *each PoI* $j \in \mathcal{P}[C_c]$ **do**
23 | | **for** *each PoI* $i \in \mathcal{P}[C_{c-1}]$ **do**
24 | | | **if** *a path exists from* i *to* j **then**
25 | | | | $path \leftarrow \text{ShortestPath}(\mathcal{G}, i, j, W)$;
26 | | | | $path\_cost \leftarrow DP[c-1][j][cost] + \text{ShortestPathCost}(\mathcal{G}, i, j, W)$;
27 | | | | $total\_cost \leftarrow total\_cost + path\_cost$;
28 | | | | *update path*;
29 | | **if** $total\_cost < min\_cost$ **then**
30 | | | $min\_cost \leftarrow total\_cost$;
31 | | | $Chosen\_PoI \leftarrow j$;
32 | | | $Best\_Path \leftarrow path$
33 | $DP[c][j] \longleftarrow \{PoI \leftarrow Chosen\_PoI, cost \leftarrow min\_cost, paths \leftarrow Best\_Path\}$

```
/* Step 4: Transition to Destinations                                        */
```
34 $last\_PoI = DP[k-1][j][PoI]$;
35 $optimal\_paths = [\ ]$;
36 $group\_path \leftarrow DP[k-1][j][path]$;
37 $shared\_cost \leftarrow DP[k-1][j][cost]$;
38 $group\_trip\_cost \leftarrow shared\_cost \times \ell$;
39 $individual\_cost\_sum \leftarrow 0$;
40 **for** *each destination* $j \in v_i^d$ **do**
41 | **if** *a path exists from* $last\_PoI$ *to* j **then**
42 | | $path \leftarrow group\_path + \text{ShortestPath}(\mathcal{G}, last\_PoI, j, W)$;
43 | | $individual\_cost \leftarrow \text{ShortestPathCost}(\mathcal{G}, last\_PoI, j, W)$;
44 | | $individual\_cost\_sum \leftarrow individual\_cost\_sum + individual\_cost$;
45 | | $optimal\_paths.append(path)$;

```
/* Step 5: Final Total Cost for Group                                        */
```
46 $total\_cost\_for\_all\_agents \leftarrow group\_trip\_cost + individual\_cost\_sum$;
47 **return** $total\_cost\_for\_all\_agents, optimal\_paths$;

from v_1 via Bus is 5, and via Train is 7. The Cost from v_2 via Car is 8, and via Ferry is 12. Next, for v_4, cost from v_1 via Car is 3, via Bus is 6. The Cost from v_2 via Ferry is 10, and via Train is 9. So, minimum cost if v_1 and v_2 want to select v_3 is $(8+5) = 13$ and if v_4 is chosen then cost will be $(10+3) = 13$. The updated value in DP table will be $DP[C_2][v_3] = 13$ and $DP[C_2][v_4] = 13$. Now, from the second category to the third category, the minimum Cost for v_3 to v_5 is 4 via Train, and v_3 to v_6 via Bus is 6. Similarly, the cost for v_4 to v_5 and v_6 are 7 and 5, respectively. So, updated value is $DP[C_3][v_5] = 17$ and $DP[C_3][v_6] = 18$. For the fourth category the updated values are $DP[C_4][v_7] = 20$ and $DP[C_4][v_8] = 21$. Finally in the last category $DP[C_5][v_9] = 25$ and $DP[C_5][v_{10}] = 24$. So, in the intermediate category, two possible minimum cost paths are $v_3 \rightarrow v_5 \rightarrow v_7$ and $v_4 \rightarrow v_6 \rightarrow v_7$. Both paths take Cost $= 20$, and these intermediate paths are common for all the agents. Hence, for agent 1 who starts journey from v_1 to v_{10} the possible path is $v_1 \rightarrow v_3 \rightarrow v_5 \rightarrow v_7 \rightarrow v_{10}$ and the cost is 29. Similarly for agent 2, possible path will be $v_2 \rightarrow v_3 \rightarrow v_5 \rightarrow v_7 \rightarrow v_9$ and travel cost is 31. So, the total group trip cost will be $(20 + (5 + 4 + 6 + 5)) = 40$.

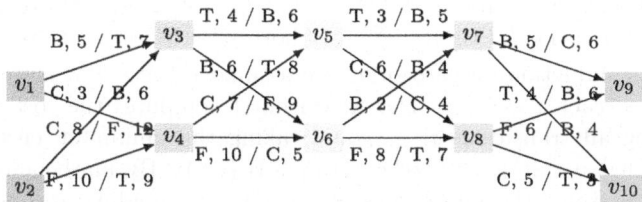

Fig. 1. Road network with categories C_1 to C_5.

Theorem 1. *The Algorithm 1 always terminates within a finite number of steps for any valid input graph $\mathcal{G}$.*

Proof. Let $|\mathcal{V}|$ be the number of PoIs and $|\mathcal{E}|$ be the number of edges in the graph $\mathcal{G}$. Each PoI and edge from the input is processed once to construct $\mathcal{G}$, which takes $\mathcal{O}(|\mathcal{V}| + |\mathcal{E}|)$. Let $C_1, C_2, \ldots, C_p$ denote the p connected components of the graph. At most $p - 1$ edges are added to connect these components, which is a finite operation. For each category $\mathcal{C} = \{C_1, C_2, \ldots, C_k\}$, the algorithm iterates through all pairs of PoIs $i \in C_{c-1}, j \in C_c$. Let $|C_c|$ denote the number of PoIs in category C_c. Then the pairwise iterations across all categories take at most $\sum_{c=2}^{k} |C_{c-1}| \cdot |C_c| \leq |\mathcal{V}|^2$. Next, the shortest path computations are performed using Dijkstra's algorithm [7] with complexity $O(|\mathcal{E}| \cdot \log |\mathcal{V}|)$ for each source-target pair and there are $|C_{c-1}| \cdot |C_c|$ such pairs, the total complexity for path computations is $\mathcal{O}(|\mathcal{V}|^2 \cdot |\mathcal{E}| \cdot \log |\mathcal{V}|)$. The dynamic programming updates for costs and paths iterate over $|C_c|$ PoIs per category, taking $O(|\mathcal{V}| \cdot |\mathcal{C}|)$. Each algorithm stage involves a finite number of operations. Thus, the Algorithm 1 terminates.

Theorem 2. *If the input graph $\mathcal{G}$ is connected, the Algorithm 1 guarantees a feasible journey from sources v_i^s to destinations v_i^d.*

Proof. Let $\mathcal{G} = (\mathcal{V}, \mathcal{E}, W)$ be the input graph. If $\mathcal{G}$ is not connected, the algorithm identifies connected components $\mathcal{C}_1, \mathcal{C}_2, \ldots, \mathcal{C}_p$, which ensures $\bigcup_{i=1}^{p} \mathcal{C}_i = \mathcal{V}$, $\mathcal{C}_i \cap \mathcal{C}_j = \emptyset$ for $i \neq j$. For each pair of disconnected components, an edge is added between a PoI $v_i \in \mathcal{C}_i$ and $v_j \in \mathcal{C}_j$. At most $p-1$ edges are added, ensuring that the graph becomes connected, i.e., $\text{diameter}(\mathcal{G}) < \infty$. For each source v_i^s, the algorithm computes the shortest path to every PoI in $\mathcal{C}_1$ using Dijkstra's algorithm, which guarantees $\exists \text{path}(v_i^s, v_j) \; \forall v_j \in \mathcal{C}_1$. By induction on the categories, we assume a path exists from v_i^s to any $v_j \in \mathcal{C}_{c-1}$. The algorithm computes the shortest path from v_j to all $v_k \in \mathcal{C}_c$, ensuring $\exists \text{path}(v_i^s, v_k) \; \forall v_k \in \mathcal{C}_c$. In the final PoI to destinations, the algorithm computes the shortest path from the last PoI in $\mathcal{C}_k$ to each destination v_i^d, ensuring $\exists \text{path}(v_i^s, v_i^d) \; \forall v_i^d \in \mathcal{V}$. Thus, a feasible journey exists for all source-destination pairs.

Theorem 3. *The Algorithm 1 computes a journey with the minimum total cost between sources v_i^s and destinations v_i^d.*

Proof. Let $\mathcal{G} = (\mathcal{V}, \mathcal{E}, W)$ be the input graph and $DP[c][j]$ represent the minimum cost to reach PoI $j \in \mathcal{C}_c$ from any source v_i^s. The recurrence relation for DP is $DP[c][j] = \min_{i \in \mathcal{C}_{c-1}} \left(DP[c-1][i] + \text{ShortestPathCost}(i,j) \right)$. For the base case $(c = 1)$, $DP[1][j] = \sum_{i \in v_i^s} \text{ShortestPathCost}(i,j)$. Now, applying mathematical induction on categories, we assume $DP[c-1][i]$ stores the minimum cost to reach $i \in \mathcal{C}_{c-1}$. The Algorithm 1 computes the cost for $j \in \mathcal{C}_c$ by evaluating all transitions $i \rightarrow j$ and taking the minimum, ensuring optimality $DP[c][j] = \min_{i \in \mathcal{C}_{c-1}}(DP[c-1][i] + W(i,j))$. By mathematical induction, $DP[c][j]$ stores the minimum cost for all paths ending at $j \in \mathcal{C}_c$. Now, for final transitions to the destinations, for each destination v_i^d, the total cost isTotalCost $= \sum_{j \in \mathcal{C}_k} DP[k][j] + \sum_{j \in v_i^d} W(j, v_i^d)$. Since all pairwise costs are minimized in the DP updates, the final cost is also minimized. Thus, the algorithm computes the globally optimal journey.

4   Experimental Evaluation

Dataset Description. We evaluate our proposed approach with two different real-world datasets. First, the networks of Switzerland, which were previously used by Sauer et al. [11,12], and the second is the road network of Helsinki[1] city of Finland. The public transit network for Switzerland and Finland is collected from the GTFS feed[2]. The Switzerland network covers the timetable of two business days, and the road networks were obtained from OpenStreetMap[3]. The Finland dataset includes travel time and distance between all SYKE (Finnish Environment Institute), calculated for walking, cycling, public transport, and car travel [16]. In Switzerland network we merged seven different feeds 'Bus', 'Train', 'Ferry', 'Funicular', 'Gondola', 'Subway' and 'Tram' into a single transport network. We categorize the PoIs into ten distinct categories 'Train Station',

---

[1] https://welcome.hel.fi/.
[2] https://gtfs.geops.ch/.
[3] https://download.geofabrik.de/.

Table 1. Description of Switzerland Dataset

Transport Medium	Base Fare (CHF)	Cost per Meter (CHF)	Cost per Minute (CHF)
Bus	2.50–4.00	0.01–0.03	0.05–0.10
Tram	2.50–4.00	0.01–0.03	0.05–0.10
Train	∼5.00	0.03–0.05	0.10–0.15
Ferry	5.00–10.00	0.05–0.10	0.15–0.25
Funicular	1.30–5.00	0.02–0.04	0.10–0.15
Gondola	5.00–15.00	0.05–0.15	0.20–0.50
Subway	2.50–4.00	0.01–0.03	0.05–0.10

'Public Square', 'City Center', 'Bridge', 'School', 'Park', 'Bus Stop', 'Airport', 'Healthcare Facility', and 'Hotel'. Further, we compute the travel cost for each medium of transport using (Travel Cost = Base Fare + (Cost per minute * Travel Time) + (Cost per meter * Travel Distance)), the information provided in Table 1. In the case of the Finland dataset, there are complete transport road networks for the city of Helsinki. However, in our problem context, we have taken a small portion of the road networks that contain 1100 unique PoIs. One point to be highlighted is that the fares given in Table 1, 2 are the approximate fares we assumed for our problem context. An overview of the networks is given in the Table 3.

Table 3. Dataset Description

Attributes	Switzerland	Finland
Stops	44557	1100
Routes	168294	4840000
Vertices	1310	1100
Edges	11370	604950

Table 2. Description of Finland Dataset

Transport Medium	Base Fare (EUR)	Cost per Meter (EUR)	Cost per Minute (EUR)
Bike	0.00–5.00	0.00–0.10	0.05–0.10
Public Transport	3.20	0.03–0.05	0.05–0.10
Private Car (Taxi)	5.90	0.01–0.05	0.74

Environment and Key Parameter Setup. The proposed and baseline methods are implemented in Python using the Jupyter Notebook Platform. All the experiments are conducted in a Ubuntu-operated desktop system with 64 GB RAM and an Xeon(R) 3.5 GHz processor. Next, we vary the number of agents ($|\mathcal{U}|$) by 5, 10, 20, 50, 100 to show the effectiveness and efficiency of the proposed solution approach. We vary the number of intermediate PoI categories (5, 10, 20) to evaluate scalability, defaulting to 10 for reporting. All experiments are averaged over three runs.

Baseline Methodologies.

- **Random PoI and Random Medium (RPRM).** In this approach, from source to destination via intermediate categories of PoI, the PoIs are selected randomly, and the transport medium between two PoIs is chosen randomly.

- **Random PoI and Random Medium (RPRM).** In this approach, from source to destination via intermediate categories of PoI, the PoIs are selected randomly, and the transport medium between two PoIs is chosen randomly.
- **Random PoI and Cheapest Medium (RPCM).** This approach selects PoIs randomly from source to destination via intermediate categories of PoIs; however, it considers the cheapest transport medium to visit one PoI to others.
- **Nearest Neighbor PoI and Cheapest Medium (NNCM).** This approach considers the nearest PoI in the first category of PoIs from the source using the cheapest transport medium, and from then on, it selects one PoI from each category, considering the same till the destination.

Goals of our Experiments. The following research questions (RQ) are our focus in this study.

- **RQ1**: Varying agents, how do the travel cost and computational time change?
- **RQ2**: Varying agents, how does the usage of transport medium change?
- **RQ3**: Varying number of PoI in each category, how do the computational time and transport medium change?

Experimental Results and Discussions. In this section, we will address the research questions posed in Sect. 4 and discuss the experimental results.

No. of Agents vs. Travel Cost. In this work, we vary the number of agents by 5 to 100 to evaluate the transport cost via the different journey mediums. In Figs. 3(a) and 3(c), travel costs rise with the increasing number of agents, significantly driven by segments from the source to the first category and from the last category to the destinations. On the Finland dataset, the proposed 'OJPA' consistently outperforms baseline methods ('NNCM', 'RPCM', 'RPRM') in terms of cost. Among the baselines, 'NNCM' achieves the lowest cost by always selecting the nearest PoI and cheapest travel medium, whereas 'RPRM' incurs the highest cost due to its random selection strategy. Similar observations are noted for the Switzerland dataset, as shown in Fig. 3(c). For instance, on the Finland dataset, with 5 agents, travel costs for 'OJPA', 'NNCM', 'RPCM', and 'RPRM' are 42.75 €, 160.765 €, 237.5 €, and 244.6 €, respectively. With 100 agents, the costs rise to 745.75 €, 2569.23 €, 4156.25 €, and 4156.25 €, respectively. For the Switzerland dataset, as the number of agents increases from 5 to 100, the travel costs for 'OJPA', 'NNCM', 'RPCM', and 'RPRM' rise sharply from $30,701.25$ CHF, $68,861.25$ CHF, $308,353.75$ CHF, and $204,588.75$ CHF to $15,005,950$ CHF, $75,782,150$ CHF, $103,046,875$ CHF, and $70,440,000$ CHF, respectively. The substantially higher costs compared to the Finland dataset result from the lack of direct paths between PoIs in Switzerland's road network, whereas such direct connections are more commonly available in Finland.

No. of Agents vs. Time. To determine the computational time required, we vary the number of agents and observe that the run time is proportional to the number of agents. We have compared our proposed approach with baseline methods like 'NNCM,' 'RPCM,' and 'RPRM'; among them, 'NNCM' takes more time than

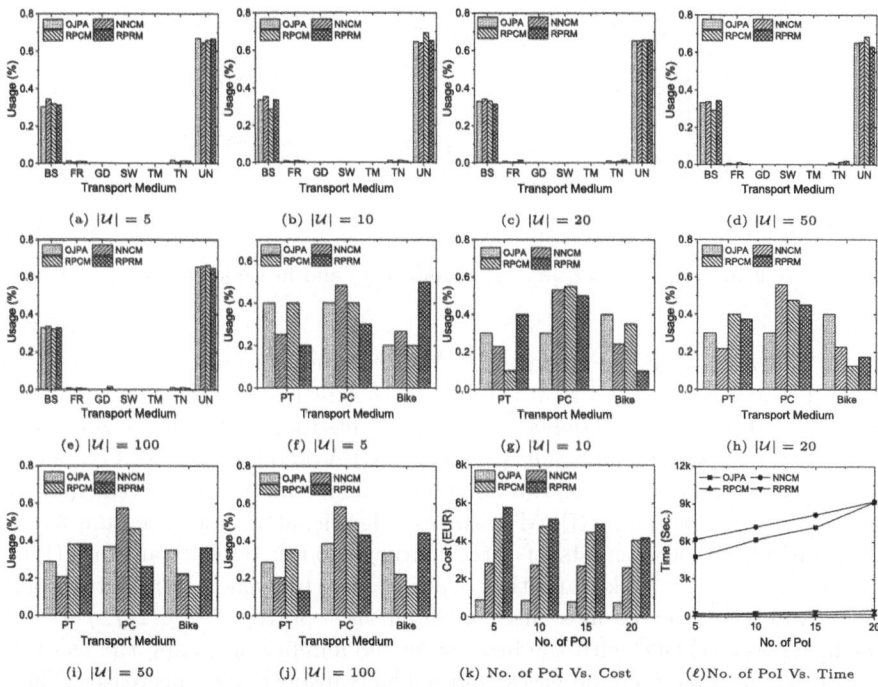

Fig. 2. Varying $|\mathcal{U}|$ Vs. Usage of transport medium in Switzerland (a, b, c, d, e) and in Finland (f, g, h, i, j), Varying No. of PoI Vs. Cost (k) and Varying No. of PoI Vs. Time (ℓ) for Switzerland dataset

the other baseline methods. This happens because 'NNCM', finds the nearest PoI from one category to the PoIs from another. It is also considered the cheapest medium of the journey, which leads to huge computational time. However, the 'RPCM' and 'RPRM' both methods use randomization to select PoIs from one category to another, leading to very less runtime, and it is negligible compared to the other methods. These observations are consistent with both the Switzerland and Finland datasets, as shown in Fig. 3(b) and 3(d), respectively. For example, in the Switzerland dataset, when we vary agents from 5 to 100, the runtime for 'OJPA', 'NNCM', 'RPCM', and 'RPRM' varies from 24, 25, 2, 2 to 546, 605, 58, 53 in seconds, respectively.

No. of Agents vs. Medium Usage. Figure 2(a) to 2(j) illustrates how the usage of different transport mediums varies with the number of agents in the Switzerland dataset, considering mediums such as bus (BS), ferry (FR), gondola (GD), subway (SW), tram (TM), and train (TN). As the number of agents increases, the usage of transport mediums also rises, but only buses, ferries, gondolas, and trams are utilized across both proposed and baseline methods. Among these, the bus is consistently the most frequently used medium, whereas the gondola is the least utilized. Figures 2(a)–(e) illustrate the average usage of travel mediums,

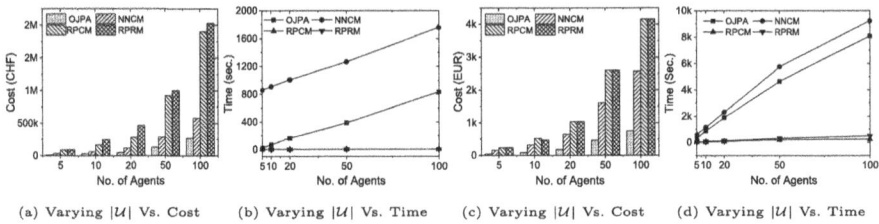

(a) Varying $|\mathcal{U}|$ Vs. Cost   (b) Varying $|\mathcal{U}|$ Vs. Time   (c) Varying $|\mathcal{U}|$ Vs. Cost   (d) Varying $|\mathcal{U}|$ Vs. Time

Fig. 3. Varying $|\mathcal{U}|$ in Switzerland (a, b) and in Finland (c, d)

clearly indicating that the Unknown (UN) medium has the highest usage. This occurs because the original road network lacks sufficient connectivity, requiring additional edges to form a connected graph. Agents frequently select these added edges for their optimal paths. The proposed 'OJPA' and baseline 'NNCM' methods show similar usage patterns for different travel mediums, whereas the random selection in 'RPCM' and 'RPRM' results in significantly varied medium usage. The Finland dataset consists of three journey modes: public transport (PT), private car (PC), and bike, and they are used very frequently. Public transport and private cars are the most used medium, as reported in Fig. $2(f, g, h, i, j)$. We have observed that with the increase in the number of agents, the usage of private cars as a medium increases, and public transport usage decreases. These observations are consistent with the Switzerland dataset.

Additional Discussion. Additionally, we have experimented with varying numbers of PoIs in each category to check the run time and usage of different media during the journey. We fixed the number of agents as 100 and varied the number of PoIs in each category by $5, 10, 15$, and 20, and the experimental results are reported in Fig. $2(k, \ell)$. We have observed that the computational time increases rapidly with the number of agents. Now, in the case of medium usage, minor changes happen in the case of the 'OJPA' and 'NNCM' approaches. However, major changes in usage occur in the 'RPCM', and 'RPRM'. We have also varied the number of intermediate categories by $5, 10, 20$ and observed that with an increasing number of categories, the computational cost increases rapidly. One point needs to be noted in the Switzerland and Finland datasets: we have considered an equal number of PoIs in each category, including the source and destination categories for all our experiments, as default settings.

5   Concluding Remarks

In this paper, we studied the GTP query problem considering multiple commuting transport mediums, a direction previously unexplored to our knowledge. We demonstrated that increasing the number of PoI categories makes the problem computationally intractable. We proposed a dynamic programming-based solution and analyzed its time and space complexity to address this. Extensive experiments on publicly available benchmark datasets revealed that our method consistently outperforms several baseline approaches. An important future research direction includes addressing fairness concerns, as minimizing the aggregated group cost may disproportionately affect individual agents.

References

1. Ahmadi, E., Nascimento, M.A.: A mixed breadth-depth first search strategy for sequenced group trip planning queries. In: 2015 16th IEEE International Conference on Mobile Data Management, vol. 1, pp. 24–33. IEEE (2015)
2. Barua, S., Jahan, R., Ahmed, T.: Weighted optimal sequenced group trip planning queries. In: 2017 18th IEEE International Conference on Mobile Data Management (MDM), pp. 222–227. IEEE (2017)
3. Ceder, A.: Public Transit Planning and Operation: Modeling, Practice and Behavior. CRC Press (2016)
4. Delling, D., Giannakopoulou, K., Wagner, D., Zaroliagis, C.: Timetable information updating in case of delays: modeling issues. Arrival Technical Report (2008)
5. Delling, D., Pajor, T., Werneck, R.F.: Round-based public transit routing. Transp. Sci. **49**(3), 591–604 (2015)
6. Delling, D., Sanders, P., Schultes, D., Wagner, D.: Engineering route planning algorithms. In: Lerner, J., Wagner, D., Zweig, K.A. (eds.) Algorithmics of Large and Complex Networks. LNCS, vol. 5515, pp. 117–139. Springer, Heidelberg (2009). https://doi.org/10.1007/978-3-642-02094-0_7
7. Dijkstra, E.W.: A note on two problems in connexion with graphs. In: Edsger Wybe Dijkstra: His Life, Work, and Legacy, pp. 287–290 (2022)
8. Hashem, T., Hashem, T., Ali, M.E., Kulik, L.: Group trip planning queries in spatial databases. In: Nascimento, M.A., et al. (eds.) SSTD 2013. LNCS, vol. 8098, pp. 259–276. Springer, Heidelberg (2013). https://doi.org/10.1007/978-3-642-40235-7_15
9. Lee, J., Park, S.: The collective trip planning query processing using G-tree index structure. In: 2020 IEEE International Conference on Big Data and Smart Computing (BigComp), pp. 173–180. IEEE (2020)
10. Li, F., Cheng, D., Hadjieleftheriou, M., Kollios, G., Teng, S.-H.: On trip planning queries in spatial databases. In: Bauzer Medeiros, C., Egenhofer, M.J., Bertino, E. (eds.) SSTD 2005. LNCS, vol. 3633, pp. 273–290. Springer, Heidelberg (2005). https://doi.org/10.1007/11535331_16
11. Potthoff, M., Sauer, J.: Efficient algorithms for fully multimodal journey planning. In: 22nd Symposium on Algorithmic Approaches for Transportation Modelling, Optimization, and Systems (ATMOS 2022). Schloss-Dagstuhl-Leibniz Zentrum für Informatik (2022)
12. Sauer, J., Wagner, D., Zündorf, T.: An efficient solution for one-to-many multimodal journey planning. In: 20th Symposium on Algorithmic Approaches for Transportation Modelling, Optimization, and Systems (ATMOS 2020). Schloss-Dagstuhl-Leibniz Zentrum für Informatik (2020)
13. Shekhar, S.: Spatial Databases. Pearson Education India (2007)
14. Solanki, N., Jain, S., Banerjee, S., Kumar S, Y.P.: Fairness driven efficient algorithms for sequenced group trip planning query problem. In: Proceedings of the 2023 International Conference on Autonomous Agents and Multiagent Systems, pp. 86–94 (2023)

15. Tabassum, A., Barua, S., Hashem, T., Chowdhury, T.: Dynamic group trip planning queries in spatial databases. In: Proceedings of the 29th International Conference on Scientific and Statistical Database Management, pp. 1–6 (2017)
16. Tenkanen, H., Toivonen, T.: Longitudinal spatial dataset on travel times and distances by different travel modes in Helsinki region. Sci. Data **7**(1), 77 (2020)

A Model-Based Approach for Simple Construction and Efficient Evaluation of Dataframes

Konstantina Zouni[1], Ioanna Moraiti[1], Sotirios Angelopoulos[1],
Damianos Chatziantoniou[1], and Verena Kantere[2(✉)]

[1] Athens University of Economics and Business, 76, 104 34 Athens, Greece
{kzouni,imoraiti,sangelopoulos,damianos}@aueb.gr
[2] University of Ottawa, 75 Laurier Ave E, Ottawa, ON K1N 6N5, Canada
vkantere@uottawa.ca

Abstract. In the dynamic field of data analytics, handling diverse, large-scale datasets poses substantial obstacles, with traditional methods, like data warehousing and manual programming, proving insufficient due to limitations in flexibility and efficiency. Dataframes, facilitated by libraries like Pandas, have gained significant popularity. However, their utility is hampered by limitations in quick construction and modification of queries by non-programming experts, as well as handling large datasets and lacking support for big data without additional libraries. To address these gaps, we introduce a novel model-based approach for dataframe analysis based on the Data Virtual Machine (DVM) framework. Building on this framework, we facilitate the construction and analysis of dataframes queries without requiring programming expertise, lowering the entry barrier for beginners and streamlining the analysis process for experts. The proposed tool facilitates easy formulation and modification of dataframe queries, efficient query evaluation, and demonstrates superior performance compared to traditional Pandas implementation as evidenced by our experimental results.

Keywords: dataframes · data virtual machines · efficient query evaluation

1 Introduction

Modern analytics environments grapple with the complexity of managing diverse datasets originating from various applications. Over the last decade, the definition of Big Data has evolved rapidly. The Three V's - Volume, Variety, and Velocity - serve as a common framework to describe Big Data, representing its high-volume, high-velocity, and high-variety nature [1]. The high-variety datasets exhibit heterogeneity not only in their semantics, but also in the data structures that store them and the systems that manage and process them possessing significant challenges [2]. To effectively address these challenges, multiple data systems and analysis platforms must coexist, integrated and federated. While data

© The Author(s), under exclusive license to Springer Nature Switzerland AG 2026
R. Wrembel et al. (Eds.): DEXA 2025, LNCS 16047, pp. 163–177, 2026.
https://doi.org/10.1007/978-3-032-02088-8_12

warehousing has been a common approach, it becomes rigid in rapidly changing data environments. In high evolving environments, creating comprehensive classical integrated schemes encompassing structured, semi-structured, and unstructured data proves to be not only highly time- and resource-intensive but often unattainable within the time constraints of analysis [4]. As an alternative, many production environments adopt a programming language (e.g., Python, R, Scala) to extract, transform, and analyze data in dataframes.

Dataframe frameworks, like Pandas, have become increasingly popular in the field of data science and analysis. They are widely used to represent, transform, and analyze data efficiently [14]. The integration of dataframes with programming languages has made them accessible to a broader audience, leading to their widespread adoption [15]. Additionally, the rise of Big Data technologies has further contributed to the popularity of dataframes. Distributed computing frameworks like Apache Spark leverage dataframes for parallel processing on large datasets, enabling faster data processing and analysis at scale [16]. At the same time, it is well-known that dataframe systems like pandas are non-interactive on moderate-to-large datasets, and break down completely when operating on datasets beyond main memory. When dealing with even moderately large datasets that exceed memory capacity, the performance of the pandas DataFrame API declines significantly [3]. Complex operations like groupby and join can also be slow on large DataFrames. The lack of Pandas in native support for parallel or distributed computing, makes it less suitable for large-scale data processing tasks. Users dealing with larger datasets may find alternative libraries like Dask, Vaex, or Apache Spark more appropriate [8]. These alternative libraries may present increased complexity. Beginners without prior experience often get overwhelmed in an attempt to get hold of these tools. Additionally, experts spend a considerable amount of time and energy daily writing the same code repeatedly to perform data analysis across different projects. Despite the widespread adoption of dataframes, there is a pressing need for a tool that empowers users lacking programming skills to conduct dataframe analysis, while also providing superior performance in handling big data and complex queries.

Contributions. Our work addresses the urgent requirement for enhanced data analysis capabilities by introducing a novel model-based approach for data- frame analysis upon the innovative Data Virtual Machine (DVM) framework. This framework revolutionizes query construction by representing queries in a graph structure, where nodes represent attribute domains and edges signify mappings between these domains. Our contributions are threefold:

i) We streamline the process of query construction and modification, democratizing with our approach data analysis for non-programmers, while also offering the possibility for easy integration with an intuitive interface to facilitate no-code query formulation.
ii) Our approach boasts efficient query evaluation process, particularly adept at handling large-scale datasets from diverse data sources and complex queries.
iii) We validate our approach by comparing it with traditional Pandas implementation in Python, demonstrating superior performance.

The subsequent sections delve into the motivation behind this paper, present the foundational concept of DVM, explore the context and semantics of Python dataframes, detail the DVM dataframe queries, describe the efficient query construction and evaluation process, present an experimental comparison between our approach and Pandas implementation, and conclude with the implications and potential future directions of our research.

2   Motivation

In contemporary data analysis practices, the widespread adoption of dataframes has surged due to their adaptability and intuitive interface. Despite their popularity, there exists a clear demand for a solution capable of seamlessly integrating with diverse data sources within organizations. This solution should empower users without programming expertise to efficiently conduct data analysis tasks using dataframes. It should cater to novice users by minimizing entry barriers while boosting the productivity of experienced analysts, enabling them to focus on analysis rather than repetitive coding tasks. At the same time this solution should offer enhanced performance in handling big datasets and complex queries and eliminate the requirement for additional libraries or manual optimization.

Motivating Example. To underscore the practical significance of our research, we offer a compelling example centered on real-world big data analysis, focusing on New York City taxi rides. Consider a scenario where a taxi company aims to optimize its operations and enhance customer satisfaction. They possess vast datasets encompassing detailed information on taxi rides, driver profiles, and passenger reviews. These datasets are illustrated in the ER model presented in Fig. 1. Our aim is to formulate a motivating query aimed at evaluating the performance of taxi drivers. Taxi driver performance can be assessed using various metrics, including trip completion rates, average trip duration, driver earnings, and customer ratings. Additionally, segmenting drivers by demographics allows for a deeper understanding of performance discrepancies.

Motivating Query. *Aggregate taxi ride data for each driver such as trip count and total distance, integrate sentiment analysis average scores from passenger reviews, and filter drivers based on age, gender, and vehicle model.*
This motivating example effectively demonstrates the research contributions. It showcases the complexity of integrating data from diverse sources, including various APIs, repositories, and data formats, reflecting real-world challenges encountered by organizations. With reference to New York City taxi rides, which surpassed 4 million in just the first two weeks of 2022, it highlights the need for enhanced performance in handling large volumes of data. Moreover, it demonstrates the approach's capability to manage complex data manipulations and analyses, such as mapping taxi rides with drivers, applying attribute-based filters and aggregating trip data. Finally, this practical use case resonates across a wide audience, ranging from researchers to data analysts and organizations involved in transportation or urban planning.

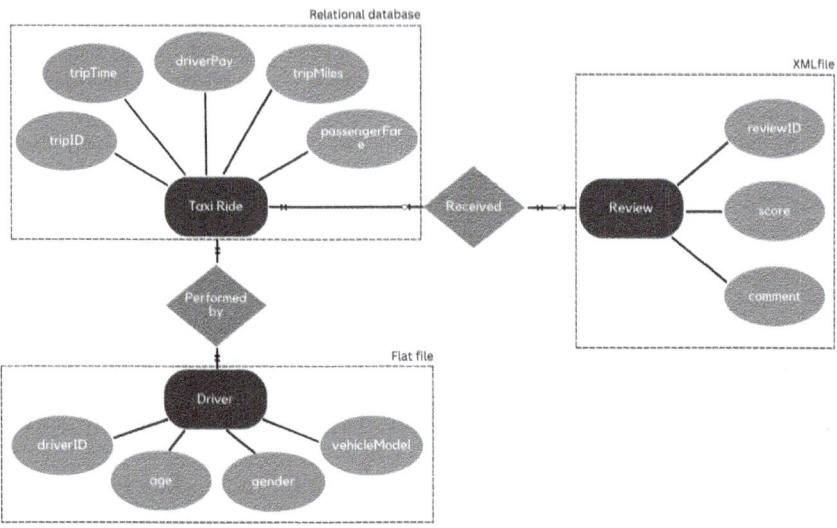

Fig. 1. NYC taxi rides ER diagram

3   Preliminaries

The DVM Conceptual Model. The DVM model provides a graphical representation of entities and attributes, akin to the Entity-Relationship (ER) model. While an ER model focuses on entities and relationships, a DVM emphasizes derived and multi-valued attributes, with entities depicted by their primary attributes. In a DVM, nodes can represent either entities or attributes based on analysis perspective. It comprises mappings as edges between attribute domains as nodes, where mappings manifest as data processes producing 2-dimensional outputs. In an ER model, attributes are directly linked to entities, and relationships are delineated separately. Conversely, in a DVM, attributes are computed and linked to entities via their primary attribute. In the DVM context, the primary key of an ER entity denotes the entity itself as a node, connected to other nodes representing its attributes. DVMs operate from a foundational approach, commencing with available data and constructing a conceptual model based on computed values. In contrast, ER models follow a top-down approach, initiating from conceptual design and progressing towards a logical model [10].

Entity-Attribute Representation. Entities within the DVM model are delineated by their primary keys, facilitating representation through key-list structures. Multi-valued attributes can manifest both as attributes and entities within the model, obviating the necessity for explicit relationships. Notably, data processing tasks yield bidirectional mappings, engendering symmetry within the model and eschewing the hierarchical constructs prevalent in traditional database schemas [10].

Data Virtual Machines. A DVM comprises a collection of mappings between attribute domains, represented as a multi-graph. Attributes serve as nodes within this graph, while data processing tasks give rise to edges, each endowed with associated key-list structures. The bidirectional nature of these mappings engenders symmetry within the DVM, obviating the need for primary keys and hierarchical relationships [10].

[Key-list Structure]. A key-list structure (KL-structure) is a set of (key, list) pairs, where each list comprises elements or is empty. Keys and elements are strings, and the structure is denoted as $K = \{(k, L_k)\}$, with $k \in keys(K)$ and $L_k \in list(k, K)$ [10].

Example. To exemplify the practical application of the DVM, we will utilize the motivating example discussed earlier. Drawing from the ER model of taxi rides outlined in the motivation section, we have constructed the corresponding DVM, showcasing its attributes, edges, and key-list structures as depicted in Fig. 2a. Additionally, we provide an illustration of the resulting key-list structures, representing the bidirectional relationship between taxi rides and drivers, as depicted in Fig. 2b.

Based on the key-list structure the dataframe query operators can be defined. These operators take as input one or more edges in a key-list structure and produce as output a single edge. Several of those operators are described in the following definitions. The operators enable us to evaluate the dataframe query [10].

[Aggregate]. The *aggregation* operator takes a key-list structure K and an aggregate function f, producing a new key-list structure K'. It constructs K' by applying f to each list L_k in K and adding the result as a single-element list L'_k associated with the key k. This operator is denoted as $Aggr(K, f)$ [10]. Common aggregate functions include min, max, average, sum, or count, applicable in any programming language.

[Filter]. The *filtering* operator takes a key-list structure K and a condition θ on a list element (a string), producing a new key-list structure K'. It constructs K' by iterating through each key k in K, adding (k, L'_k) to K', where L'_k contains all $x \in L_k$ satisfying $\theta(x)$. This operator is denoted as $Filter(K, \theta)$ [10].

[Map]. The *mapping* operator takes a key-list structure K and a function f with signature `string f(x:string)`, producing a new key-list structure K'. It constructs K' by iterating through each key k in K and adding (k, L'_k) to K', where L'_k replaces each $x \in L_k$ with $f(x)$. This operator is denoted as $Map(K, f)$ [10]. An example could be transforming customer reviews into sentiment scores using sentiment analysis.

[RollupJoin]. The *RollupJoin* operator combines two key-list structures K_1 and K_2, yielding a new key-list structure K. It constructs K by iterating through each key k in K_1 and adding (k, L_k) to K, where L_k is the concatenation of lists from K_2 corresponding to the elements in $list(k, K_1)$. This operator is denoted as $rollUpJoin(K_1, K_2)$ [10].

[ThetaCombine]. Given:

- a list O of key-list structures $K_1, K_2, ..., K_n$, called the output list (can be empty),
- a list S of key-list structures $K_{0_1}, K_{0_2}, ..., K_{0_m}$, called the selection list (can be empty),
- a boolean expression $\theta(k, l_1, l_2, ..., l_m)$ involving k (an atomic value), $l_1, l_2, ..., l_m$ (lists of values), called the selection condition,

We define an operator called *thetaCombine*, denoted as thetaCombine$(O; S; \theta)$, which returns a key-list structure K constructed as follows:

$$\forall k \in \text{keys}(K_1) \cap ... \cap \text{keys}(K_n) \cap \text{keys}(K_{0_1}) \cap ... \cap \text{keys}(K_{0_m}) :$$

If $\theta(k, \text{list}(k, K_{0_1}), ..., \text{list}(k, K_{0_m}))$ is true, add a pair (k, L_k) to K, where:

- if O is empty then $L_k = [\,]$ (the empty list),
- else $L_k = \oplus_{i=1}^{n} \text{list}(k, K_i)$.

This operator is defined to take the intersection of the keys of the involved key-list structures as the set of keys of the result, resembling a kind of inner join. Different versions of this operator can be defined, such as taking the union of the keys of key-list structures in O.

4   Pandas Dataframes

Pandas dataframes were used to compare our proposed tool due to its widespread popularity and robust functionality, which has made it a standard in data science for tasks such as data manipulation, statistical computing, and visualization. Pandas DataFrames have gained a significant user base across various industries and research fields. For example, Modin, a parallel dataframe system based on Pandas, has attracted over 1 million users from different industries [3]. Dataframes were initially introduced in the S programming language at Bell Laboratories in 1990 to facilitate statistical computations. The concept of dataframes was presented by Chambers, Hastie, and Pregibon at the Computational Statistics conference. They described dataframes as a class of objects in S that can conveniently organize variables relevant to specific analyses [5]. In 2011 Wes McKinney introduced Pandas, a Python library designed to work with structured data sets, providing rich data structures and tools for data manipulation and analysis. It aimed to be a foundational layer for statistical computing in Python and complement the existing scientific Python stack. Pandas addressed the need for rich data structures and metadata handling through its dataframe object, which allowed for flexible and intuitive manipulation of labeled data sets, ensured automatic data alignment, and supported hierarchical indexing for advanced representation of higher-dimensional data within a 2D dataframe [6].

Semantics. Pandas library documentation defines dataframes as "Two- dimensional, size-mutable, potentially heterogeneous tabular data". The data structure

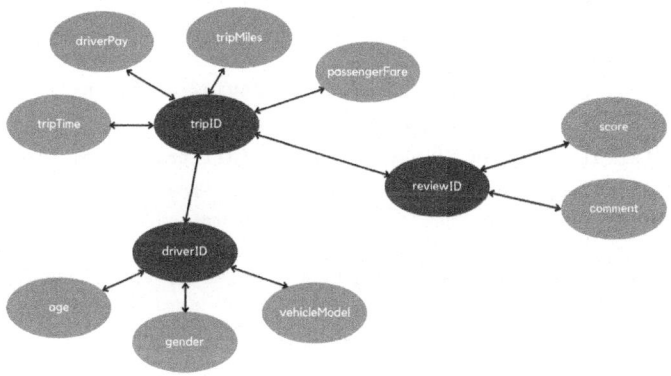

(a) Taxi rides DVM

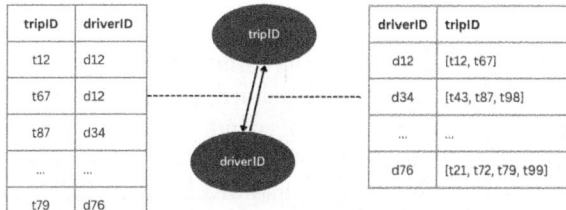

(b) Taxi ride and driver edges key list structures

Fig. 2. Taxi rides DVM illustrations

contains labeled axes (rows and columns). The arithmetic operations align on both row and column labels. The Dataframe can be thought of as a dict-like container for Series objects. The definition relies on Series, which are "One-dimensional ndarray with axis labels (including time series)". The index of a Series is used to label and identify each element of the underlying data. It can be regarded of as an immutable ordered set (technically a multi-set, given it may contain duplicate labels), and is used to index and align data [21].

The authors of "Towards Scalable Dataframes" provide a Dataframe definition based on the APIs of Pandas and R. A dataframe consists of elements from a known set of domains, denoted as $Dom = \{dom1, dom2, ...\}$. Commonly used domains include strings (Σ^*), integers, floats, booleans, and categories. Each domain includes a distinguished null value (NA) and a parsing function $p_i : \Sigma^* \to dom_i$ to interpret cell values. Unlike the relational model, a key aspect of dataframes is that the domains of its columns can be determined from data post hoc. This is facilitated by a schema induction function $S : \Sigma^* \to Dom$ that assigns an array of strings to a domain [3,7].

```
# Prior to this, read and load data into dataframes
# for taxi trips, reviews and drivers
trip_driver_data = pd.merge(trip_data, driver_data, on='driverID')
all_trip_data = pd.merge(trip_driver_data, review_data, on='tripID')
all_trip_data[all_trip_data['gender'] == 'Female']}
all_trip_data[(all_trip_data['age'] >= 20) &(all_trip_data['age'] <= 30)]
all_trip_data[all_trip_data['vehicleModel'] == 'Toyota']}
# Prior to this, define a function that performs
# sentiment analysis on a given column with text
all_trip_data['comment'] = all_trip_data['comment'].apply(sentiment)
# Prior to this, remove the columns that aren't usefull for the analysis
result = all_trip_data.groupby('driverID').agg({
    'value': ['count', 'sum', 'mean', 'min', 'max']
}).reset_index()
```

Listing 1: Pandas implementation of motivational query

Acknowledging the absence of the formal dataframe semantics, the authors of "Towards Scalable Dataframes" defined the following set of operators that encapsulates the extensive APIs of pandas and R. **SELECTION:** Eliminates rows based on conditions, akin to filtering in pandas/R. **PROJECTION:** Removes specific columns, like SELECT in SQL or column indexing. **UNION:** Combines two dataframes, useful for dataset merging. **DIFFERENCE:** Computes elements unique to one dataframe. **CROSS PRODUCT / JOIN:** Combines dataframes by matching elements. **DROP DUPLICATES:** Ensures data integrity by removing duplicate rows. **GROUPBY:** Groups rows based on columns and aggregates others. **SORT:** Orders rows lexicographically. **RENAME:** Changes column names. **WINDOW:** Applies functions using a sliding window. **TRANSPOSE:** Swaps rows and columns. **MAP:** Applies a function to each row. **TOLABELS:** Projects one column as new row labels. **FROMLABELS:** Inserts row labels as a new column. **TRANSPOSE (in schema):** Handles schema changes induced by transposing [3].

Limitations. Although pandas is praised for its extensive API, it also contains significant redundancies among its operators, leading to varied performance implications. This complexity burdens users who must manually select the appropriate pandas API calls. For example, a single task can be done using various APIs with performance ranging from very fast to slow. Users can choose between openpyxl, xlrd, or pyxlsb to read an Excel file, each offering different performance results. Consequently, many users opt to use only a small subset of operators, avoiding the bulk of the API. The intricacies of the API and its evaluation semantics make it challenging to apply traditional query optimization techniques. Each operator within a pandas "query plan" is executed entirely before subsequent operators, lacking extensive optimization, reordering, or pipelining, unless explicitly instructed by the user using .pipe. Furthermore, when dealing with even moderately large datasets that exceed memory capacity, the performance of the pandas.DataFrame API declines significantly. This is due to pandas' eager evaluation approach, where intermediate data sizes often surpass memory limits and require paging [3]. Another major concern is its memory usage, as it stores the entire dataset in memory, causing issues with datasets that exceed available memory and leading to performance bottlenecks. Pandas' memory usage is a widely recognized drawback attributed to its internal memory requirements [9]. An additional drawback is the lack of native support for parallel or distributed computing, which makes it less suitable for large-scale data processing tasks.

Some additional considerations of the data analysis approach with Pandas include the following. End-users who are not proficient in programming may find it challenging to explore data directly, as they would with graphical data exploration tools. Moreover, the lack of reusability in the approach can lead to duplication of effort and maintenance challenges when creating new datasets or performing different analyses. When data sources undergo changes or new data sources are introduced, it becomes challenging to update the code efficiently to accommodate these modifications. This difficulty arises due to the need for manual intervention and adjustments in the existing codebase, which can be time-consuming and error-prone. As data structures and formats evolve, the code must be adapted to handle these alterations, and when new data sources are integrated, the codebase may require significant reworking or rewriting.

Recent Advancements. Dataframes have indeed seen significant progress in recent years, particularly in the context of big data processing and analysis. The development of scalable dataframe systems has been a key focus. For instance, MODIN, a scalable dataframe system, has demonstrated performance 30 times faster than traditional systems like pandas [3]. Additionally, there has been a growing emphasis on parallelism approaches, such as those used in the Julia programming language, which has shown promise in improving the efficiency of numerical computing tasks [11]. Furthermore, the integration of dataframe systems with cloud computing technologies has been explored, with initiatives like NumS presenting scalable array programming for the cloud, indicating a shift towards leveraging cloud infrastructure for dataframe operations [12]. In the realm of big data processing, Apache Spark has indeed emerged as a prominent platform, offering distributed computing capabilities for dataframe operations. The advancements in Spark have contributed to the efficient handling of large-scale datasets and complex analytics tasks.

Example: We will use the motivation query to demonstrate the capabilities of the Pandas library in creating Dataframes. Listing 1 presents the Python code snippet that produces the Dataframe of our motivation query. First we load the data by leveraging Python libraries such as Pandas for reading CSV files, xml.etree.ElementTree for parsing XML files, and psycopg2 for connecting to the database. The data loading part is not in the code as we want to emphasize the query analysis per se. For the motivational query analysis, we begin by merging the `trip` and `driver` data first and then merging the `passengerReview` data to create a dataframe with all the data. We keep the rows where the driver's gender is female and then filter out the rows of data where the driver's age is lower than 20 years old and higher than 30 years old, removing all rows that don't meet those criteria. We conduct sentiment analysis on the `passengerReview` column using the TextBlob library, transforming textual reviews into numerical sentiment scores. Finally, we group the data based on the `driverID` and apply the desired aggregators.

5   DVM Dataframes

In this section, we illustrate how dataframe queries can be formulated and analyzed over DVM, offering a streamlined approach compared to the intricacies involved in Pandas dataframe analysis.

[**Dataframe Queries over DVM**]. A dataframe query over a DVM is formally defined as a tree structure, wherein each node corresponds to an attribute of the DVM, and edges represent relationships between these attributes. Transformations and selection conditions are applied to each node and edge to specify the desired output [10]. Given a DVM $G = \{A, S\}$, a dataframe query is a tree structure Q, in which:

- each node N of Q has a unique name and a label
- for each edge $N \rightarrow N'$ in Q, there exists an edge: $label(N) \rightarrow label(N')$
- each edge e of Q is annotated with a list of transformations, called the transformations string
- each node N of Q is annotated with the selection condition

The starting point (root) of a dataframe query within the DVM can be any attribute, implicitly symbolizing the entity under analysis. It could represent the trip or driver entities (such as tipID or driverID) or attributes like gender or age. Notably, nodes and edges of the DVM may appear multiple times within a dataframe query, even within the same path. Visualizing dataframe queries over DVMs can be done with intuitive and straightforward ways by either connecting a user interface or using a declarative method as analyzed in the subsequent section.

Example: To illustrate our motivating query within the DVM tree structure, we refer to Fig. 3. We leverage the DVM operators outlined in Sect. 2 to construct our query tree. Beginning with `driverID` as the root node, we employ *thetaCombine* with `driverAge` and `gender` composing our Output list. For the first node's θ case, we specify `driverAge` between 20 and 30 years old. Similarly, for the `gender` node, we specify the θ case as equality to the string value "FEMALE" and for the `vehicleModel` node, we specify the θ case as equality to the string value "Toyota Corolla".

Next, we traverse to the `tripID` node, matching it with the selected `driverID`s using *rollUpJoin*. We then apply *thetaCombine* with an output list comprising `driverPay`, `tripTime`, and `tripMiles` corresponding to the filtered drivers. Here, we omit specifying a θ case, selecting only the desired nodes.

Subsequently, we apply various aggregators including average, count, sum, min, and max to the attributes `tripTime`, `driverPay`, `tripMiles` and `passenger- Fare`. Following this, we employ *rollUpJoin* to match the `passengerReview` attribute from the XML file using the `tripID` node. Finally, we utilize the *map* operator to apply sentiment analysis scoring to `passengerReview` using a Python script, and subsequently, apply the average, count, sum, min, and max aggregators.

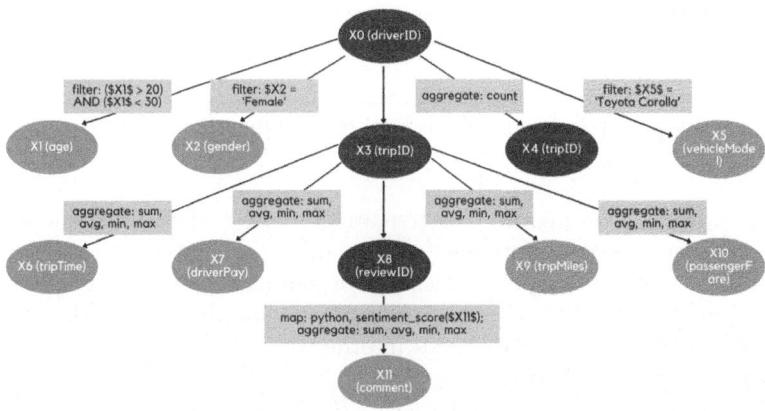

Fig. 3. DVM tree structure of motivating query.

6   Implementation and Experiments

6.1   Implementation

Query Construction and Modification. As previously stated in Sect. 3 and Sect. 5, a DVM is essentially a graph-based model, with dataframe queries over DVMs represented as tree structures. This tree architecture encapsulates the complexity of a query, allowing non-data experts to easily define queries in a natural manner rather than learning the extensive API of tools such as Pandas, as seen in Table 1.

The construction and modification of queries can be simplified by introducing GUI tools that allow for the visual formulation of such queries. As demonstrated in Fig. 3, users can select nodes from the DVM to be included in the output dataframe by specifying them as query tree nodes, and data transformation tasks can be incorporated into the domain attributes, by designating the usage of any of the dataframe query operators mentioned in Sect. 3. Under the hood, the dataframe query tree can be expressed in a declarative manner by using a format such as XML, as demonstrated in Listing 2.

```
<query>
  <rootnode>X0</rootnode>
  <node>
    <label>X0</label>
    <onnode>driverID</onnode>
    <children>X4</children>
  </node>
  <node>
    <label>X4</label>
    <onnode>tripID</onnode>
    <transformations>aggregate:count</transformations>
  </node>
</query>
```

Listing 2: Example XML Dataframe Query over DVM

Efficient Query Evaluation. The tree architecture of dataframe queries over DVMs offers enhanced performance during query evaluation. In essence, the DVM query evaluation pipeline consists of initially parsing the query tree bottom-up and then, for each edge, utilizing the specification of the DVM to access the corresponding data sources in order to load the specified data while also executing any data processing tasks.

A DVM is a virtual schema and as such, the key-list structures are not materialized and data is loaded incrementally, beginning with the edges that stem from the leaf nodes and progressing to the root node. This tree architecture of dataframe queries over DVMs offers enhanced performance during query evaluation due to the partitioning of the tree structure, which allows for the parallel execution of tasks. Partitioning is a common strategy of breaking up a problem into smaller subproblems that can be solved concurrently [17]. When evaluating a query, multiple threads can be used to concurrently load data and run computations on edges that are at the same level. Each thread processes one or more edges and waits for all the threads in the layer to complete before moving up to the next layer. At each step, an edge's computation is determined by the results of its children and their accompanying edges.

Having mentioned that existing dataframe systems break down completely when operating on datasets that exceed the constraints of the main memory, we need to involve persistent memory in the evaluation step. Even though storing data in persistent memory introduces new challenges considering the performance implications of relying on I/O as opposed to reading in-memory data, it is possible to balance memory space and data analytics performance [20]. We adopt techniques similar to MapReduce [19] - in order to reduce main memory usage - by using intermediate files for offloading data after a layer of the tree is complete.

6.2   Experiment

To evaluate the performance of our implementation against Pandas, we utilize our motivating example to construct a query and measure the performance. In Sects. 4 and 5, we presented an initial graphical representation of the motivating query by using Pandas in listing 1 and the DVM operators in Fig. 3.

Experimental Setup. For our experiments, we utilized a Debian server with 72.6 GB of RAM and 8 CPUs. For the relational database we used the PostgreSQL version 13.11 database running locally at the Debian server. For Pandas experiments with Python, we used the Python version 3.9.2. The DVM implementation was built with Maven version 3.6.3 and compiled with Java version 17.0.7.

Data Description. As described in Sect. 1, our data comprises taxi rides in New York City in 2022, augmented with relevant data from taxi driver records. This dataset includes attributes detailing the taxi rides, such as `tripID`, `driverPay`, `tripTime`, `tripMiles`, `PU-Location`, `DOLocation`, `PUDate- time`, `requestDatetime`, `passengerFare`, `airportFee`, and `driverTip`. All those attributes were sourced from this dataset published by the NYC Gov website. Additional attributes such as the `passengerReview` attribute, along with `driverID`, `gender`, and `driverAge`, were incorporated from two Kaggle datasets containing customer reviews and driver information for Uber drivers.

Since most of our data was in CSV format, we refactored it into varying data source types, including Excel, XML, and relational database tables, to test our query. Initially, the dataset containing taxi rides was transformed into a PostgreSQL relational database with four tables: `trip_details`, `trip_time`, `trip_speed`, and `trip_location`. The attributes were distributed among these tables with `tripID` serving as the primary key. The total size of our database was 3,094,897 rows across all four tables. The passenger review data were converted to XML format, while the taxi trip data remained in CSV format. It should be noted that the datasets contained additional attributes that were not utilized in our experiments. We have presented only those fields that contribute the most to the explanation of our experiments. Additionally, the attribute names might not align precisely with those in the datasets. This was done to ensure more consistent and understandable naming conventions across the various data sources in our conceptual framework.

Results. The results of our experiments are presented in Table 1. We observed a 22% decrease compared to Pandas Dataframes. This improvement in performance can be attributed to the distribution of the workload to multiple threads as was discussed in Sect. 6 – *Efficient Query Evaluation* and Python's relative inefficiency as an interpreted language compared to the compiled language used to develop the tool for this experiment [18].

Table 1. Experiment results

Implementation	Time (minutes)	Pandas Difference
Pandas	20:01	–
DVM	15:27	-22%

7   Conclusions and Future Work

In conclusion, in this paper we present an approach that facilitates the efficient construction, modification and evaluation of big data queries. Our approach provides a simplified process for building dataframe queries from multiple big data sources. Our model-based approach seeks to make dataframe analysis approachable to a wider audience including non-programming experts. In comparison to Pandas - the most widely used tool for dataframe analysis - the results of our experiment not only indicate the efficiency in terms of execution time, but also in the ease of use. By looking at the different implementations of our motivating query it is evident that the creation of queries using our approach abstracts the need for technical expertise and enables the user to perform queries more intuitively. Additionally, the connection of complex data sources, such as databases, is only performed once and the connection of database tables is done automatically. Finally, our approach offers a decrease of 22% in query execution time, with prospects of even greater efficiency in the future through further optimization of the query evaluation.

Looking ahead, this discovery opens up a number of research directions. More complex queries with multiple root nodes for analysis should be investigated further. The query evaluation and the current operators could be optimized further using concurrency and parallelization during data loading and nodes transformation. Furthermore, this approach could be extended with a user interface to enhance user experience.

References

1. Gandomi, A., Haider, M.: Beyond the hype: big data concepts, methods, and analytics. Int. J. Inf. Manag. **35**(2), 137–144 (2015). ISSN 0268-4012, https://doi.org/10.1016/j.ijinfomgt.2014.10.007
2. De Mauro, A., Greco, M., Grimaldi, M., Nobili, G.: Beyond data scientists: a review of big data skills and job families. J. Inf. Process. Manag. (2016)
3. Petersohn, D., et al.: Towards Scalable Dataframe Systems, arXiv preprint, eprint: 2001.00888, archivePrefix: arXiv, primaryClass: cs.DB, 2020
4. Abadi, D.: The beckman report on database research. Commun. ACM **59**, 692–99 (2016)
5. Chambers, J., Hastie, T., Pregibon, D.: Statistical Models in S. In: Momirović, K., Mildner, V. (eds.) Compstat. pp, pp. 317–321. Physica-Verlag HD, Heidelberg (1990)

6. McKinney, W.: Pandas: a foundational python library for data analysis and statistics. Python High Perform. Sci. Comput. (2011)
7. Abiteboul, S., Hull, R., Vianu, V.: Foundations of Databases, vol. 8. Addison-Wesley, Reading, Boston (1995)
8. Nelluri, R.: What are the limitations of pandas?, Insights and Data. https://insightsndata.com/what-are-the-limitations-of-pandas-35d462990c43. Accessed 10 Apr 2021
9. Sinthong, P., Carey, M.J.: AFrame: Extending DataFrames for Large-Scale Modern Data Analysis (Extended Version), arXiv preprint, eprint: 1908.06719, archivePrefix: arXiv, primaryClass: cs.DB, 2019
10. Chatziantoniou, D., Kantere, V., Antoniou, N., Gantzia, A.: Data virtual machines: simplifying data sharing, exploration & querying in big data environments. In: IEEE International Conference on Big Data (Big Data), 2022
11. Bezanson, J., Edelman, A., Karpinski, S., Shah, V.: Julia: a fresh approach to numerical computing. SIAM Rev. **59**(1), 65–98 (2017). https://doi.org/10.1137/141000671
12. Elibol, M., et al.: Nums: scalable array programming for the cloud (2022). https://doi.org/10.48550/arxiv.2206.14276
13. Amović, M., Govedarica, M., Radulović, A., Jankovic, I.: Big data in smart city: management challenges. Appl. Sci. **11**(10), 4557 (2021). https://doi.org/10.3390/app11104557
14. Rehman, M.S., Elmore, A.: Fuzzydata: a scalable workload generator for testing dataframe workflow systems. In: Proceedings of the 2022 Workshop on 9th International Workshop of Testing Database Systems, 2022, Association for Computing Machinery, New York, NY, USA, pp. 17–24, 2022. https://doi.org/10.1145/3531348.3532178
15. Macdonald, C., Tonellotto, N., MacAvaney, S., Ounis, I.: Pyterrier: declarative experimentation in python from BM25 to dense retrieval. In: Proceedings of the 30th ACM International Conference on Information & Knowledge Management (CIKM '21), 2021, Association for Computing Machinery, New York, NY, USA, pp. 4526–4533, https://doi.org/10.1145/3459637.3482013
16. Uta, A., Ghit, B., Dave, A., Rellermeyer, J., Boncz, P.: In-memory indexed caching for distributed data processing. In: Proceedings of the 2022 IEEE International Parallel and Distributed Processing Symposium (IPDPS), pp. 104–114, 2022. https://doi.org/10.1109/IPDPS53621.2022.00019
17. Joseph, J.: An Introduction to Parallel Algorithms. Addison-Wesley Publishing Company, Boston (1992)
18. Stepanek, H.: Thinking in Pandas, pp. 37–39. Springer, Cham (2020). https://doi.org/10.1007/978-1-4842-5839-2
19. Dean, J., Ghemawat, S.: MapReduce: simplified data processing on large clusters, pp. 107–113. Association for Computing Machinery, 2008. https://doi.org/10.1145/1327452.1327492
20. Gai, K., Qiu, M., Liu, M., Xiong, Z.: In-memory big data analytics under space constraints using dynamic programming. Future Gener. Comput. Syst. 219–227 (2018). https://doi.org/10.1016/j.future.2017.12.033
21. Pandas documentation — pandas 2.2.1 documentation. https://pandas.pydata.org/docs/. Accessed 30 Mar 2024

Energy and Performance Evaluation of Serverless and Serverful Models on Spark for Database Join Operations

Phan-An-Truong Tran[1]([✉]), Laurent D'orazio[2], Thuong-Cang Phan[3], and Le Gruenwald[4]

[1] Vinh Long University of Technology Education, Vinh Long, Vietnam
truongtpa@vlute.edu.vn
[2] Univ Rennes, CNRS, IRISA, Rennes, France
laurent.dorazio@univ-rennes.fr
[3] Can Tho University, Can Tho, Vietnam
ptcang@cit.ctu.edu.vn
[4] School of Computer Science, The University of Oklahoma, Norman, USA
ggruenwald@ou.edu

Abstract. The demand for environmentally friendly cloud computing is on the rise, leading cloud service providers to focus on reducing carbon emissions by using renewable energy sources and energy-efficient computing models. This study assesses the performance and energy consumption of serverless and serverful architectures, specifically looking at join operations using Apache Spark for big data processing in a private cloud combined with Kubernetes. By using the TPC-DS benchmark, we examine the impact of cold-start and warm-start phases in the serverless environment, as well as the auto-scaling capabilities of Spark in serverless environments within the private cloud. The results show that the efficient and flexible resource management in serverless environments in private clouds leads to more optimal processing times and energy consumption compared to serverful architectures, especially in warm-start scenarios. These findings offer valuable insights for organizations seeking to streamline their big data infrastructure while also making a positive environmental impact within the IT industry.

Keywords: Serverless · Serverful · Warm-start · Cold-start · Energy consumption

1 Introduction

Big data processing has become essential for organizations to maximize value from information. The global big data analytics market, valued at $307.51 billion in 2023, is projected to reach $924.39 billion by 2032 with 13.0% CAGR [1]. In the field of big data analytics and processing, Apache Spark stands out as a powerful and flexible platform when compared with other tools such as Apache Hadoop,

R. Wrembel et al. (Eds.): DEXA 2025, LNCS 16047, pp. 178–193, 2026.
https://doi.org/10.1007/978-3-032-02088-8_13

Flink, and Storm [8]. Spark's advantages include in-memory data processing capabilities, support for diverse data types, and ease of application development. Apache Spark can be deployed in both traditional on-premise environments and cloud services as virtual machines. However, operating Apache Spark in these environments faces significant limitations. In terms of cost, both on-premise and cloud-based solutions require substantial investments, especially when computational needs are intermittent but a certain number of virtual servers must be maintained. Additionally, Spark encounters difficulties in performance optimization due to unpredictable workloads. In this context, the serverless environment emerges as a potential solution, with advantages such as no need for server infrastructure management, simplified configuration, and a Pay-as-you-go pricing model. Services like Amazon EMR serverless, Google Cloud Dataproc serverless, and Azure HDInsight on Azure Kubernetes Service are leading this trend in big data processing with Apache Spark [5,9]. Energy efficiency is increasingly important as data centers consume 1–2% of global electricity, driving cloud providers toward renewable energy adoption and carbon neutrality by 2050 [4]. However, there is a lack of comprehensive studies comparing performance and energy consumption between Apache Spark models in serverful (on-premise or virtual servers) and serverless environments. This creates difficulties for organizations in selecting the optimal environment for their specific needs.

To address this issue, we propose the study of Energy and Performance of serverless and serverful Models on Spark for Database Join Operations. This research makes the following contributions: (1) An analysis of factors influencing the performance of Apache Spark during join operations on TPC-H benchmark datasets of varying sizes. The analysis focuses on the impact of cold-start latency, worker node configurations, and resource scalability in both serverless and serverful deployment models; (2) A comparative study of performance and energy consumption between serverless and serverful environments during join-intensive workloads. The study highlights how deployment characteristics—such as the initialization state of serverless environments—affect execution behavior; and (3) An evaluation of the role of auto-scaling mechanisms in serverless environments, analyzing how dynamic resource allocation influences both performance and energy efficiency when processing large-scale data workloads with variable resource demands. The results of this study provide diverse insights for organizations considering transitioning or optimizing their big data processing infrastructure while contributing to the collective effort to minimize the environmental impact of the information technology industry.

The remainder of this paper is structured as follows. Section 2 reviews the related work on the performance and energy consumption of serverless and serverful architectures, particularly in the context of big data processing. Section 3 provides the background on the technical characteristics of Apache Spark in both serverless and serverful environments. Section 4 describes the monitoring system used for tracking performance and energy metrics. Section 5 presents the experimental setup and workloads, followed by detailed energy and performance evaluations in Sect. 6. Finally, Sect. 7 discusses the key findings from the experiments and provides suggestions for future research directions.

2   Related Work

With the increase in big data, processing tools like Apache Hadoop and Apache Spark, along with deployment models such as serverless and serverful, are advancing rapidly. Some research [2,16,17] has been conducted to explore the performance evaluation and energy consumption of Spark in both the serverless and serverful environments. Changpeng et al. [17] evaluated the performance of Spark on a bare metal environment and a Kubernetes environment with several algorithms, such as SQL Join, Kmeans, and Wordcount and the study showed that Spark on Kubernetes performs better than the serverful environment in certain cases, such as memory-intensive stages (e.g., reduce stage in Sort), when using cache efficiently, or when applying custom task scheduling to improve data locality and executor utilization. Qi Zhang et al. [16] conducted a study comparing the performance of Spark when deployed on containers versus virtual machine environments. The research showed that Apache Spark's performance on containers is higher than that of virtual machine environments. Alhindi et al. [2] conducted a study to evaluate the energy consumption of the serverless Function as a Service (FaaS) model on the OpenFaaS platform based on Docker and Kubernetes. The study proposed a method to estimate energy usage based on mathematical formulas of idle energy levels and execution energy levels. The OpenFaaS platform with Faas was shown to have an energy level of 12–17.5% lower when compared to the Docker model.

Previous studies lack a comprehensive evaluation of executing Spark on a serverless model versus on a serverful model across factors like data size, nodes, and auto-scaling. Our study fills this gap by comparing the cold-start time and power consumption in the two serverless environments (cold-start and warm-start) and the serverful environment, and benchmarking Spark performance and energy efficiency across data sizes, cluster sizes, auto-scaling configurations, and serverless start states. Our key contributions include experimentally comparing Spark on serverless and serverful environments across multiple scenarios for database join operations. The results offer insights into performance differences, guiding appropriate architecture selections.

3   Background

This section introduces the architecture, characteristics, advantages, and limitations of serverless and serverful systems in the context of executing join operators on Apache Spark. This section explains the differences in how to execute applications on serverless and serverful environments in terms of initialization, startup, processing data, and clean-up. We also explain how to measure energy consumption on Spark.

3.1   Serverless and Serverful Environments

Serverful computing represents the traditional model where applications are deployed manually or semi-automatically on configured physical or virtual servers, with developers and administrators responsible for provisioning, maintaining, and scaling the infrastructure [13]. In contrast, *serverless computing* is a cloud paradigm where providers manage all server infrastructures and automatically allocate computational resources, allowing developers to focus solely on application code [13]. The serverless model operates through stateless containers triggered by events, offering key advantages including on-demand computation, automatic scaling, and usage-based payment that enhance business flexibility and cost efficiency.

Within serverless environments, cold-starts and warm-starts significantly impact application performance [6]. *A cold-start* occurs when a container initializes for the first time or after extended inactivity, requiring complete runtime environment setup including container image retrieval, resource allocation, and application initialization, which introduces considerable latency. Conversely, *warm-starts* utilize recently active containers, dramatically reducing response time. This distinction between cold and warm starts substantively affects query response time and overall serverless application performance, particularly in scenarios with irregular workload patterns or sudden traffic spikes, making effective management of these states crucial for optimizing both performance and cost in serverless architectures.

3.2   Spark Serverless and Serverful

In the context of deploying Spark serverless, both public cloud and private cloud provide solutions that make application deployment easier for developers: Major public cloud providers such as AWS, Azure, and Google Cloud have developed specialized Spark serverless services like Amazon EMR serverless, Google Cloud Dataproc serverless, and Azure HDInsight on Azure Kubernetes Service [3]. These services offer automatic scaling capabilities, resource management, and pay-per-use billing based on actual usage. In private cloud serverless environments, Spark serverless deployment typically involves using Kubernetes (K8s) in combination with container management tools like Docker [10,12,15].

In this study, we chose a private cloud environment with the Kubernetes platform combined with the MinIO object storage system as the experimental environment for Spark on serverless due to their many similarities [7]. Figure 1 compares the differences between Spark applications when processing data in serverful and serverless environments. After a job is submitted to Spark, the scheduling begins before the application is executed. This process includes two main parts: DAG Scheduler and Task Scheduler. The DAG Scheduler is used to schedule and manage jobs as a directed, acyclic graph. This graph represents the dependencies between RDDs (Resilient Distributed Datasets) and transformations on them in a Spark application. The Task Scheduler manages and schedules tasks on executors on worker nodes in a Spark cluster. Its task is to distribute tasks to executors to perform parallel data processing. When the data processing is finished, the results are aggregated at the Spark Driver and returned to the user. Before termination, cleanup is performed and resources are freed [14] (Fig. 1.a).

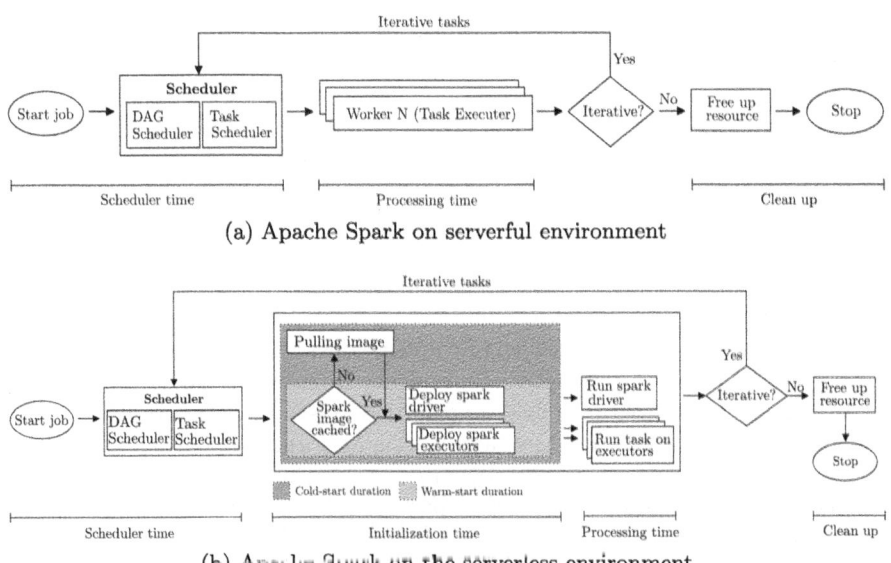

(a) Apache Spark on serverful environment

(b) Apache Spark on the serverless environment

Fig. 1. Comparing Spark serverless and serverful environments

When comparing the processing of a Spark application on serverless and serverful environments, there are some differences in the initialization process, especially in the serverless environments with cold-start and warm-start phases. The cold-start phase in the serverless environment is the first stage when a Spark application is executed. During this phase, Spark has not yet initialized the environment. Therefore, it takes a specific time to do some work, such as downloading the Spark image to all workers and then initializing the *Spark driver* and *Spark workers* (Fig. 1.b). After the environment has been initialized, the Spark application on the serverless environment begins to execute. From the second run onwards, the serverless environment is considered *warm-start* since Spark images already exist on the workers. It just needs to initialize the *Spark driver* and *Spark workers* and start executing the application. This phase is called *warm-start* (Fig. 1.b).

Spark in serverless and serverful environments have different initialization characteristics. Spark in the serverless environment helps automate the initialization process, reclaim resources after each execution to help save costs, and auto-scale. Meanwhile, Spark must manually initialize and reclaim resources in the serverful environment.

4   Proposed Monitoring System

Monitoring the performance and energy consumption of Apache Spark in both serverless and serverful environments presents a major challenge in private cloud settings due to the complexity of multi-server hardware and distributed processing. To address this, we developed GUESS [11], a monitoring system that collects key metrics such as resource utilization, query execution time, and power consumption using Prometheus, Grafana, Spark History Server, and OpenManage Enterprise Power Manager. Figure 2 illustrates the *GUESS* model, designed to track Spark's performance and energy usage across different deployment environments.

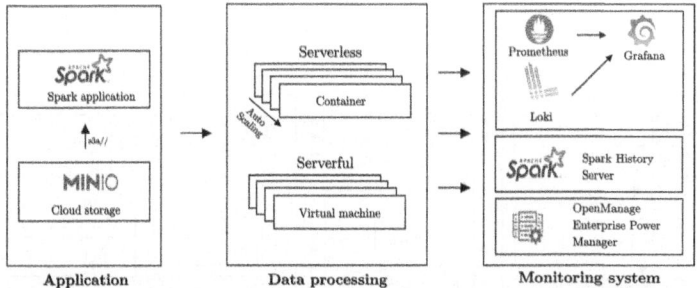

Fig. 2. GUESS: Monitoring Join query Execution in serverless and serverful Spark [11]

Energy consumption is measured at the hardware level using Dell iDRAC with 1-s granularity. Performance metrics including CPU, memory, I/O, and network utilization are collected through Prometheus and visualized via Grafana dashboards [11]. Total energy is calculated by aggregating power usage across all cluster nodes during the complete execution period from job submission to cleanup, enabling accurate comparison between serverless and serverful environments.

5   Experimental Setup and Workloads

In this section, we outline the experimental environment and workloads utilized to assess the performance and energy consumption of Apache Spark during join operations in both serverless and serverful architectures.

5.1   Experimental Setup

Figure 3 illustrates the proposed experimental architecture, deployed on a private cloud infrastructure consisting of 5 physical Dell PowerEdge servers, each hosting 7 virtual machines running Ubuntu Server 22.04. Each VM is allocated 6vCPU,

16 GB memory and 128 GB storage, connected via 10Gbps Ethernet network. Network bandwidth and latency are monitored to ensure consistent performance across experiments. In serverful environments, Spark is installed manually on each virtual machine. In the serverless environment, Spark is deployed on a Kubernetes cluster in the private cloud with the horizontal pod autoscaling (HPA) mode, which is the typical environment we use in experiments due to its similarities with Spark serverless on public cloud services. The HPA is configured with 70% CPU target utilization, scaling 2 pods per 30 s (up) and 1 pod per 60 s (down), with 5–20 executor range. Cold-start time is measured from pod creation to first task execution, while warm-start time begins from executor reuse to task execution, tracked through Kubernetes events and Spark logs. Using the MinIO cluster for storage as an object storage saves the data produced by our experiments.

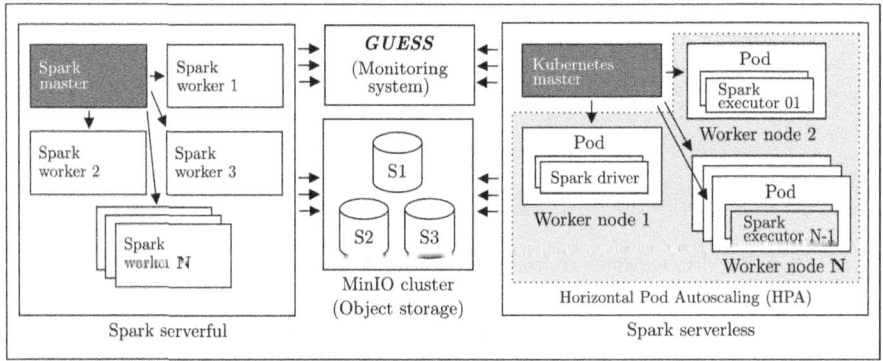

Fig. 3. The proposed experimental architecture

5.2   Database Size

To evaluate Spark performance across serverless and serverful environments, we employed the TPC-DS database benchmark, a sophisticated workload specifically designed for assessing big data processing capabilities. This benchmark is particularly valuable for our study due to its complex query patterns across 24 database tables, diverse analytical scenarios represented by 99 distinct query types, and extensive join operations that effectively stress distributed processing frameworks. Our experiments consisted of three database scenarios: *Case Study 1 (small database - 10* GB*):* Examines initialization delay from cold-start in serverless environments using a small dataset where startup time dominates execution time, isolating container image pulling and environment setup impacts; *Case Study 2 (medium database - 25* GB*):* Evaluates performance and energy consumption across all three deployment environments (serverful, serverless cold-start, and serverless warm-start) using a balanced workload that allows comparison of processing efficiency without overwhelming infrastructure constraints; and *Case Study 3 (large database - 50 GB)*: Stressed automatic scaling capabilities with our largest database (50 GB), allowing serverless environments to scale between 10–20 nodes dynamically. While 50 GB may appear modest for 10–20 nodes, this size was chosen to demonstrate auto-scaling behavior under controlled conditions and to ensure completion within reasonable experimental timeframes. Future work will examine larger datasets up to multi-terabyte scales.

Each environment maintained consistent base configurations (12 GB driver memory, 6 executor cores) with the primary distinction being the ImagePullPolicy setting between serverless cold-start (*Never*) and warm-start (*Always*). For the heavy workload scenario, we enabled dynamic allocation in serverless environments (max: 20 executors, min: 5 executors), a capability not available in the serverful environment.

6   Energy and Performance Evaluation

In this section, we present the results of our experiments, focusing on the energy consumption and performance of Apache Spark in both serverless and serverful environments.

6.1   Case Study 1: The Impact of "Cold-Start" on Join Query Performance

Our Case Study 1 aims to quantify the time duration of the cold-start process in the serverless environment compared to the serverful environment. In the experiment, we use the *PullPolicy* parameter in Kubernetes to control how container images are pulled. Specifically, we compare two values of *PullPolicy*: *Never* (corresponding to the cold-start serverless environment) and *Always* (corresponding to the warm-start serverless environment). With *Never*, the container image is only pulled if it does not exist locally, while *Always* forces Kubernetes to pull a new image from the registry every time a container is started. By changing *PullPolicy*, we can evaluate the impact of the image-pulling process on the cold-start time in the serverless environment. The Spark images are stored on *Docker Hub* with different tags (image digests) to allow for multiple tests.

The experiment is conducted on a compute cluster consisting of 1 master node and 10 worker nodes. Each node is configured with 12 GB of memory, 6 vCPU executor cores, and 10 worker nodes for each environment (serverful, cold-start serverless, and warm-start serverless). We use the TPC-DS benchmark with a database size of 10 GB and execute a total of 31 different queries, with each environment being tested 5 times consecutively to obtain the most accurate results.

The results in Table 1 show that the serverless environment in the cold-start state the serverful environment, while in the warm-start state, the serverless

Table 1. Average processing time (Seconds) by each stage in Case Study 1

Stage	Serverless cold-start	Serverless warm-start	Serverful
Stage 1: Initialization			
Scheduling	3.3	3.3	2.0
Pulling images	35.5	0.0	0.0
Creating	11.0	11.0	0.0
Stage 2: Processing			
Running the app	382.7	374.2	370.0
Stage 3: Clean-up			
Killing	2.5	2.5	1.0
Total Time	**435.0**	**391.0**	**373.0**

Total time is calculated as the sum of Stage 1 (Initialization), Stage 2 (Processing), and Stage 3 (Clean-up).

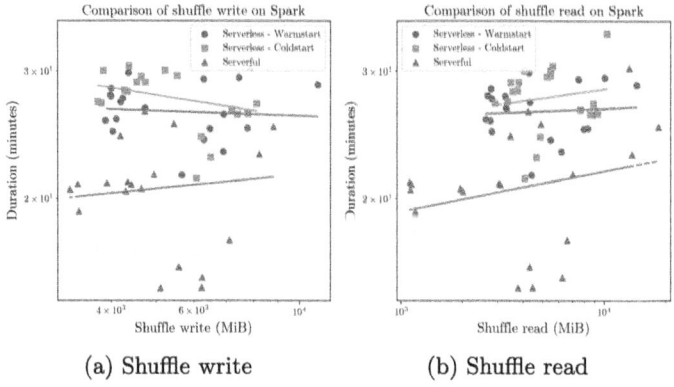

(a) Shuffle write   (b) Shuffle read

Fig. 4. Comparison of the data shuffling process between serverless and serverful environments

environment has 4.8% higher total average processing time than the serverful environment. The reason for this difference lies in the cold-start phase of the serverless model, where the startup process requires more preparation steps than the serverful model, leading to longer startup times. Specifically, when a serverless cluster begins processing, the workers may need to download container images from a registry (such as Docker Hub or Amazon ECR) to set up the execution environment. However, this process does not occur synchronously and is influenced by several factors, such as network bandwidth, disk read/write speed, and other issues.

For Apache Spark in a serverless environment, when the workers cannot pull images from the registry synchronously, some workers have to wait until all others complete initialization before starting the tasks, directly affecting the performance of distributed operations, especially complex operations like Join. A Join operation requires the simultaneous distribution and processing of data across workers, but if some workers are not ready, the process will be delayed, increasing latency and causing imbalance in the computation. Furthermore, Spark can encounter data skew issues, where some workers have to process significantly more data than others, leading to bottlenecks and reduced overall cluster performance. Figures 4.a and 4.b illustrate that the read and write latencies of the serverful environment (Triangle points) are lower than those of serverless (Circle and Square points). To mitigate these negative impacts, solutions such as warm-start, caching optimization, and the use of *Adaptive Query Execution (AQE)* can be applied to improve performance and reduce the latency caused by cold-start.

6.2   Case Study 2: Comparing the Performance and Energy Consumption

In this experiment, we evaluate the performance and energy consumption of query operations in Spark in three environments: serverless with cold-start, serverless with warm-start, and serverful, using a TPC-DS database with a size of 25 GB. The experiment was conducted with 10 worker nodes to compare data processing capabilities and resource optimization across these environments, aiming to identify differences in performance and energy consumption in various deployment scenarios. We used 31 sample queries, primarily focusing on data Join operations. Each query in every environment was executed 5 times, and the experimental results were averaged over the 5 runs.

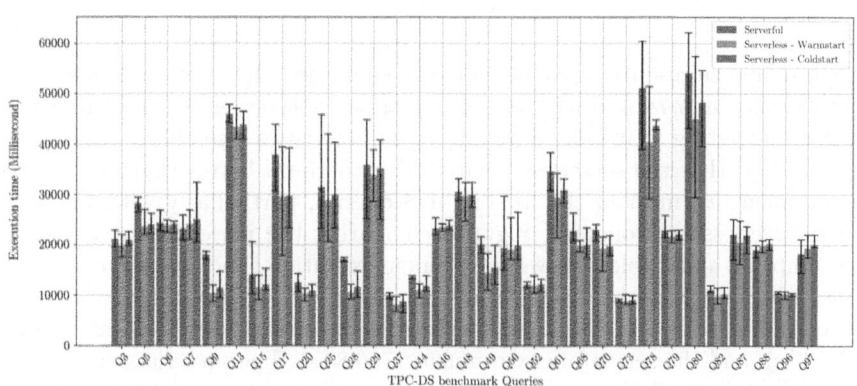

Fig. 5. TPC-DS benchmark queries on dataset of size is 25 GB

Analyzing the error bars in the average execution time chart of 31 TPC-DS queries (Fig. 5) reveals that a serverless environment has higher stability than the serverful environment. The shorter error bars in the serverless environment indicate lower fluctuations and better predictability. This stability is due to the serverless environment's isolation and flexible resource allocation capabilities,

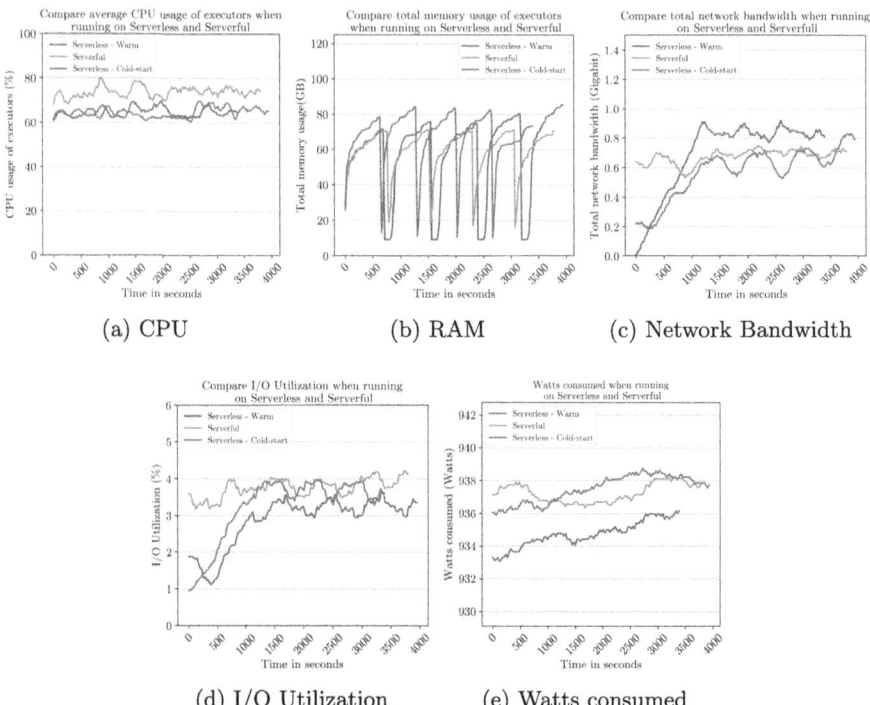

(a) CPU   (b) RAM   (c) Network Bandwidth

(d) I/O Utilization   (e) Watts consumed

Fig. 6. Compares performance and energy consumption in Case Study 2

which minimize resource contention between queries. However, for some specific queries, such as *q78* and *q80*, the error bars on the serverless environment are longer, suggesting that individual query characteristics also affect stability. The results show that the serverless environments perform better than the serverful environment for most of the 31 TPC-DS queries, particularly with the warm-start serverless environment. Figure 6 comparing the performance and energy consumption and Table 2 shows the average execution time across 5 consecutive runs, which reveals that the warm-start serverless environment has the lowest execution time (651.90 s), followed by the cold-start serverless environment (677.49 s), and the serverful environment (736.83 s). The serverless environment also demonstrates better energy consumption than the serverful environment. The warm-start serverless environment consumes the least energy (1200 Wh), followed by the cold-start serverless environment (1380 Wh), and the serverful environment (1330 Wh).

The efficiency in performance and energy consumption of the serverless environment, especially in the warm-start scenario, can be explained by several factors. First, the serverless architecture allows for flexible resource allocation and scaling, ensuring that resources are optimally used for each query, particularly in complex *join* operations within the TPC-DS database. Second, the initialization

Table 2. Comparison of execution time and energy consumption in Case Study 2

Environment	Execution time (s)	Total energy usage (Wh)
Serverless cold-start	677.49	1380
Serverless warm-start	651.90	1200
Serverful	736.83	1330

process in the serverless environment, including container creation and resource allocation, is optimized to minimize overhead, especially in warm-start scenarios. This reduces waiting time and allows the system to quickly become ready to handle workloads. Moreover, the nature of the TPC-DS workload, with short-lived join queries and frequent changes in resource requirements, aligns well with the flexible scaling capabilities of serverless. In join queries, *data locality* plays a crucial role: serverless intelligent resource scheduling ensures that tasks are processed closer to the data, reducing the need for network data transfer and thus improving overall performance. In contrast, serverful environments with fixed resources can lead to resource underutilization during low demand and resource contention during peak loads, negatively impacting both performance and energy efficiency. Serverful environments often fail to optimize *data locality* and cannot dynamically adjust resources, resulting in lower processing efficiency, especially in large-scale join operations such as those in TPC-DS.

6.3   Case Study 3: Impact of Auto-Scaling of Resources

In this experiment, our focus is to study the impact of the automatic scaling capabilities of the serverless environment when processing TPC-DS queries with a larger database size of 50 GB. We evaluate the effectiveness of automatically adjusting the number of worker nodes in a serverless environment based on the workload. We compare the resulting query processing performance and energy consumption with those in a serverful environment. The experiment was conducted on a compute cluster consisting of 1 master node and a varying number of worker nodes ranging from 10 to 20 nodes. For the serverless environment, we use the configured auto-scaling parameters *spark.dynamicAllocation.enabled*, *spark.dynamicAllocation.maxExecutors*, and *spark.dynamicAllocation.minExecutors* to allow Spark to automatically adjust the number of executors based on resource demand. We measured and compared the cold-start time, total processing time, and power consumption in the serverless environment (in both cold-start and warm-start states) and the serverful environment.

When we examine the cold-start time results in Table 3, we can see that enabling auto-scaling actually increases the cold-start time in both serverless cold-start and warm-start states. Specifically, with a configuration of 1 master node and 10 worker nodes, the cold-start time increases from 87.50 s to 98.50 s for the cold-start serverless environment, and from 19.50 s to 30.50 s for the warm-start serverless environment when auto-scaling is enabled. We see a similar trend with the configuration of 1 master node and 20 worker nodes. This increase occurs because of the initialization and configuration processes of the components relating to auto-scaling, such as the Kubernetes Cluster Autoscaler and Horizontal Pod Autoscaler, which require additional time for setup and synchronization. However, in the serverful environment with fixed resource allocation, enabling or disabling auto-scaling has no effect on the cold-start time. Even though enabling auto-scaling increases the cold-start time in the serverless environment, it can be thought of as an acceptable trade-off to achieve flexible scalability and more efficient resource utilization while processing queries on big data.

Table 3. Comparing serverless and serverful performance with auto-scaling in Case Study 3

	1 master, 10 workers			1 master, 20 workers		
	Cold-start	Warm-start	Serverful	Cold-start	Warm-start	Serverful
I. Without Auto-Scaling						
Cold-start time	87.5	15.0	2.0	150.0	30.0	2.0
Processing time	1134.6	1146.0	1150.0	1024.8	960.0	1025.0
Total (s)	**1222.1**	**1161.0**	**1152.0**	**1174.8**	**990.0**	**1027.0**
Power (Wh)	**1671**	**1600**	**1650**	**2500**	**2250**	**2450**
II. With Auto-Scaling						
Cold-start time	98.5	15.0	2.0	150.0	30.0	2.0
Processing time	1100.0	1140.0	1150.0	990.0	951.0	1025.0
Total (s)	**1198.5**	**1155.0**	**1152.0**	**1140.0**	**981.0**	**1027.0**
Power (Wh)	**1655**	**1569**	**1650**	**2468**	**2187**	**2450**

For the processing time in Table 3, enabling auto-scaling significantly improves both the serverless environments with cold-start and warm-start states. With a configuration of 1 master node and 10 worker nodes, the processing time slightly decreases from 1134.60 s to 1100.00 s for the cold-start serverless environment and from 1146.00 s to 1140.00 s for the warm-start serverless environment when auto-scaling is applied.

When comparing power consumption in Table 3, it is evident that applying auto-scaling results in the use of energy more efficiently, especially in the serverless environments. With a configuration of 1 master node and 10 worker nodes, the power consumption decreased from 1600 Wh to 1569 Wh (a 1.97% reduction). Similarly, with a configuration of 1 master node and 20 worker nodes, enabling auto-scaling reduced power consumption from 2250 Wh to 2187 Wh (a 2.88% reduction) in the warm-start serverless scenario. These results highlight the role of auto-scaling in optimizing resource utilization and minimizing energy waste. In contrast, power consumption in the serverful environment showed no difference between applying and not applying auto-scaling (with energy consumption being 4.90% and 10.73% higher compared to warm-start serverless, respectively), indicating that serverful architectures may be less flexible in optimizing resources and saving energy compared to serverless environments.

Although the energy savings may seem small, at large-scale data centers, auto-scaling can lead to significant reductions in energy consumption. This is crucial for zero-carbon cloud computing, where providers aim to cut emissions through renewable energy and efficient computing models. As data and computational demands grow, auto-scaling and serverless architectures will be essential for sustainable cloud computing.

7 Conclusion and Future Research

This study highlights the fundamental differences between serverless and serverful environments in terms of performance, energy consumption, and resource management. Serverless architectures excel in handling intermittent and unpredictable workloads by enabling automatic initialization, elastic scaling, and timely resource deallocation, which contributes to greater energy efficiency. In contrast, serverful models offer consistent performance and lower latency for continuous workloads but are generally less efficient due to static resource allocation. Our experimental findings show that, under warm-start conditions, serverless systems outperform serverful setups in both execution speed and energy usage. However, cold-start latency remains a critical drawback for serverless deployments, limiting their suitability for real-time applications. Overall, the decision to adopt a serverless or serverful architecture depends on multiple factors, including workload pattern, data volume, operational costs, and system management capabilities.

Future studies should further explore the applicability of serverless Spark in public cloud ecosystems such as AWS Lambda, Google Cloud Functions, and Azure Functions, where service constraints and runtime environments vary. A deeper investigation into cold-start mitigation techniques—especially in relation to interarrival times—could provide key strategies for improving responsiveness. Moreover, examining hybrid deployment models that integrate both serverless and serverful paradigms may yield solutions that balance flexibility and performance. Evaluating energy efficiency, cost trade-offs, and scalability in real-world cloud environments will also help to establish practical best practices for deploying Apache Spark in production workloads.

References

1. Big data analytics market forecast, 2024–2032 (2024). https://www.fortunebusinessinsights.com/big-data-analytics-market-106179. Accessed 25 Mar 2025
2. Alhindi, A., Djemame, K., Heravan, F.B.: On the power consumption of serverless functions: an evaluation of OpenFaaS. In: IEEE International Conference on Utility and Cloud Computing (UCC) (2022). https://doi.org/10.1109/UCC56403.2022.00064
3. Borra, P.: Serverless computing: the future of scalability and efficiency with AWS, Azure, and GCP. Int. J. Adv. Res. Sci. Commun. Technol. (2025). https://doi.org/10.48175/IJARSCT-23373
4. Cao, Z., Zhou, X., Hu, H., Wang, Z., Wen, Y.: Toward a systematic survey for carbon neutral data centers. IEEE Commun. Surv. Tutorials (2022). https://doi.org/10.1109/COMST.2022.3161275
5. Darius, P.S., et al.: From data to insights: a review of cloud-based big data tools and technologies. In: Big Data Computing: Advances in Technologies, Methodologies, and Applications (2024). https://doi.org/10.11648/j.ajist.20240803.11
6. Ghorbian, M., Ghobaei-Arani, M.: A survey on the cold start latency approaches in serverless computing: an optimization-based perspective. Computing (2024). https://doi.org/10.1007/s00607-024-01335-5
7. Govind, H., González–Vélez, H.: Benchmarking serverless workloads on kubernetes. In: IEEE/ACM International Symposium on Cluster Computing and the Grid (CCGRID) (2021). https://doi.org/10.1109/CCGrid51090.2021.00085
8. Luengo, J., García-Gil, D., Ramírez-Gallego, S., García, S., Herrera, F.: Big data: technologies and tools. Big Data Preprocess. (2020). https://doi.org/10.1007/978-3-030-39105-8_2
9. Mittal, V., et al.: Mu: an efficient, fair and responsive serverless framework for resource-constrained edge clouds. In: SoCC 2021: Proceedings of the ACM Symposium on Cloud Computing (2021). https://doi.org/10.1145/3472883.3487014
10. Perez, A., Risco, S., Naranjo, D.M., Caballer, M., Molto, G.: On-premises serverless computing for event-driven data processing applications (2019). https://doi.org/10.1109/CLOUD.2019.00073
11. Phan, A.T., d'Orazio, L., Phan, T., Gruenwald, L.: GUESS: monitoring join query execution in serverless and serverful spark. In: International Conference on Database Systems for Advanced Applications (DASFAA), Gifu, Japan, July 2024 (2024). https://doi.org/10.1007/978-981-97-5575-2_45
12. Sethy, K.K., Singh, D., Biswal, A.K., Sahoo, S.: Serverless implementation of data wizard application using azure kubernetes service and docker. In: 1st IEEE International Conference on Industrial Electronics: Developments & Applications (ICIDeA) (2022). https://doi.org/10.1109/ICIDEA53933.2022.9970103
13. Shafiei, H., Khonsari, A., Mousavi, P.: Serverless computing: a survey of opportunities, challenges, and applications. ACM Comput. Surv. (2022). https://doi.org/10.1145/3510611
14. Shanahan, J.G., Dai, L.: Large scale distributed data science using apache spark. In: 21th ACM SIGKDD International Conference on Knowledge Discovery and Data Mining (2015). https://doi.org/10.1145/2783258.2789993
15. Vahidinia, P., Farahani, B., Aliee, F.S.: Cold start in serverless computing: current trends and mitigation strategies. In: International Conference on Omni-Layer Intelligent Systems (COINS) (2020). https://doi.org/10.1109/COINS49042.2020.9191377

16. Zhang, Q., Liu, L., Pu, C., Dou, Q., Wu, L., Zhou, W.: A comparative study of containers and virtual machines in big data environment. In: IEEE International Conference on Cloud Computing, CLOUD (2018). https://doi.org/10.1109/CLOUD.2018.00030
17. Zhu, C., Han, B., Zhao, Y.: A comparative study of spark on the bare metal and kubernetes. In: International Conference on Big Data and Information Analytics (BigDIA) (2022). https://doi.org/10.1109/BigDIA51454.2020.00027

Graph Applications

The Missing Link: Joint Legal Citation Prediction Using Heterogeneous Graph Enrichment

Lorenz Wendlinger[1]([✉]) [iD], Simon Alexander Nonn[1] [iD],
Abdullah Al Zubaer[1] [iD], and Michael Granitzer[1,2] [iD]

[1] Universität Passau, Passau, Germany
[2] Interdisciplinary Transformation University Austria, Linz, Austria
{lorenz.wendlinger,simon.nonn,
abdullahal.zubaer,michael.granitzer}@uni-passau.de

Abstract. Legal systems heavily rely on cross-citations of legal norms as well as previous court decisions. Practitioners, novices and legal AI systems need access to these relevant data to inform appraisals and judgments. We propose a Graph-Neural-Network (GNN) link prediction model that can identify Case-Law and Case-Case citations with high proficiency through fusion of semantic and topological information. We introduce adapted relational graph convolutions operating on an extended and enriched version of the original citation graph that allow the topological integration of semantic meta-information. This further improves prediction by 3.1 points of average precision and by 8.5 points in data sparsity as well as showing robust performance over time and in challenging fully inductive prediction. Jointly learning and predicting case and norm citations achieves a large synergistic effect that improves case citation prediction by up to 4.7 points, at almost doubled efficiency.

Keywords: Legal Tech · Link Prediction · Graph Neural Networks

1 Introduction

To prepare any legal argument and especially an appraisal, the relevant norms for the specific circumstances of the case must be found, checked, and referenced with concurring or dissenting opinions in the literature and court judgments to support one's (legal) argumentation. For experienced legal practitioners, who specialize in one area of the law, this context information is available through convention and memorization or by consulting the relevant commentaries. However, for novice legal practitioners and students, the relevant references can be challenging to research; especially the initial norms for a case can be hard to find. Some conventions are not obvious without intimate knowledge of the relevant law and legal opinions as well as court decisions in a specific domain.

While legal reference extraction [14,18] can make this data available, it cannot extrapolate missing references. The prediction of references exceeds the pure

© The Author(s), under exclusive license to Springer Nature Switzerland AG 2026
R. Wrembel et al. (Eds.): DEXA 2025, LNCS 16047, pp. 197–211, 2026.
https://doi.org/10.1007/978-3-032-02088-8_14

retrieval of similar documents, as it utilizes and extends the reference corpus of a legal text that is inexorably linked to its outcome and logical framework. It can help legal students and practitioners by suggesting suitable references or indicating incorrect citations for their argumentation. Furthermore, the so-extracted context can be used for further processing in generative models, making link prediction an essential task for building effective legal tech systems.

Recent advances in LLMs make this problem tractable with generative models, which are more versatile than discriminative models, but can be less robust and tend to hallucinate. Recent work [24] suggests that they require resource-intensive compute and natural language inference, making them unsuited to resource-constrained and real-time applications.

We solve this task by predicting legal norm and case references for documents with missing references based on semantics and topology to provide a missing link between these data. This fusion enables us to include rich meta-information into document representation by modelling them in a joint citation graph. This meta-data implicitly encodes relevant details about the relevant law domain and makes this important context available even if other data is missing. Specifically, our contributions are:

- We effectively and efficiently predict case and law references in a large citation network composed of real cases and laws by leveraging semantic text-based and topological information
- We incorporate semantically relevant categorical meta-information via graph enrichment without the need for separate feature extraction
- We modify the original relational graph convolution with a general residual for better representation learning in large sparse graphs
- We jointly learn and predict two types of references, drastically improving efficiency and results on smaller graphs
- We propose a realistic evaluation method that is temporally cohesive and tailored to a wide range of possible applications

2   Related Work

There is a large catalogue of work on graph learning and related tasks. While many rule-based approaches to link prediction have been explored in the past [2], they are not easily adaptable to heterogeneous graphs. We therefore focus on deep learning techniques that can be extended to handle such data, which are commonly known as Graph Neural Networks (**GNN**).

A link prediction model can be considered a graph auto-encoder as it learns to reconstruct the original graph from representations generated by the encoder. A popular choice for the decoder is the dot product for its computational efficiency. This makes the decoder completely static, while the encoder is learned via back-propagation on the reconstruction task, i.e. binary link classification.

Most GNN encoders are built around the graph convolution (**GCN**) developed by Kipf et al. in [9]. They use topological information to compute node-level features in each layer by propagation along graph edges, effectively leveraging the representations of neighbouring nodes.

The Variational Graph Auto Encoder (**VGAE**) [10] is an extension that predicts the target distribution and is trained using the KullbackâĂŞLeibler divergence in conjunction with the reconstruction loss.

Veličković et al. [19] propose the graph attention mechanism (**GAT**) which computes attention scores along edges. This includes a learnable notion of node importance into the neighbourhood aggregation.

Schlichtkrull et al. introduce the relational **RGCN** [17] for relational graphs that propagates all relations using separate learned weights and then take their normalized sum as the node representation.

However, Wang et al. [21] show that, in a fair comparison, GCN outperforms these three adaptations in the most common link prediction task. This partially negates that purported progress of link prediction models and emphasizes the importance of realistic evaluation.

Wang et al. explore Heterogeneous Attention Networks (**HAN**) for node classification on small heterogeneous graphs in [20]. They operate on reachable graphs induced on the original network via tuples of relations, so-called metapaths. This method is not directly adaptable to link prediction, as it removes the original edges, losing valuable information.

Dhani et al. explore legal knowledge graphs for link prediction and document similarity in [5]. Their employ an RGCN with LegalBERT [3] features on 2 286 Indian law documents. In their link prediction experiments, document embeddings outperform both hand-crafted and bag-of-words node features.

Palmer et al. [13] compare various semantic embedding methods for to Case-Case citation prediction. They rephrase this task on a paragraph level and show that it is challenging to solve.

3   Joint Link Prediction with Heterogeneous Graph Enrichment

Our Heterogeneous Graph Enrichment Model (**HGE**) expands on previous work regarding link prediction on relational data and utilization of meta-features.

We focus on citation networks, a heterogeneous graph structure $\mathcal{G} = (\mathcal{V}, \mathcal{E})$, composed of Nodes $\mathcal{V}$ that represent the topology of documents and the citations $\mathcal{E}$ between them. Specifically, $\mathcal{V} = \{\mathcal{V}_C, \mathcal{V}_L\}$ contains both *Cases* $\mathcal{V}_C$ and *Laws* $\mathcal{V}_L$, with relations, or edge types, $T_e = \{r_{CC}, r_{CL}\}$. Cases can reference cases via $r_{CC} : \mathcal{V}_C \rightarrow \mathcal{V}_C$ and laws via $r_{CC} : \mathcal{V}_C \rightarrow \mathcal{V}_L$. An example is given in Fig. 2.

3.1   Heterogeneous Graph Convolution

Like the relational Graph Convolution of [17], we use separate weight matrices, i.e. GCN layers, for each relation type $r \in T_e$. For node i in layer l, the original RGCN with activation σ and normalization constant $c_{i,j}$ computes

$$h_i^{(l)} = \sigma\left(\sum_{r \in T_e} \sum_{j \in \mathcal{N}_i^r} \frac{1}{c_{i,r}} W_r^{(l)} h_j^{(l)} + W_0^{(l)} h_i^{(l-1)} \right), \qquad (1)$$

where W_r is a learnable parameter for each relation and $\mathcal{N}_i^r$ is the neighbourhood of i under r. Unlike [17], we do not model the residual self-loop separately, but instead add a general residual as the self-loop is contained in the enriched graph relation types T_e' in our model:

$$h_i^{(l)} = \sigma\left(h_i^{(l-1)} + \sum_{r \in T_e'} \sum_{j \in \mathcal{N}_i^r} \frac{1}{c_{i,r}} W_r^{(l)} h_j^{(l)} \right), \qquad (2)$$

Aggregation is still performed for each layer as the sum over all relevant relation types for each node type, c.f. Fig. 1. This can help derive useful features for sparse graphs, such as those of a scale-free nature, by making short feature paths available through residual modelling alongside the long dependencies computed by stacking multiple GNN layers, while alleviating the issue of over-smoothing that plagues deeper GNNs.

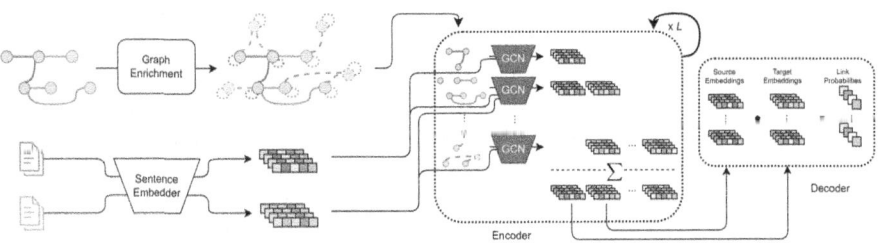

Fig. 1. Heterogeneous Graph Enrichment model with L layers including and node embedding for *Cases* and *Laws*. All learned parameters are contained in the encoder. Asymmetric decoding, activation and residuals are omitted for visual clarity.

3.2   Joint Modelling

In contrast to previous work, we investigate link prediction of two distinct edge types. By jointly learning prediction for both edge types, we can condition the model to produce representations that are informative for both tasks, as Case nodes are used in both relations. In many graphs, but in our multi-faceted case especially, shortcuts can lead to suboptimal training results that do not generalize well. Joint learning achieves a regularizing effect that prevents collapse to the trivial solution.

3.3   Graph Enrichment

We perform metafeature-based graph enrichment to accurately model the relationship between cases and laws. Like in [17], we also add reverse edges and self-loops, though they are processed as regular relations, c.f. Eq. 2.

Exposed Meta-features. Descriptive and structural meta-data can provide relevant information about the data itself, as document semantics go beyond just the text. Especially for a complex legal system, the level of jurisdiction, type of case, or area of law can drastically change how and which references are made. We want to include these additional semantics in the node representation, though, unlike the initial text-based node features, this meta-information is categorical and therefore not natively compatible with the node feature space. However, we can include the meta-information about nodes into their representation via the citation network topology by exposing it as discrete nodes for each feature expression. For a meta-feature $m : \mathcal{V} \rightarrow \{0,1\}^M$ with M possible expressions, we modify the relational graph $\mathcal{G} = (\mathcal{V}, \mathcal{E})$ such that

$$\mathcal{V}' = \mathcal{V} \cup \{f : \exists u \in \mathcal{V} : m(u) = f\}$$
$$\mathcal{E}' = \mathcal{E} \cup \{(u, m(u)) \; \forall u \in \mathcal{V}\} \tag{3}$$

This does not require additional features for the categories, e.g. courts or law areas. Moreover, it directly models how nodes that share meta-information are related. In this extensional model, a meta-feature is exclusively defined by the nodes that express it, such as an area of law is defined by the laws that govern it via a specific law book, c.f. Fig. 2. In contrast to latent aggregation over relevant node representations, we propose dynamically generating it via the same GNN propagation rule. They are finally aggregated via the same sum rule for heterograph convolution, c.f. Eq. 2.

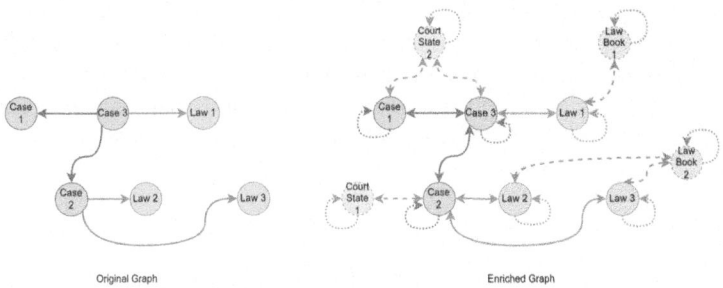

Original Graph   Enriched Graph

Fig. 2. Original and fully enriched graph with dynamic *Case* and static *Law* nodes and directed original edges. The enriched graph adds *Law Book* and *Case Court State* nodes with corresponding feature edges and self-loops for all nodes.

Intuitively, this serves two purposes: it collects salient representations for exposed features from their respective nodes and propagates them back to the original nodes for contextualizing. We experiment with multiple ways of exposing this information, such as metapath modelling (c.f. [20]) and direct normalization, but we find that including them as discrete nodes offers the most versatility.

We choose the meta-features `Case Court Type`, `Law Code` and `Case Court State` for graph enrichment. The `Case Court Type` is especially informative, as it is composed of both the court level of appeal and jurisdiction. Together, these meta-features encode information about the most important legal context, i.e. the type of case, area of the law and the local regulations, and make it available explicitly through our graph enrichment.

3.4   Initial Node Features

Based on the work by Milz et al. [11], who find strong links between graph and text expression, we propose a fusion of topological and semantic information by integrating textual features into our graph neural network. We use a BERT-based [4] sentence transformer model to generate latent embeddings for the text-based Case and Law nodes. All legal references are removed via pattern-matching to prevent shortcutting and data leakage. More specifically, documents are split into context-window size chunks with a small overlap via recursive splitting on white space and then aggregated via their mean. This ensures maximum context size while maintaining consistency over breakpoints. We use Jina V2 Embeddings [6] for their large context window of 8192 and availability of German-specific `jina-v2-base-de` embeddings[1].

We compare the effect of different overlaps and the multilingual Siamese BERT encoder `all-mpnet-base-v2`[2] [15,16] in Subsect. 5.5.

3.5   Training Procedure and Parameter Setting

To generate link likelihoods from node representations, we use the asymmetric inner product decoder from [23]. HGE is trained by optimizing the cross-entropy reconstruction loss via adaptive momentum [8] gradient descent for 200 epochs with a learning rate of 10^{-4}. Random dropout [7] is applied to the node representations to prevent co-adaptation and promote robustness with a probability of 0.2. We use 3 layers of size 256 for all models and 2 attention heads for GAT.

4   Datasets

We empirically verify our method on legal citation graphs from [11,12], and label them according to their *Case* counts as **OLD36k**[3] and **OLD201k**[4].

Ostendorff et al. [12] propose the **O**pen **L**egal **D**ata platform for facilitating open access to legal data. They collect data for more than 250k German laws and court decisions up to 12/2022 by crawling government websites and trusted services. They perform reference extraction, named entity recognition and topic extraction for some of them and publish the resulting graph, OLD36k.

Milz et al. [11] analyse a German citation network based on data from [12]. They use an improved reference extraction and linking method to construct the

---

[1] https://huggingface.co/jinaai/jina-embeddings-v2-base-de.

[2] https://huggingface.co/sentence-transformers/all-mpnet-base-v2.

[3] https://static.openlegaldata.io/dumps/de/refs/; March 30, 2025.

[4] https://osf.io/8d2v4/; March 30, 2025.

Table 1. Dataset statistics for OLD36k and OLD201k. *: not directly used, only for graph enrichment, c.f. Subsect. 3.3, †: edge counts differ from those reported [11], which were generated on preliminary data and have been updated (see Footnote 4).

Type	Name	OLD36k	OLD201k	meta-information
Node	Case	36 113	201 823	type, date
	Law	10 304	50 814	law book code, law book title, section
	Court*	-	1 119	name, type, slug, city, description state, jurisdiction, level of appeal
Edge	**Case-Case**	17 065	90 189†	-
	Case-Law	424 862	971 625†	-
	Case-Court*	-	201 823†	-

larger OLD201k citation network. They further show their data to be scale-free and find node similarity to correlate with text similarity. They identify reference linking as a pertinent issue for Case-Case citations, as only 16.3% of extracted references could be added to the network due to missing data.

As OLD36k and OLD201k are extracted from Open Legal Data dumps, they have full text available for all cases and laws. In addition, both datasets contain meta-information about the cases, laws and courts (directly modelled for OLD201k, indirectly for OLD36k).

Because both graphs are generated from different data bases with different extraction method, and none of the included identifiers is fully unique, losslessly mapping between the two citation networks is not possible and we treat them as separate samplings from the same domain.

As OLD201k is slightly newer with a cutoff date of 2020-12-10, while OLD36k was created on 2019-02-19, and because it is more complete (c.f. Table 1), we use it for the bulk of our experiments. OLD36k is an almost subgraph of OLD201k with a mismatch of only 28 Cases and elucidates behaviour in scenarios with more limited reference sets or missing data.

Their data distributions are similar (c.f. Fig. 3), though the case library of OLD201k stretches back much further, with some cases from 1970. OLD201k also contains slightly longer Cases and Laws, with both distributions showing less positive skew. Consequently, 85% of OLD201 cases fit within one embedding context window, c.f. Fig. 3b, but 88.9% of OLD36k cases do.

5   Experiments

We first report full link prediction results on OLD201k and OLD36k in Subsect. 5.2 and contrast them with separate prediction in Subsect. 5.3. An ablation study investigates the effect of various model components in Subsect. 5.4. The influence of initial node embedding, i.e. text embedding models and chunking parameters,

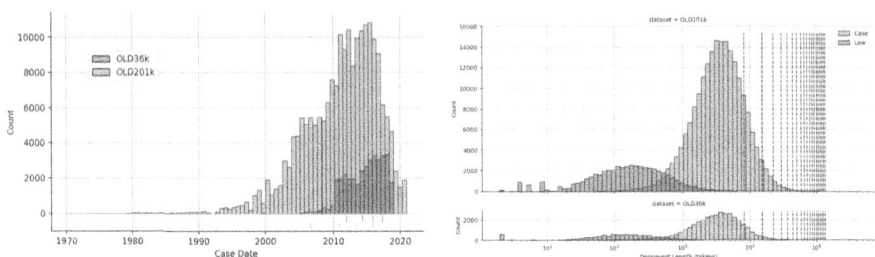

(a) Case date with cutoffs for date splitting indicated for OLD201k and OLD36k

(b) Document Length with embedding context window marked

Fig. 3. OLD201k and OLD36k Dataset Distributions.

are explored in Subsect. 5.5. Finally, Subsect. 5.6 investigates model robustness and sensitivity in the context of time-based influences and data availability.

We implement all models in DGL 2.4.0[5] using pytorch 2.6.0[6] compiled with CUDA 12.4 and run experiments on an Nvidia A100 GPU. All code, experiment setups, final and intermediate results are available at https://github.com/wendli01/missing_link.

5.1 Methodology

Graph splitting is challenging as graphs are a cohesive data structure from which samples cannot be drawn independently. For link prediction, two main scenarios can be differentiated. In transductive settings, the node set is the same for training and testing, i.e. all nodes are known at training time, whereas for inductive splitting, predictions are made for new nodes.

Naive random splitting may introduce artifacts by removing nodes that leave their neighbours in an invalid state. To control this effect, we employ time-based splitting, which ensures a valid training graph for each split, as suggested in [22]. This produces a training graph, a semi-inductive inference graph including training data that representations for the test predictions are generated on, and the indicator graph that contains node pairs for prediction, c.f. Fig. 4b.

For OLD36k we treat laws as static nodes, while cases are split into equal-sized splits via a cut-off date based on their associated date, c.f. Fig. 4b. This automatically results in valid Case-Case edges, as only cases that are already published can be referenced.

By default, we perform cumulative training and testing by training on all old nodes and evaluating on all new nodes, c.f. Fig. 4a. This varies the train/test proportion for each split and adds an estimate of data efficiency as well as robustness to concept drift over time. We evaluate these effects separately in Sect. 5.6.

---

[5] https://www.dgl.ai/, March 30, 2025.
[6] https://pytorch.org/, March 30, 2025.

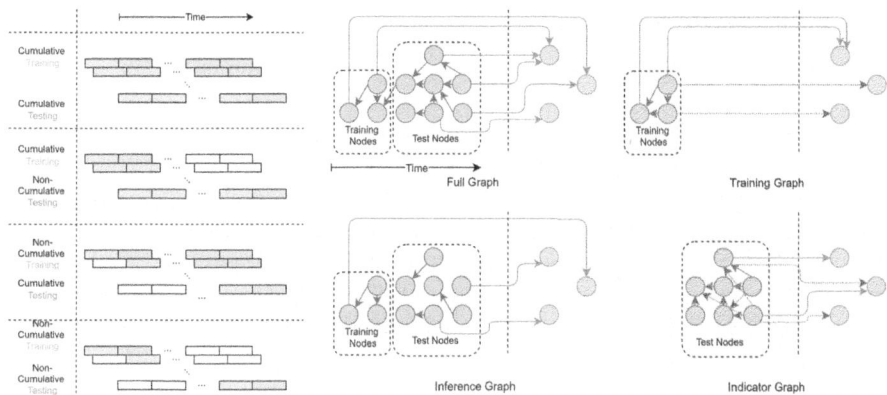

(a) Time-based splitting with (non-)cumulative training and testing.

(b) Time-based graph splitting with dynamic Case and static Law nodes including <u>real</u> edges and training / test <u>negatives</u>.

Fig. 4. Time-based splitting modes and graph splitting.

By subsampling on both edge types, we can create an incomplete inference graph and a test graph. This represents a scenario in which complete training data is available, but the current data is missing references. By default, we create 5 such test splits with a test ratio of 90% for each time split. Varying the test ratio allows us to gauge how much information the model can recover without requiring retraining, which we exploit to assess the effects of data sparsity in Sect. 5.6. For simplicity, we employ balanced negative sampling for both training and testing. We use uniform per-source sampling, which generates a negative edge $(u, v') \notin \mathcal{E}$ for each real edge $(u, v) \in \mathcal{E}$.

For each experiment we report the average precision (AP) as well as the area under the Receiver-Operator-Characteristic curve (AUC-ROC). While AUC-ROC mostly measures the ranking quality of predictions, AP puts a larger emphasis on the positive performance. These measures are more impactful than metrics that evaluate at a threshold of 0.5, as they assess performance across all possible threshold values. This is especially relevant in our study, as applications might focus on top-end performance, like recommending the top-n references to a user, or across the whole spectrum, i.e. for retrieval-augmented generation, or at the bottom end for rejecting erroneous user citations.

Scores are calculated separately for the two edge types as well as combined into a micro and macro score. For imbalanced datasets, macro-averaging ensures that performance on all edge types is adequate, instead of the score being dominated by the majority class, i.e. Case-Law citations, especially as the majority type is usually easier to predict. This is representative of the application case, where a case citation can be as impactful as a law citation.

5.2   Link Prediction

We evaluate the joint link prediction capabilities of our model and previous approaches on both OLD versions in Table 2. Performance is generally high, owed to the potent fusion of semantic and topological information in our GNN models. A purely semantic SGD classifier operating only on text embeddings cannot compete. While for OLD201k homogeneous methods still perform adequately, they cannot recover from the information loss incurred on the homogenized OLD36k. Their performance is also inconsistent across the two datasets, with VGAE showing promising results on OLD201k but not on OLD36k.

Among the heterogeneous models, our novel HGE out-performs RGCN by a large margin on OLD201k, 2.2 and 3.1 points of average precision and AUC-ROC respectively. Only HGE can effectively utilize heterogeneous information for improved ranking, while RGCN does not offer AUC-ROC improvements over VGAE. In the data sparsity of OLD36k, the margin is higher still at 7.2 and 8.5 points respectively. At the same time, HGE does not add a large performance penalty, with run-time comparable to RGCN[7], as this method also processes each edge type separately, and faster than VGAE, which requires extra standard deviation estimation and loss propagation. As HGE scales well with increased graph size, it is comparatively more efficient for OLD201k.

Table 2. Link prediction results in 5 folds with date-based splitting and evaluation on 5 90% test splits. Results $\pm\sigma$ are reported in % with macro-averaging over the edge types, for per-edgetype results see Table 3. GAT, GCN and VGAE operate on the fully homogeneous bi-directed graph.

Data	OLD201k				OLD36k			
		AUC-	time(s)			AUC-	time(s)	
	AP	ROC	test	train	AP	ROC	test	train
SGD	$72.1_{\pm0.88}$	$72.6_{\pm1.11}$	$4.2_{\pm1.25}$	$21.8_{\pm16.3}$	$64.4_{\pm0.22}$	$64.0_{\pm0.29}$	$1.5_{\pm0.6}$	$14.9_{\pm6.6}$
GAT	$79.6_{\pm2.75}$	$81.2_{\pm2.04}$	$2.3_{\pm1.1}$	$45.0_{\pm30.3}$	$73.0_{\pm0.83}$	$73.0_{\pm0.72}$	$0.7_{\pm0.9}$	$19.9_{\pm2.5}$
GCN	$80.4_{\pm1.8}$	$81.7_{\pm1.62}$	$2.3_{\pm1.1}$	$44.9_{\pm30.5}$	$73.3_{\pm0.95}$	$73.3_{\pm0.94}$	$0.8_{\pm01.0}$	$20.5_{\pm2.6}$
VGAE	$82.9_{\pm1.04}$	$82.7_{\pm1.21}$	$3.3_{\pm1.1}$	$111.8_{\pm69.4}$	$71.0_{\pm0.85}$	$71.7_{\pm1.0}$	$1.0_{\pm1.0}$	$30.3_{\pm3.9}$
RGCN	$85.9_{\pm1.46}$	$\underline{82.8}_{\pm1.99}$	$0.3_{\pm0.1}$ (see Footnote 7)	$39.2_{\pm15.9}$	$\underline{80.3}_{\pm1.4}$	$\underline{77.3}_{\pm1.27}$	$0.1_{\pm0.0}$ (see Footnote 7)	$27.1_{\pm4.9}$
HGE	$\mathbf{88.1}_{\pm1.33}$	$\mathbf{85.9}_{\pm1.55}$	$3.3_{\pm0.9}$	$76.7_{\pm33.9}$	$\mathbf{87.5}_{\pm1.05}$	$\mathbf{85.8}_{\pm1.34}$	$1.3_{\pm1.1}$	$46.4_{\pm5.3}$

The promising AP and AUC-ROC scores of HGE on both OLD graphs show that our relational modelling and enrichment is effective and make it suitable for link prediction across a variety of possible thresholds and application scenarios.

5.3   Joint Prediction

As we predict jointly by default, we compare the efficacy and effectiveness to separate prediction here. The input data is the same for both scenarios, only the

---

[7] The large test time difference is due to the optimized RGCN implementation in DGL.

prediction target changes - separate models therefore have access to all edges for propagation. For joint prediction, *CC* and *CL* are learned and predicted by the same model, while separate models are just that.

Table 3. Edge type scores for HGE in joint vs. separate prediction.

dataset	training	Average Precision		ROC-AUC		time (s)	
		CC	CL	CC	CL	train	test
OLD201k	joint	$\mathbf{78.2}_{\pm 2.63}$	$98.0_{\pm 0.1}$	$\mathbf{73.8}_{\pm 3.03}$	$98.0_{\pm 0.12}$	$\mathbf{76.7}_{\pm 34}$	$\mathbf{3.3}_{\pm 0.9}$
	separate	$77.6_{\pm 3.12}$	$98.0_{\pm 0.1}$	$73.4_{\pm 3.18}$	$98.0_{\pm 0.1}$	$137.4_{\pm 63}$	$7.1_{\pm 1.1}$
OLD36k	joint	$\mathbf{79.8}_{\pm 2.03}$	$95.2_{\pm 0.18}$	$\mathbf{76.8}_{\pm 2.56}$	$94.7_{\pm 0.25}$	$\mathbf{46.4}_{\pm 5.3}$	$\mathbf{1.3}_{\pm 1.1}$
	separate	$75.1_{\pm 1.2}$	$95.1_{\pm 0.22}$	$69.4_{\pm 1.36}$	$94.7_{\pm 0.26}$	$60.3_{\pm 3.4}$	$1.6_{\pm 1.2}$

Regardless of dataset, *CC* edges are more challenging to predict than *CL*, though for OLD201k the difference is more pronounced. While this may have inherent reasons, such as higher heterogeneity and time-sensitivity of case citations, it can also be explained by the significant linking issues for CC edges reported in [11], c.f. Sect. 4.

We find that joint prediction offers regularization that prevents short-cutting and substantially improves results for the minority edge type, *CC*. This is especially pronounced for OLD36k, where fewer data is available and joint learning increases AP and ROC-AUC by 4.7 and 7.4 points, c.f. Table 3. For OLD201k the improvement is still measurable at 0.6 and 0.4 points. More importantly, due to the joint nature, training is sped up by a factor of 1.79x and 2.15x respectively, with testing scaling similarly.

5.4   Ablation Studies

We also evaluate the precise impact of different HGE components on link prediction on OLD201k in Table 4, while holding all other hyper-parameters constant.

We find that type information has the greatest impact on average precision, though HGE operating on the homogeneous graph still out-performs all other approaches, including the RGCN (Table 2), as it can recover a portion of the type information through the semantic node features. The effect of reverse edges is comparatively small, and at 0.5 points macro AP, comparable to the 0.7 points macro ROC-AUC difference of omitting residuals (c.f. Table 4). Not exposing meta-features lowers AP and ROC-AUC scores by 0.7 and 1 point(s).

We can deduce that reverse edges and type information mostly benefit the identification of positives, while skip connections and exposed meta-features also help with finding representations that can encode negatives.

With everything else equal, as per the HGE results without exposed features, our adapted heterograph convolution still outperforms the original RGCN by 4.6 and 2.9 points of AP and ROC-AUC respectively (c.f. Tables 4 and 2).

Table 4. HGE ablation study for OLD201k link prediction in 5 folds with date-based splitting and evaluation on 5 90% test splits.

	Average Precision		ROC-AUC		time (s)	
	micro	macro	micro	macro	test	train
HGE	$97.1_{\pm0.12}$	$88.1_{\pm1.32}$	$97.0_{\pm0.12}$	$85.9_{\pm1.54}$	$3.2_{\pm0.9}$	$74.7_{\pm31.4}$
w/o reverse edges	$96.1_{\pm0.27}$	$87.6_{\pm1.69}$	$95.8_{\pm0.28}$	$85.9_{\pm2.31}$	$2.8_{\pm0.9}$	$49.8_{\pm20.4}$
w/o skip connections	$96.9_{\pm0.1}$	$87.6_{\pm1.44}$	$96.8_{\pm0.13}$	$85.2_{\pm1.84}$	$3.2_{\pm0.9}$	$75.4_{\pm32.5}$
w/o exposed features	$96.9_{\pm0.09}$	$87.4_{\pm1.5}$	$96.8_{\pm0.15}$	$84.9_{\pm1.8}$	$2.9_{\pm1.8}$	$51.5_{\pm24.6}$
homogeneous	$97.0_{\pm0.07}$	$85.7_{\pm1.43}$	$97.0_{\pm0.06}$	$85.9_{\pm0.96}$	$2.6_{\pm0.9}$	$34.5_{\pm10.6}$

5.5   Node Embeddings

We investigate the impact of the initial node embedding method, as it generates the representation space that nodes are mapped into for all our experiments.

Table 5. Node Embedding study on OLD201k using the full HGE.

	context	overlap	Average Precision		ROC-AUC	
			micro	macro	micro	macro
all-mpnet-base-v2 [16]	384	16	$96.6_{\pm0.2}$	$86.8_{\pm1.47}$	$96.5_{\pm0.22}$	$84.3_{\pm1.96}$
jina-V2-base-de [6]	4096	512	$97.1_{\pm0.08}$	$88.0_{\pm1.4}$	$97.0_{\pm0.09}$	$85.7_{\pm1.67}$
	2048	256	$97.0_{\pm0.12}$	$88.0_{\pm1.44}$	$96.9_{\pm0.13}$	$85.7_{\pm1.79}$
	8192	1024	$97.1_{\pm0.09}$	$88.1_{\pm1.34}$	$97.0_{\pm0.12}$	$85.9_{\pm1.56}$

Overall, we observe the impact of the semantic node embedding method to be small, with chunking parameters showing no significant effect (c.f. Table 5). The over-smoothing usually encountered when aggregating over many embeddings is not detrimental. Even with the substantially smaller embedding and context window of all-mpnet-base-v2, most relevant information can be recovered by HGE. While we did experiment fine-tuning with Attention over Sentence Embeddings [1], it did not improve results and could not easily be scaled to our data. This confirms our decision of using latent initial node embeddings.

5.6   Robustness and Sensitivity

In the time-based evaluation scenario we have chosen, multiple aspects can be varied to gauge robustness and sensitivity to temporal and data changes.

While performance does improve over time, c.f. Fig. 5a, this is mostly due to the low data quality of the earliest cases that include only very few extracted citations. Conversely, this low-quality data does not poison the cumulative training set, as non-cumulative training performance is worse, and HGE does not experience degradation due to drifts. Non-cumulative testing confirms that there is some temporal dependency in the data, though it is not well-behaved.

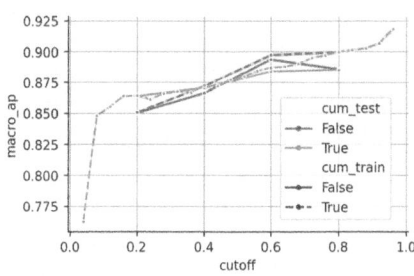 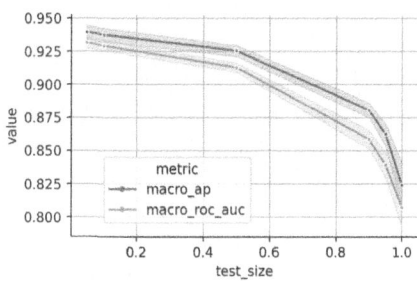

(a) Time and Quantity via date splitting from 1970 to 2022 with (non) cumulative training and testing.

(b) Data Sparsity via test graph edge ratio. At 1.0 prediction is fully inductive.

Fig. 5. Robustness of HGE on OLD201k over time and varying data availability.

We estimate the model to be somewhat sensitive to data sparsity in the availability of inference edges, c.f. Fig. 5b. With lower test sizes than the default of 0.9, and correspondingly more inference edges available, model performance is improved. In the fully inductive setting without any inference links (i.e. test size at 100%), HGE can still recover information with scores above 80%, by leveraging the topology induced through our meta-feature graph enrichment.

6   Conclusion and Future Work

We solve the task of legal reference prediction on two large real-world citation networks with a semantic-topology fusion GNN approach that outperforms semantic methods by a large margin. We further present a novel method for heterogeneous citation prediction that can effectively and efficiently exploit meta-information and synergies in joint learning. It is robust over a large temporal range and handles data sparsity well. While it operates on full text features, these are latent and the model is discriminative, which results in promising scalability.

We believe that this can be an effective tool for the writing and assessing of legal documents for both legal practitioners and novices by lowering the barrier of entry for finding norm and case citations for a relevant case. Moreover, it can help identify even obscure relevant precedent, which is an important step in building legal tech systems.

Though our system is effective, robust and scalable, there are limitations to its applicability, such as the availability of new legal data. Integrating new knowledge requires embedding cases, extracting references and re-training the model, which we estimate at a runtime of less than one hour (c.f. Table 3). Predicting the references of one case requires encoding it and querying all possible targets, i.e. computing the dot-product with all other nodes, which we measure to take less than 5 s for OLD201k, though with an optimized implementation, or reference set partitioning, this could be reduced even further.

Both considered citation graphs have missing links due to imperfect reference extraction and linking, which, we believe, could be improved by applying new Natural Language Inference methods and limiting the reference set to disjoint graph components. Identifying them is feasible along jurisdiction and geographic lines with the help of legal experts and would also improve link prediction effectiveness and efficiency by limiting the considered reference pool.

Our reference prediction is limited to the document level throughout, while for legal practitioners, it might be valuable to know which part of their opinion should reference which legal paragraph or case detail, as this can drastically change the outcome of a legal appraisal. Re-framing this along the lines of [13] would require either a separate fine-grained prediction model or modelling as paragraph nodes, with corresponding sentence-level attention, and might only be tractable with a limited reference set, as mentioned above.

As a simplification, we have treated the law as a static document set, though in practice it is ever-changing. Collecting and integrating previous versions into our reference set could improve performance on and transfer from historic cases.

Acknowledgements. The paper has been partially funded by COMET K1- Competence Center for Integrated Software and AI Systems (INTEGRATE) within the Austrian COMET Program and by the German Federal Ministry of Education and Research (BMBF) within the project DeepWrite (Grant. No. 16DHBKI059).

References

1. Abdaoui, A., Dutta, S.: Attention over pre-trained sentence embeddings for long document classification. arXiv preprint arXiv:2307.09084 (2023)
2. Arrar, D., Kamel, N., Lakhfif, A.: A comprehensive survey of link prediction methods. J. Supercomput. **80**(3), 3902–3942 (2024)
3. Chalkidis, I., Fergadiotis, M., Malakasiotis, P., Aletras, N., Androutsopoulos, I.: LEGAL-BERT: the Muppets straight out of law school. arXiv preprint arXiv:2010.02559 (2020)
4. Devlin, J., Chang, M.W., Lee, K., Toutanova, K.: BERT: pre-training of deep bidirectional transformers for language understanding. In: Burstein, J., Doran, C., Solorio, T. (eds.) Proceedings of the 2019 Conference of the North American Chapter of the Association for Computational Linguistics: Human Language Technologies, Volume 1 (Long and Short Papers), pp. 4171–4186. Association for Computational Linguistics, Minneapolis, Minnesota, June 2019
5. Dhani, J.S., Bhatt, R., Ganesan, B., Sirohi, P., Bhatnagar, V.: Similar cases recommendation using legal knowledge graphs. arXiv preprint arXiv:2107.04771 (2021)
6. Günther, M., et al.: Jina embeddings 2: 8192-token general-purpose text embeddings for long documents (2024). https://arxiv.org/abs/2310.19923

7. Hinton, G.E., Srivastava, N., Krizhevsky, A., Sutskever, I., Salakhutdinov, R.R.: Improving neural networks by preventing co-adaptation of feature detectors. arXiv preprint arXiv:1207.0580 (2012)
8. Kingma, D.P., Ba, J.: Adam: a method for stochastic optimization. arXiv preprint arXiv:1412.6980 (2014)
9. Kipf, T.N., Welling, M.: Semi-supervised classification with graph convolutional networks. arXiv preprint arXiv:1609.02907 (2016)
10. Kipf, T.N., Welling, M.: Variational graph auto-encoders (2016). https://arxiv.org/abs/1611.07308
11. Milz, T., Granitzer, M., Mitrović, J.: Analysis of a German legal citation network. In: Proceedings of the 13th International Joint Conference on Knowledge Discovery, Knowledge Engineering and Knowledge Management (IC3K 2021) - Volume 1: KDIR, pp. 147–154. INSTICC, SciTePress (2021)
12. Ostendorff, M., Blume, T., Ostendorff, S.: Towards an open platform for legal information. In: Proceedings of the ACM/IEEE Joint Conference on Digital Libraries in 2020, JCDL 2020, pp. 385–388. ACM, August 2020. https://doi.org/10.1145/3383583.3398616
13. Palmer Olsen, H., Garneau, N., Panagis, Y., Lindholm, J., Søgaard, A.: Re-framing case law citation prediction from a paragraph perspective. In: Legal Knowledge and Information Systems, pp. 323–328. IOS Press (2023)
14. Palmirani, M., Brighi, R., Massini, M.: Automated extraction of normative references in legal texts. In: Proceedings of the 9th International Conference on Artificial Intelligence and Law, pp. 105–106 (2003)
15. Reimers, N., Gurevych, I.: Sentence-BERT: sentence embeddings using Siamese BERT-networks. In: Proceedings of the 2019 Conference on Empirical Methods in Natural Language Processing. Association for Computational Linguistics (2019)
16. Reimers, N., Gurevych, I.: Making monolingual sentence embeddings multilingual using knowledge distillation. In: Proceedings of the 2020 Conference on Empirical Methods in Natural Language Processing. Association for Computational Linguistics (2020). https://arxiv.org/abs/2004.09813
17. Schlichtkrull, M., Kipf, T.N., Bloem, P., van den Berg, R., Titov, I., Welling, M.: Modeling relational data with graph convolutional networks. In: Gangemi, A., et al. (eds.) ESWC 2018. LNCS, vol. 10843, pp. 593–607. Springer, Cham (2018). https://doi.org/10.1007/978-3-319-93417-4_38
18. Shulayeva, O., Siddharthan, A., Wyner, A.: Recognizing cited facts and principles in legal judgements. Artif. Intell. Law $25(1)$, 107–126 (2017). https://doi.org/10.1007/s10506-017-9197-6
19. Veličković, P., Cucurull, G., Casanova, A., Romero, A., Lio, P., Bengio, Y.: Graph attention networks. arXiv preprint arXiv:1710.10903 (2017)
20. Wang, X., et al.: Heterogeneous graph attention network. In: The World Wide Web Conference, pp. 2022–2032 (2019)
21. Wang, X., Vinel, A.: Benchmarking graph neural networks on link prediction (2021). https://arxiv.org/abs/2102.12557
22. Yang, Y., Lichtenwalter, R.N., Chawla, N.V.: Evaluating link prediction methods. Knowl. Inf. Syst. 45, 751–782 (2015)
23. Yu, Y., Wang, X.: Link prediction in directed network and its application in microblog. Math. Probl. Eng. $2014(1)$, 509282 (2014)
24. Zhang, K., Yu, W., Dai, S., Xu, J.: CitaLaw: enhancing LLM with citations in legal domain (2025). https://arxiv.org/abs/2412.14556

Graph Patterns in Fine-Grained Access Control for Graph-Structured Data

Daniel Schmid[1], Aya Mohamed[1,2,3], Dagmar Auer[1,2(✉)],
Bahara Muradi[1,2], and Josef Küng[1,2]

[1] Institute for Application-Oriented Knowledge Processing (FAW), Linz, Austria
[2] LIT Secure and Correct Systems Lab (SCSL), Linz Institute of Technology (LIT),
Johannes Kepler University (JKU) Linz, Linz, Austria
daniel@wwwmaster.at,
{dagmar.auer,bahara.muradi,josef.kueng}@jku.at
[3] University of Applied Sciences Upper Austria, Hagenberg, Austria
aya.mohamed@fh-hagenberg.at

Abstract. More and more data, one of organizations' most important assets, are graph-structured today and require advanced access control. Graph patterns are a promising approach to specify and enforce sophisticated, fine-grained authorization policies for graph-structured data. However, within the context of authorization policy specification, the term *graph pattern* has been subject to varying interpretations. Therefore, a consistent definition of graph patterns for access control models and their implementations is required. This work contributes to a more precise and consistent understanding of graph patterns within fine-grained authorization models for graph-structured data. We examine how graph patterns are conceptualized and implemented by two representative access control approaches. Both models apply graph patterns for fine-grained access control on property graphs. Although graph patterns are specified in the authorization policy and enforced independently of the datastores, these access control models differ by whether the approach is *pattern-first*, i.e., patterns are at the core of the model, or the graph pattern is an *extension*. We study the differences in pattern definition, enforcement, and implementation through a comparative analysis.

Keywords: Access Control · Authorization Policy · Graph Pattern · Graph-structured Data · Fine-grained

1 Introduction

Data are one of the most important assets of organizations, if not the most important. This requires sophisticated data protection, with access control playing an important role. In recent years, there has been a strong focus on fine-grained access control models, both in research and increasingly in practice. Today, more

R. Wrembel et al. (Eds.): DEXA 2025, LNCS 16047, pp. 212–227, 2026.
https://doi.org/10.1007/978-3-032-02088-8_15

and more of this data is graph-structured. Global Market Insights[1] for example forecast continuous robust growth in the graph database market over the next years. For graphs, both vertices and edges are considered first-class entities. Therefore, not only vertices, but also edges should be considered in access control for graphs. Edges can be used to describe semantic relationships between immediate neighboring vertices, but also between ones connected via paths of variable length. Focusing on isolated vertices and ignoring the interconnected nature of graphs can pose a data security risk. We consider the integration of this contextual information to be important, and therefore we focus on graph patterns in this paper to specify such paths.

We address the following research questions in the context of *graph patterns* for *fine-grained access control* to *graph-structured data*.

RQ1 What are the core concepts of graph patterns?
RQ2 What are the current access control approaches to graph-structured data using patterns?
RQ3 What are the main differences concerning graph patterns in these access control approaches?

In [10], we have compared access control approaches to protect graph-structured data without considering context in the data graph. As graph patterns are an intuitive way to describe this context, we identify and describe access control models to answer RQ2, and further study and compare them in detail for RQ3.

The main contributions of our work to the area of access control for graph-structured data are in defining and highlighting the concept of graph patterns for fine-grained access control. In addition to our study of the underlying concepts, we compare two access control models and their implementations with respect to graph patterns. XACML4G [9] is an *extension* of XACML to support graph patterns, while AReBAC [11] is entirely pattern based and therefore follows a *pattern first* approach. To enable a fair comparison, we had to overcome the limitations of Rizvi's AReBAC implementation [11] and re-implemented the AReBAC core model and algorithms in an open source project.

Our research is based on the results of our comparative study on access control models for graph-structured data in [10]. We start by studying the core concepts *graph pattern*, *property graph* and *fine-grained access control for graph-structured data*, and then search for corresponding access control models to be studied in detail, which meet the following constraints: (1) authorization definition and enforcement are considered and (2) a detailed description of the theory and a (prototypical) implementation are provided. The comparison considers the core aspects of graph patterns and describes some of them according to a test scenario.

The rest of the paper is structured as follows. In Sect. 2, we discuss the core terms property graph and fine-grained access control for graph-structured data and formally define our understanding of graph patterns. We give insight into related work in Sect. 3 and provide an access control scenario in Sect. 4 to support

---

[1] https://www.gminsights.com/de/industry-analysis/graph-database-market.

the discussion of specification, enforcement, and application of graph patterns for fine-grained access control in Sect. 5. In Sect. 6, we compare and analyze the different approaches. The paper concludes with a summary and an outlook on future work in Sect. 7.

2   Graph Pattern Concepts

Depending on the domain and the graph model, the understanding of *graph patterns* may differ. In the following, we explain how we understand graph patterns that support fine-grained access control for graph-structured data, i.e., graph patterns in authorization specification and enforcement, and clarify relevant key concepts.

Property Graph. We assume that the data to be protected can be modeled as a property graph and thus focus on approaches applied to this graph model.

Since mathematics has been dealing with graph theory for almost 300 years, there is no standard vocabulary. Vertices are often called nodes or points, whereas relationships, lines, and arcs are synonyms of edges [4]. In the following, we use the terms *vertex* and *edge*. We solely employ *node* and *relationship* in the description of the data layer. Depending on the source, terms are used as synonyms or defined differently. To gain insight into the variety of *property graph* definitions, we discuss the following characteristics [2, 4, 5]:

- *Directed/undirected edges.* Gutierrez et al. [5] and Angles [2] define property graphs as directed graphs, whereas Green et al. [4] emphasize their hybrid edge structure, i.e., edges can be directed or undirected.
- *Multigraph.* Multiple edges between two vertices are highlighted in [2, 4, 5].
- *Pseudograph.* Only Gutierrez et al. [5] stress self-referencing edges.
- *Labeled.* Labels on vertices and edges are considered by [2, 5], but not by [4].
- *Properties/attributes.* Both terms are used synonymously in various publications. All consider properties on vertices and edges fundamental to property graphs. A property is a pair of key and value. Vertices and edges can have a finite set of properties. The term attributed graph is often used synonymously with property graph (e.g. [13]), but also for graphs with properties on the vertices [1]. It is also used to distinguish property graphs from graphs that have properties only on the vertices [1].

We define property graph as a directed labeled multigraph with properties on vertices and edges. We consider the term attributed graph as a synonym.

Fine-Grained Access Control for Graph-Structured Data. The finest granularity of the protection level is a single vertex or a single edge. We do not consider the protection of specific attributes. However, attributes and labels should be taken into account in the definition and enforcement of authorizations on vertices and edges. Since subjects can be modeled as part of the overall graph, attribute-based constraints apply not only to the protected objects, but also to the requesting subjects. In our research [8–10] and also in relationship-based access control research [11,12], it is argued that due to the interconnected nature of a graph, we need to consider not only isolated resources, but also their context to make access decisions. Graph patterns seem to be an intuitive way to describe this context.

Graph Pattern. In general terms, a graph pattern is a structure that defines a specific set of vertices and edges that serve as a template for matching sub-structures within a larger graph. Graph patterns are used in different areas, such as graph query languages, graph pattern recognition, or graph analytics. Our domain is protecting graph-structured data that are retrieved via queries. Thus, our focus is on pattern specification in query languages, in particular on the recently published Graph Query Language (GQL) standard[2].

Taking the GQL standard into account, we consider a graph pattern to be a set of path patterns. A path pattern matches a path, i.e., an odd-length sequence of graph elements, starting with a vertex and followed by any number of edge vertex pairs. Moreover, patterns on single elements can be specified to match a single vertex or edge. In our context, such element patterns can incorporate both the labels of graph elements and their associated properties. According to Green et al. [4], property graphs (in the context of GQL) have mixed edges. We assume property graphs to have directed edges. However, undirected edges can be used in the specification of patterns for queries and authorizations. The grammar in Listing 1 defines the structure of a graph pattern (in EBNF notation).

Listing 1. Graph pattern structure

```
GraphPattern = PathPattern { PathPattern } . // no duplicates
PathPattern = VertexPattern { EdgePattern VertexPattern } .
VertexPattern = ident↑vertexId { ident↑vertexLabelName } { Attribute } .
EdgePattern = ident↑edgeId [ident↑edgeLabelName] Direction { Attribute } .
Attribute = ident↑name ident↑value .
Direction = "-->" | "<--" | "--" .
```

In this section, we have discussed the basic characteristics of graph patterns for property graphs with focus on query specification. We also presented our understanding of these essential concepts.

3   Related Work

This paper is based on Mohamed et al. [10], where we compared authorization and access control approaches for graph-structured data in terms of access control approach, authorization policy definition, and enforcement. The approaches were selected according to the following criteria: protecting graph-structured

---

[2] ISO/IEC 39075:2024 Information technology—Database languages—GQL, https://www.iso.org/standard/76120.html.

data, enforcing fine-grained access control, and application in graph datastores. Graph patterns were not considered.

In this paper, we focus on access control approaches based on graph patterns. Starting with our previous results, we have to exclude the work of Bereski et al. [3] which extended the *role-based access control (RBAC)* model in Neo4j to support attributes. The authorization policy relies on the Neo4j access control model (i.e. statically defined rules) with an extension to support fine-grained conditions, but does not consider paths, which are a core part of graph patterns. The general query rewriting approach (GQRA) proposed in Hofer et al. [6,7] enforces fine-grained access control by rewriting Cypher queries. Although authorizations can be implemented as a proprietary set of rules using paths with placeholders for runtime values, GQRA focuses on the enforcement, which is not sufficient for our current study. The remaining models, XACML4G [9] and AReBAC [11], represent two basic approaches to use graph patterns in access control: extending an access control model by integrating graph patterns (*extension*) and a model that is fundamentally based on graph patterns (*pattern first*). These two models will be studied in detail in Sect. 5 supported by a demonstration case introduced in Sect. 4.

4   Access Control Scenario

We use a common access control scenario as a running example to demonstrate the specification and enforcement of graph patterns. The scenario relies on a subset of the Airbnb dataset provided by Inside Airbnb[3]. We only consider listings, hosts, reviewers, and reviews with their relationships. A host can offer listing(s) for renting (e.g., rooms, flats, houses). A reviewer writes review(s) after staying in one or more listed accommodations. Figure 1 shows a part of the graph abstracted from the concrete data.

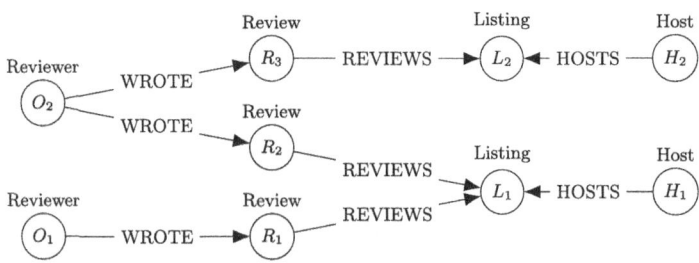

Fig. 1. Sample graph containing reviewers, reviews, listings, and hosts

The Host H_1 is connected to the Listing L_1 via the HOSTS relationship. There are two reviews R_1 by O_1 and R_2 by O_2 for this listing. In our authorization scenario, only the reviewer who wrote the review or the host of the reviewed

---

[3] https://insideairbnb.com/get-the-data/.

listing is allowed to read the reviews. For example, the reviewer O_2 can read the review R_2, but not the review R_1, although there is a path between O_2 and R_1, as O_2 is not the reviewer of R_1. Graph patterns can be used to define much more complex authorizations with a wide variety of context restrictions.

5   Access Control Based on Graph Patterns

In this section, we discuss different ways to specify and enforce graph patterns for fine-grained access control in *XACML for Graphs (XACML4G)* [9] and in *Attribute-supporting Relationship-Based Access Control (AReBAC)* [11]. Since the provided AReBAC implementation used a test dataset that was no longer available, it was not feasible to reproduce the reported results or apply the implementation to alternative scenarios required for our comparison. Thus, we implemented the pattern-related parts of AReBAC to compare both approaches based on the access control scenarios defined in Sect. 4.

5.1   XACML4G

XACML for Graphs is a fine-grained datastore-independent authorization and access control approach to graph-structured data [8,9]. It is based on *extensible access control markup language (XACML)*[4], which relies on the *attribute-based access control (ABAC)* model and supports fine-grained constraints for the basic entities (e.g., subject, resource, action and environment conditions). However, XACML lacks the natural specification of the relationships between the subject and the resource. Therefore, XACML4G extends the policy language model introducing *pattern* and the reference architecture to apply concepts from the *relationship-based access control (ReBAC)* model.

Pattern Specification. Each rule in the authorization policy can optionally define a graph pattern. The pattern element is composed of a recursively defined path as a vertex, a connecting edge, and either another vertex or an entire path. In theory, the base case is a single vertex, but practically there is no need for such a pattern, as it can be simply matched and evaluated with the standard XACML policy. Listing 2 provides the structure and attributes of *pattern*, *vertex*, and *edge* as defined in our XML schema definition (XSD).

Listing 2. XACML4G pattern structure and attributes

```
Pattern  →   ε  |  <xacml4g:pattern> Path </xacml4g:pattern>
Path  →   <xacml4g:path> Vertex Edge (Vertex | Path) </xacml4g:path>
Vertex  →   <xacml4g:vertex> Attr_vertex> AnyOfSequence </xacml4g:vertex>
Attr_vertex→  (vertexId)? (label)? (category)?
Edge  →   <xacml4g:edge> Attr_edge> AnyOfSequence </xacml4g:edge>
Attr_edge→  (edgeId)? (type)? (category)? (direction)?
            (length)? (minLength)? (maxLength)?
AnyOfSequence  →   (<xacml:anyOf>...</xacml:anyOf>)*
```

---

[4] https://docs.oasis-open.org/xacml/3.0/xacml-3.0-core-spec-os-en.html.

Each path element (i.e., vertex and edge) can have constraints at the attribute level like in XACML (i.e., a sequence of *anyOf*) along with an identifier, a vertex label/edge type and a category as attributes ($Attr_{Vertex}$ and $Attr_{Edge}$). The vertex category indicates whether it is a subject, a resource, or belongs to the path, whereas an edge has only the resource or path category. The edge element can have additional attributes, such as direction (i.e., from, to, or any) and length-related properties (i.e. *length*, *minLength*, and *maxLength*) for flexible patterns by specifying a range for some parts of the path.

The element to describe pattern-related conditions for a policy rule is called *pattern condition*. Based on the identifier of vertices and edges, the pattern condition can join and compare pattern elements (as a whole or individual attributes) in addition to specifying further constraints using supported functions, e.g., for comparing strings.

In XACML4G, requests can also have a path instead of attributes for the basic entities only. For this, the request is extended to differentiate between action and path attributes. Path attributes are structured into a sequence of attributes, each representing a vertex of any category, except for the resource, which can also be an edge. As the subject and resource elements in the path are identified by their category, there can be any number of vertices in between or even beyond. To handle the different naming conventions for the *id* attribute in different data sources, the identifier name and value of a vertex or a resource edge can be specified in the request attribute value.

Applying the access control scenario provided in Sect. 4, two different subject vertices (i.e., *Host* and *Reviewer*) are allowed to access the *Review* vertex as a resource. Thus, a policy having the resource and action in the *target* element is defined with two rules, each of them specifying the subject and path in the *target* and *pattern* elements respectively. The *target* vertices are defined using their labels without additional constraints at the attribute level. For this, the attribute xacml4g:1.0:subject:subject-vertex-label in XACML4G is used to define the vertex label for the subject category. The same applies to the resource vertex label. Both rules have *permit* as *effect*, indicating that access is authorized when the policy is matched and the respective rule is successfully evaluated. The path of reviewers to their reviews is direct via the WROTE relationship, while the pattern from hosts to reviews of their listings requires a recursive path as described in Listing 3.

Listing 3. XACML4G pattern for access control scenario

```
<xacml4g:Pattern PatternId="hostToReviewPattern">
  <xacml4g:Path>
    <xacml4g:Vertex Category="urn:oasis:names:tc:xacml:1.0:subject-category:access-subject"/>
    <xacml4g:Edge Category="xacml4g:1.0:path-category:edge" Type="HOSTS" Direction="from"/>
    <xacml4g:Path>
      <xacml4g:Vertex Label="Listing" Category="xacml4g:1.0:path-category:vertex"/>
      <xacml4g:Edge Category="xacml4g:1.0:path-category:edge" Type="REVIEWS" Direction="to"/>
      <xacml4g:Vertex Category="urn:oasis:names:tc:xacml:3.0:attribute-category:resource"/>
    </xacml4g:Path>
  </xacml4g:Path>
</xacml4g:Pattern>
```

Enforcement. The XACML conceptual components are extended to apply the XACML4G pattern along with its conditions, and evaluate them against request paths. An access request is first processed by the context handler, which also extracts the path attributes to be used in policy matching and pattern evaluation. XACML4G patterns are currently handled as a XACML condition, which is dynamically added during the evaluation of the matched policy rule(s) in the policy decision point (PDP). Accordingly, a query is generated from the path elements and pattern-related conditions in the policy in addition to path attributes representing the input path in the request. It searches for the intersection of the rule pattern from the matched policy and the request pattern. This query and other intermediate ones needed for fetching information for policy matching (e.g., vertex label and edge type) are executed in the extended policy information point (PIP). If the query returns true, the pattern evaluation is successful. If the generated query for the condition does not yield a result, an *indeterminate* decision is returned.

In addition to our proprietary implementation supporting the policy specification in JSON [8], we extended the XSD for the XACML policy language and an open-source implementation of XACML called *Balana*[5] to support the specification and enforcement of elements related to the graph pattern. Since XACML4G follows the permit-deny access control model approach [10], the result is a decision: permit, deny, not applicable (i.e., no matching policy) or indeterminate (i.e., error during policy evaluation).

Recalling the scenario in Sect. 4, the policy is matched according to the defined subject vertex with label `Host` and `Review` as a resource vertex. The following condition in Listing 4 is added to the conditions' list of the matched policy rule, which checks whether the query associated with the pattern identifier returns true or not.

Listing 4. Condition for pattern evaluation

```
<Condition>
  <Apply FunctionId="urn:oasis:names:tc:xacml:1.0:function:string-equal">
    <Apply FunctionId="urn:oasis:names:tc:xacml:1.0:function:string-one-and-only">
      <AttributeDesignator Category="urn:oasis:names:tc:xacml:1.0:subject-category:access-
          subject" AttributeId="hostToReviewPattern" DataType="http://www.w3.org/2001/
          XMLSchema#string" MustBePresent="true"/>
    </Apply>
    <AttributeValue DataType="http://www.w3.org/2001/XMLSchema#string">true</AttributeValue>
  </Apply>
</Condition>
```

The Cypher query in Listing 5 is generated and executed in the PIP based on the constraints in the pattern (refer to Listing 3). The placeholders in the query are the path attributes of the request. Attributes related to the subject and resource are also added to the policy pattern (`p1`) to narrow down the result. Upon successful matching of the policy and evaluation of the pattern, the rule effect *permit* is returned.

Listing 5. Cypher query generated for the XACML4G pattern

```
MATCH p2 = ({host_id:$hostId})--({listing_id:$listingId})--({review_id:$reviewId})
MATCH p1 = (:Host{host_id:$hostId})-[:HOSTS]->(:Listing)<-[:REVIEWS]-
```

---

[5] https://github.com/wso2/balana.

```
        (:Review{review_id:$reviewId})
WHERE ALL (x IN nodes(p2) WHERE x IN nodes(p1)) AND
      ALL (x IN relationships(p2) WHERE x IN relationships(p1))
RETURN p1 IS NOT NULL AS result
```

5.2   AReBAC

The *Attribute-supporting Relation-Based Access Control* model in Rizvi and
Fong [11] extends the ReBAC model and combines it with ABAC by incorporating attribute-based constraints. It allows access decisions based on the paths
between the subject and the resource considering edge labels as well as attributes
on vertices and edges. Both queries and policies are graph patterns that specify a graph and any number of constraints. Graph patterns restrict the query
result, therefore, the datastore must contain a subgraph corresponding to the
graph pattern where all constraints must match. Query and policy are combined
in a process called *Weaving* resulting in a single graph pattern that can be evaluated against a datastore. AReBAC graph patterns are database- and query
language-independent and consist of the following components:

- A set of vertices.
- A set of directed edges including edge labels.
- A set of mutual exclusion constraints each containing two vertices.
- Sets of vertex and edge attribute requirements for fine-grained constraints.
- A set of *returned vertices* specifying the result set, which is a subset of the
  overall set of vertices.
- A category (with relevant actors for authorization) used for policy matching.
- A mapping from the actors in the category to the corresponding vertices in
  the graph pattern. This mapping is used to combine graph patterns.

To compare XACML4G and AReBAC based on the implemented scenarios, we
extended AReBAC to support labeled vertices.

Pattern Specification. The AReBAC model is designed for database- and
query language-independent graph queries that are specified as graph patterns.
Their essential properties correspond to a subset of Cypher[6] called Nano-Cypher,
which is why graph patterns can be translated to Nano-Cypher and vice versa.
A Nano-Cypher query consists of MATCH statements specifying requested vertices and labeled edges connecting them, WHERE statements constraining these
vertices and edges, and a RETURN statement listing all vertices from MATCH statements that should be included in the result. Listing 6 shows a Nano-Cypher query
containing an edge with the label WROTE between a reviewer (O) and review (R).
With this query, the reviewer is expected to contain a specified identifier with
the review being returned.

Listing 6. Simple Nano-Cypher query

```
MATCH (O)-[: WROTE]->(R)
WHERE O.id=$reviewerId
RETURN R
```

---

[6] https://opencypher.org/.

A Nano-Cypher query can have an arbitrary number of MATCH statements as shown in Listing 7 which queries reviewers with reviews on multiple distinct listings of a given host using a mutual exclusion constraint (<>).

Listing 7. Nano-Cypher query with multiple MATCH statements

```
MATCH (reviewer)-[: WROTE]->(review1)-[: REVIEWS]->(listing1)<-[: HOSTS]-(host)
MATCH (reviewer)-[: WROTE]->(review2)-[: REVIEWS]->(listing2)<-[: HOSTS]-(host)
WHERE host.id=$hostId AND review1 <> review2
RETURN reviewer
```

Given that both queries and policies are represented as graph patterns, Nano-Cypher can be used for specifying not only queries but also policies.

Returning to the example in Fig. 1, consider a policy that allows hosts to view the reviews of their own listings. This policy can be used on queries like the one in Listing 6. This query can be translated to the graph pattern shown in Fig. 2 (a). The actors are specified above the corresponding vertices. The authorization policy in Fig. 2 (b), also specified as a graph pattern, requires the subject to host the listing for the requested review.

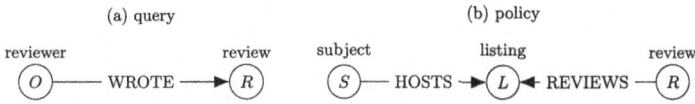

Fig. 2. Graph of a query corresponding to Listing 6 and a policy allowing hosts to access reviews on their listings

In the AReBAC model, each request comes with a category that specifies the actors accessed. These categories can have associated policies. Furthermore, *category hierarchies* are supported, where a category is *superior* to another if the former category includes all actors of the latter. All restrictions of the *inferior* categories are propagated to the superior ones. To get the matching policy for a request, one needs to first find all categories matching the request and then combine the matching policies using the *Weaving* algorithm. If a policy is specified for the superior category, it is also included in the Weaving process along with policies from inferior categories.

Enforcement. To perform access control, the request and policy graph patterns are combined into a single pattern. The *Weaving* algorithm can combine an arbitrary number of graph patterns into a single one, including all constraints from the input patterns (e.g., subject must be a specific user). The combined pattern matches a result if and only if all input patterns match. If one actor is assigned to vertices in different input patterns, these vertices ultimately result in a single vertex. Figure 3 shows the result of combining the graph patterns (query and policy) in Fig. 2.

In this example, the actor review is part of both input patterns and is therefore combined to a single vertex for that actor. The combined pattern matches a subgraph of the datastore if and only if the query and the policy match.

Fig. 3. Graph of combined pattern allowing hosts to access reviewers of their listings

As a Nano-Cypher query can be translated to a graph pattern and vice versa, the overall graph pattern can be either evaluated by Neo4j (reperesented in a Nano-Cypher query) or directly in AReBAC using the *GP-Eval* algorithm proposed by Rizvi and Fong [11]. When a graph pattern is evaluated against a datastore, nodes from that datastore are included in the result if and only if all of the following conditions are satisfied:

- Each vertex in the graph pattern must be assigned to a node in the datastore.
- Each edge in the graph pattern must have a corresponding relationship in the datastore.
- If a mutual exclusion constraint exists in the graph pattern, the corresponding nodes in the datastore must be different.
- Any vertex or edge attribute requirement in the graph pattern must be satisfied by the corresponding node or relationship in the datastore.
- A node from the datastore is included in the result if and only if it corresponds to one of the returned vertices in the graph pattern.
- Any distinct combination of returned datastore nodes can be included in the result only once, i.e., no duplicate results are allowed.

Recall the combined graph pattern in Fig. 3 for the graph shown in Fig. 1. When the subject H_1 executes the authorized query with the reviewer being set to O_2, only review R_2 is returned. R_1 is excluded because it is written by another reviewer. On the other hand, R_3 is excluded because the policy does not allow access to reviews for listings of other hosts.

The GP-Eval implementation provided by Rizvi[7] does not include the Weaving algorithm. It is also tightly coupled to Neo4j and written specifically for their test cases and benchmarks. Therefore, we provide a datastore-independent implementation[8] of the Weaving and GP-Eval algorithms, which allows custom adapters for arbitrary datastores based on property graphs. We included adapters for embedded Neo4j databases and in-memory graphs and provide tests and benchmarks. Our implementation can be used for further analysis of the AReBAC model that might not necessarily use Neo4j. In contrast to Rizvi and Fong [11], we use labeled vertices and only consider a mapping of actors to vertices instead of categories, as we only implemented the Weaving and GP-Eval algorithms, but no policy matching.

---

[7] https://github.com/szrrizvi/arebac/.

[8] https://github.com/danthe1st/AReBAC/.

6   Discussion

After discussing the notion of graph patterns in XACML4G and AReBAC in terms of specification and enforcement, we compare their concepts and implementations. XACML4G and AReBAC both use graph patterns to define fine-grained constraints at the attribute level taking relationships between entities into consideration. However, the definition and structure of the patterns differ and are applied differently. XACML4G is defined in an XML Schema (XSD) based on XACML's XSD [9], while AReBAC is formally defined in [11]

In XACML4G, which is based on the XACML policy language and reference architecture, authorization policies are hierarchically structured and matched with the access request based on the defined target in terms of subject, resource, action, and other attributes (e.g., environment). The pattern is part of the rule that belongs to the matched policy. In AReBAC, access requests and authorization policies are described entirely in graph patterns. To enforce the policy, it is combined with the request pattern using the *Weaving* algorithm and then evaluated by the GP-Eval algorithm. The combined graph pattern can also be translated to a Nano-Cypher query which is evaluated in Neo4j. Thus, graph patterns are an extension in XACML4G, while they are the core of AReBAC. Another fundamental difference, independent of the use of graph patterns, is that XACML4G is a *permit-deny access* approach that returns the access decision, while AReBAC is a *filtered results* approach only returning authorized nodes.

Pattern Specification. An overview of the differences in pattern specification is given in Table 1. In XACML4G, patterns are specified in XML or JSON format, while AReBAC relies on its proprietary graph pattern definition or Nano-Cypher, a subset of the graph query language Cypher.

In contrast to AReBAC, XACML4G not only protects vertices but also edges. A practical example is a public Address vertex which could be related to a Person. If the aim is to protect sensitive information about whose address it is and not the public Address vertex, the edge should be protected rather than the vertex. However, AReBAC could be extended to protect edges and even single attributes by adapting the policy matching logic.

Table 1. Comparison with respect to graph pattern specification

Specification aspect	XACML4G	AReBAC
Language	XML/JSON	Graph Pattern/Nano-Cypher
Protected resource	Vertices and edges	Vertices only
Structure	Recursive path patterns	Arbitrary graph patterns
Definition	Flexible paths	Fixed paths
Validation	XSD	Not considered

Concerning their structure, graph patterns in XACML4G are path patterns limited to one edge between two given vertices, due to their recursive structure. AReBAC graph patterns, on the other hand, do not restrict the number of edges for a given vertex, but arbitrary patterns can be defined.

However, the two approaches require different levels of information about the underlying data graph to specify the graph patterns. While in AReBAC all vertices and edges of the pattern must be defined, XACML4G supports, for example, the specification of a minimum and maximum number of intermediate vertices between two specific vertices, which allows more flexibility in pattern specification.

XACML4G is entirely defined in XML Schema (XSD), which allows to check, if a policy defined in XML, complies with XACML4G. No comparable means are available for AReBAC, where incorrect patterns lead to empty result sets in the evaluation, without any indication of the cause.

Enforcement. In AReBAC, authorization policies are enforced by first combining all policy and request patterns with the Weaving algorithm and then evaluating the resulting pattern with either the GP-Eval algorithm or translating the pattern to Nano-Cypher and executing the query in Neo4j. The GP-Eval algorithm maps one or more fixed vertices (e.g. subject) to nodes in the datastore, defining a subgraph which contains the result nodes. XACML4G enforces access control by matching the policy and then generating and executing queries based on pattern constraints and conditions, resulting in a decision whether access is granted.

XACML4G like XACML supports negative permissions, which can lead to conflicts at the rule or policy level and must be resolved using the rule/policy combining algorithm. In AReBAC, policies (e.g., with attribute requirements in graph patterns) are used like filters on the result set; thus, no conflicts occur. Table 2 concludes the discussion of graph pattern enforcement.

Table 2. Comparison with respect to enforcement of graph patterns

Enforcement aspect	XACML4G	AReBAC
Enforcement basis	Matching	Weaving
Evaluation	Query-based	GP-Eval or Query-based
Response	Access decision	Authorized nodes
Conflict resolution	Combining algorithm	Conflict-free by design

Finally, the performance evaluation for XACML4G in Mohamed et al. [9] showed promising results. Since the benchmarking results of Rizvi and Fong [11] could not be reproduced, we carried out preliminary benchmarks on our implementation of AReBAC to provide a basis for future performance comparison with other access control models in different datastores.

7 Conclusion

With the continuous growth of graph-structured data, there is an increasing need for access control models that provide fine-grained and context-aware data protection. Protecting single elements in a graph ignoring the related context can result in security violations. Accordingly, graph patterns specifying authorizations as paths or subgraphs are required for graph-structured data. In this work, we presented a detailed discussion on graph patterns for specifying and enforcing fine-grained access control for graph-structured data. Addressing RQ1, we have defined the core concepts of graph patterns, especially in the context of property graphs, where both vertices and edges may have labels and attributes relevant for authorization decisions.

With regard to RQ2, we studied two representative approaches, XACML4G and AReBAC. Both models use graph patterns. However, their concepts and technical implementations differ. While XACML4G *extends* the well-established XACML standard, with recursive path-structured patterns that are evaluated based on queries, AReBAC proposes a *pattern first* approach extending the ReBAC model. AReBAC combines graph patterns with the Weaving algorithm and evaluates the resulting pattern using the GP-Eval algorithm.

We addressed RQ3 through a comparative analysis of pattern specification, enforcement, and implementation details. Our AReBAC implementation allowed us to test and demonstrate both approaches (XACML4G and AReBAC) based on an access control scenario derived from a real-world Airbnb dataset. Our comparison shows that XACML4G offers flexibility through recursive path definitions, pattern validation via XSD, and supports edges as resources. In contrast, AReBAC provides a formal definition of the graph pattern-based model and the two core algorithms - pattern combining (Weaving) and pattern evaluation (GP-Eval).

In future work, we consider integrating the strengths of XACML4G and our implementation of the AReBAC concepts to enhance access control for graph-structured data. Furthermore, we plan to focus on generalizing graph pattern definitions for broader applicability, particularly in the context of knowledge graphs and other linked data systems. This includes extending validation mechanisms for patterns to ensure policy compatibility, and conducting further performance evaluations across different datastores and access control models.

Acknowledgments. This work has been supported by the LIT Secure and Correct Systems Lab funded by the State of Upper Austria and the Linz Institute of Technology, and by the COMET-K2 Center of the Linz Center of Mechatronics (LCM) funded by the Austrian federal government and the federal state of Upper Austria.

Disclosure of Interests. The authors have no competing interests to declare that are relevant to the content of this article.

References

1. Akbas, E., Zhao, P.: Attributed graph clustering: an attribute-aware graph embedding approach. In: Proceedings of the 2017 IEEE/ACM International Conference on Advances in Social Networks Analysis and Mining 2017, ASONAM 2017, pp. 305–308. Association for Computing Machinery, New York, NY, USA (2017). https://doi.org/10.1145/3110025.3110092

2. Angles, R.: The property graph database model. In: Olteanu, D., Poblete, B. (eds.) Proceedings of the 12th Alberto Mendelzon International Workshop on Foundations of Data Management. CEUR (2018). https://ceur-ws.org/Vol-2100/paper26.pdf. https://ceur-ws.org/

3. Bereksi Reguig, A.A., Mahfoud, H., Imine, A.: Towards an effective attribute-based access control model for Neo4j. In: Mosbah, M., Kechadi, T., Bellatreche, L., Gargouri, F. (eds.) Model and Data Engineering, pp. 352–366. Springer, Cham (2024)

4. Green, A., Guagliardo, P., Libkin, L.: Property graphs and paths in GQL: mathematical definitions (2021). https://doi.org/10.54285/ldbc.TZJP7279. https://ldbcouncil.org/publication/ldbc-tr-tr-2021-01/

5. Gutierrez, C., Hidders, J., Wood, P.T.: Graph Data Models, pp. 1–6. Springer, Cham (2018). https://doi.org/10.1007/978-3-319-63962-8_81-1

6. Hofer, D., Mohamed, A., Auer, D., Nadschläger, S., Küng, J.: Rewriting graphDB queries to enforce attribute-based access control. In: Strauss, C., Amagasa, T., Kotsis, G., Min Tjoa, A., Khalil, I. (eds.) Database and expert systems applications. LNCS, vol. 14146, pp. 431–436. Springer, Cham (2023). https://doi.org/10.1007/978-3-031-39847-6_34

7. Hofer, D., Mohamed, A., Nadschläger, S., Auer, D.: An intermediate representation for rewriting cypher queries. In: Kotsis, G., et al. (eds.) DEXA 2023, vol. 1872,, pp. 86–90. Springer, Cham (2023). https://doi.org/10.1007/978-3-031-39689-2_9

8. Mohamed, A., Auer, D., Hofer, D., Küng, J.: Extended authorization policy for graph-structured data. SN Comput. Sci. **2**(5) (2021). https://doi.org/10.1007/s42979-021-00684-8

9. Mohamed, A., Auer, D., Hofer, D., Küng, J.: XACML extension for graphs: flexible authorization policy specification and datastore-independent enforcement. In: Proceedings of the 20th International Conference on Security and Cryptography, pp. 442–449. SCITEPRESS - Science and Technology Publications (2023). https://doi.org/10.5220/0012090000003555

10. Mohamed, A., Auer, D., Hofer, D., Küng, J.: Comparison of access control approaches for graph-structured data. In: di Vimercati, S.D.C., Samarati, P. (eds.) Proceedings of the 21st International Conference on Security and Cryptography SECRYPT - Volume 1, vol. 1, pp. 576–583. Science and Technology Publications (SciTePress) (2024). https://doi.org/10.5220/0012861500003767

11. Rizvi, S.Z.R., Fong, P.W.L.: Efficient authorization of graph database queries in an attribute-supporting ReBAC model. In: Zhao, Z., Ahn, G.J., Krishnan, R., Ghinita, G. (eds.) CODASPY 2018, pp. 204–211. The Association for Computing Machinery, New York, New York (2018). https://doi.org/10.1145/3176258.3176331

12. Rizvi, S.Z.R., Fong, P.W.L., Crampton, J., Sellwood, J.: Relationship-based access control for an open-source medical records system. In: Weippl, E., Kerschbaum, F., Lee, A.J. (eds.) Proceedings of the 20th ACM Symposium on Access Control Models and Technologies - SACMAT 2015, pp. 113–124. ACM Press, New York, New York, USA (2015). https://doi.org/10.1145/2752952.2752962
13. Wang, Y., Li, Y., Fan, J., Ye, C., Chai, M.: A survey of typical attributed graph queries. World Wide Web **24**(1), 297–346 (2020). https://doi.org/10.1007/s11280-020-00849-0

An Efficient Point-of-Interest Placement Method Based on Betweenness Centrality

Ryuta Shiraishi[1]([⊠]), Ryusei Ohtani[1], Yuko Sakurai[1][iD],
and Satoshi Oyama[2,3][iD]

[1] Nagoya Institute of Technology, Nagoya, Japan
{r.shiraishi.749,r.otani.638}@stn.nitech.ac.jp, sakurai@nitech.ac.jp
[2] Nagoya City University, Nagoya, Japan
oyama@ds.nagoya-cu.ac.jp
[3] RIKEN Center for Advanced Intelligence Project, Chuo, Japan

Abstract. The problem of optimally placing points of interest (POIs) in public spaces such as urban streets or parks to enhance user accessibility has been a central topic in network science and artificial intelligence. However, because this task constitutes a combinatorial optimization problem, finding exact solutions becomes computationally intractable for large-scale instances. Assuming that users move from their staying locations toward exits or key destinations within the environment, we propose an approximation algorithm based on the concept of betweenness centrality from graph theory. By prioritizing locations along the most critical paths between users and destinations, our method provides a practical and scalable approach to POI placement. The simulation results demonstrate that our approach achieves high-quality solutions with significantly reduced computation time.

Keywords: POI placement · Graph · Betweenness Centrality

1 Introduction

In various domains including urban planning, tourism promotion, and disaster prevention, optimizing the placement of public facilities is essential for the effective utilization of limited resources. In urban areas, facilities such as public transportation transfer points, emergency shelters, and tourist information centers are directly related to the convenience and safety of users. The appropriate placement of these facilities, or Points of Interest (POI), in public spaces such as streets and parks has long been recognized as an important research Problem, as their locations greatly influence the overall efficiency of the urban network [4].

The POI placement problem has been widely studied as a combinatorial optimization problem, with numerous approaches proposed based on graph theory and artificial intelligence techniques [5–7], particularly in the contexts of urban planning and transportation network optimization. In particular, approaches that utilize betweenness centrality to place POIs based on the structure of the network, as well as methods that apply machine learning to dynamically place

© The Author(s), under exclusive license to Springer Nature Switzerland AG 2026
R. Wrembel et al. (Eds.): DEXA 2025, LNCS 16047, pp. 228–233, 2026.
https://doi.org/10.1007/978-3-032-02088-8_16

POIs in response to predicted user demand [9], have attracted considerable attention.

In this study, we consider a practical instance of this problem: placing trash cans in parks. Specifically, after events such as cherry blossom viewing, waste is often generated. Currently, many parks in Japan follow a "carry-in carry-out" policy, providing no trash bins within the premises. As a result, litter frequently accumulates at park entrances, exits, and nearby train stations. We address the question of where trash cans should be placed in the park, assuming that people move from their point of stay to the park's entrance or exit. Selecting multiple optimal locations to minimize the total cost of people's travel constitutes a combinatorial optimization problem, which is computationally difficult to solve exactly within a reasonable time.

To address this problem, we propose an approximate algorithm based on the concept of betweenness centrality. Betweenness centrality, defined by Freeman [3], quantifies how often a given node appears on the shortest paths between pairs of other nodes. This measure provides a way to identify nodes that play an important role in the flow of information or people within a network. Numerous studies have investigated centrality measures. However, a key issue with betweenness centrality is that its computational cost becomes prohibitively large as the network size increases. To address this, Brandes proposed an efficient algorithm for computing betweenness centrality [1]. Furthermore, Erdős et al. developed an enhanced version of this approach called Brandes++ [2]. In addition, Xiaohuan et al. pointed out that conventional centrality measures assume uniform interactions between all node pairs, and proposed a new index called SIBC (Spatial Interaction incorporated Betweenness Centrality), which incorporates spatial interactions into the computation [8].

We propose an approximate algorithm for garbage bin placement based on betweenness centrality, with the objective of minimizing the total movement cost of individuals under the assumption that people move from their stay locations within the park toward the exits. This algorithm focuses on agents' shortest paths and computes betweenness centrality accordingly. Through simulation experiments, we demonstrate that this centrality-based approach can efficiently obtain high-quality solutions while significantly reducing computation time. These results contribute to the field of POI placement from the perspective of designing practically applicable algorithms.

2   Problem Setting

This study focuses on a problem setting that models the placement of trash cans in a park. Each agent is assumed to carry trash from their location to the park exit. The park operator intends to place m trash cans so that all agents must pass at least one trash can. At this time, the capacity of the trash cans is unlimited, and the number m is fixed. Additionally, considering that agents experience psychological stress and physical constraints while carrying trash, we assume that additional costs are incurred during the movement to reach the trash cans. This assumption introduces asymmetry in movement costs between states with and without trash into the model. Therefore, the objective is to determine the locations of m relay points (i.e., trash bins) that minimize the total movement cost for all agents, subject to the constraint that each agent must pass through at least one of the selected relay nodes along the path.

To formalize this, let $N = \{1, 2, \ldots, n\}$ be the set of agents, and let $G = (V, E, w)$ be a weighted directed graph. For any edge $(v, v') \in E$, the weight $w(v, v') \in R_{\geq 0}$ denotes the cost of moving from v to v'. Each agent $i \in N$ has a starting node s_i and a destination node t_i. In addition, each agent has a cost coefficient $k_i (\geq 1)$ representing the increased cost of traveling with trash to a relay node v. The total cost of traveling from s_i to t_i via v is given by $k_i \cdot d(s_i, v) + d(v, t_i)$, where $d(v, v')$ denotes the shortest path distance between nodes v and v', which can be computed using Dijkstra's algorithm or a similar method. The objective is to find a set V^* of m relay nodes that minimizes the total cost over all agents:

$$V^* = \arg \min_{\{R \subset V, |R| = m\}} \sum_{i \in N} \min_{v \in R} (k_i \cdot d(s_i, v) + d(v, t_i)) \tag{1}$$

3   POI Placement Based on Betweenness Centrality

Betweenness centrality is one of the fundamental metrics for quantifying the importance of nodes in network analysis [3]. It measures the extent to which a particular node lies on the shortest paths between pairs of nodes in a network. The betweenness centrality $C_b(v)$ of a node v is defined as:

$$C_b(v) = \sum_{s,t \in V \setminus \{v\}} \frac{\sigma(s, t|v)}{\sigma(s, t)} \tag{2}$$

Here, $\sigma(s, t)$ denotes the total number of shortest paths between nodes s and t, and $\sigma(s, t|v)$ is the number of those paths that pass through node v. In other words, the importance of node v is quantified by summing the ratios of how frequently v appears on the shortest paths between all other node pairs.

To efficiently compute betweenness centrality, the Brandes algorithm [1] is widely used. It constructs a shortest-path subgraph rooted at a source node s and computes the dependency $\delta_s(v)$ in reverse order from the terminal nodes back to the source. By repeating this process for all nodes as source nodes, the betweenness centrality for all nodes can be obtained.

In our setting, it is not necessary to compute the betweenness centrality for all node pairs. Instead, we calculate a restricted betweenness centrality that considers only the pairs of starting and ending nodes (s_i, t_i) for each agent i. Specifically, the betweenness centrality score $C_b(v)$ is computed as follows:

$$C_b(v) = \sum_{(s_i, t_i) \in V_{st}} \delta_{s_i, t_i}(v) \tag{3}$$

$$\delta_{s_i,t_i}(v) = \sum_{t':v\in P_{s_i,t_i}(t')} \frac{\sigma(s_i,v)}{\sigma(s_i,t')}(1+\delta_{s_i,t_i}(t')), \delta_{s_i,t_i}(s_i) = \delta_{s_i,t_i}(t_i) = 0 \qquad (4)$$

Here, $P_{s_i,t_i}(t')$ denotes the set of parent nodes of node t' in the shortest path subgraph where s_i is the root and t_i is the leaf. Also, V_{st} is the set of all pairs of starts and goals of all agents.

Algorithm 1. Overall Framework for Relay-Point Selection and Cost Evaluation

---

Require: Directed graph $G = (V, E, w)$, agent set $N = \{(s_i, t_i)\}_{i=1}^n$, cost multipliers $\{k_i\}_{i=1}^n$, number of bins m
Ensure: Total travel cost
1: **Step 1:** Compute betweenness centrality scores $C[v]$ for all $v \in V$ using Eq.(3)
2: **Step 2:** Select candidate relays $R \leftarrow$ top m nodes by $C[v]$ in descending order
3: totalCost $\leftarrow 0$
4: **for all** $i = 1, \ldots, n$ **do**
5:     $(s_i, t_i) \in N$, best$_i \leftarrow \infty$
6:     **for all** relay $v \in R$ **do**
7:         $c_1 \leftarrow$ dist$(s_i, v) \times k_i$, $c_2 \leftarrow$ dist(v, t_i)
8:         best$_i \leftarrow$ min$\big($best$_i$, $c_1 + c_2\big)$
9:     totalCost $\leftarrow$ totalCost $+$ best$_i$
10: **return** totalCost

We will follow Algorithm 1. Compute betweenness centrality from the graphs and agents given as input. At this time, the cost coefficient k_i is ignored to improve the efficiency of calculation. Then, the nodes with the top m betweenness centrality scores are determined as the locations of the trash cans. Then, the agent selects the trash can with the smallest movement cost among the placed trash cans. At this time, the cost coefficient k_i is considered. Therefore, only promising candidate locations are considered without searching all nodes, which significantly reduces the computation cost and execution time.

Example 1. We consider the case where two garbage bins are to be placed. Agent 1 moves from s_1 to t_1, Agent 2 moves from s_2 to t_2, and Agent 3 moves from s_3 to t_3. First, when determining the betweenness centrality, the additional cost coefficient k_i to the trash can is ignored. Therefore, the shortest paths for the agents are $A \rightarrow D \rightarrow F \rightarrow I$, $B \rightarrow E \rightarrow H \rightarrow J$, and $C \rightarrow E \rightarrow G \rightarrow I$. Using Eq. (3), we compute the betweenness centrality scores of the intermediate nodes. For Agent 1, node F is assigned a score of 1, and node D is assigned a score of 2. For Agent 2, node H receives a score of 1, and node E receives a score of 2. For Agent 3, node G receives a score of 1, and node E again receives a score of 2. As a result, the total scores are as follows: node E has a total score of 4, node D has a score of 2, nodes F, H, and G each have a score of 1, and all other nodes have a score of 0. Therefore, the top two nodes in terms of betweenness centrality score—nodes E and D—are selected as the locations for garbage bin placement (Fig. 1).

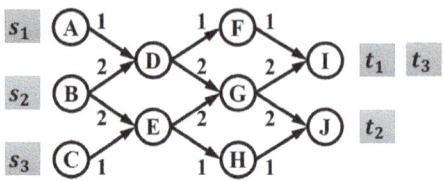

Fig. 1. Example

4   Experimental Results

We conducted computational experiments to evaluate the performance of the proposed algorithm based on betweenness centrality. We used a workstation with two AMD EPYC 7763 CPUs (2.45 GHz, 64 cores, 128 threads) and 2 TB of memory, running Ubuntu Desktop Linux 22.04 LTS and Python 3.7. The experiments were performed on randomly generated weighted directed graphs. Specifically, for graphs with $|V| \in \{10, 30, 50, 100, 200\}$ nodes, directed edges between each pair of nodes were generated with a probability of 0.3. Each edge weight was assigned as an integer uniformly sampled from the range $[1, 100]$. The number of agents was fixed at 100. For each agent i, the starting node s_i and the destination node t_i were randomly selected. The number of relay points was set to $m = 3$ and the cost coefficient k_i was different for each agent, chosen uniformly at random from the set $\{1, 2, 3\}$.

We generated and evaluated 100 random instances for each graph size. Three performance metrics were measured: (1) *feasibility*—whether it is possible to select m relay nodes such that each agent's path passes through at least one of them, (2) *total travel cost*, and (3) *CPU time*. For small graphs ($|V| \in \{10, 30, 50\}$), optimal solutions were computed exactly. For larger graphs ($|V| \in \{100, 200\}$), exact computation was infeasible due to high computational complexity, so performance was estimated using nonlinear regression based on observed data.

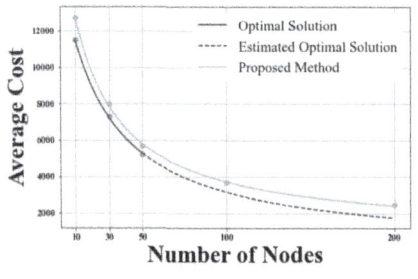

Fig. 2. Average Travel Cost

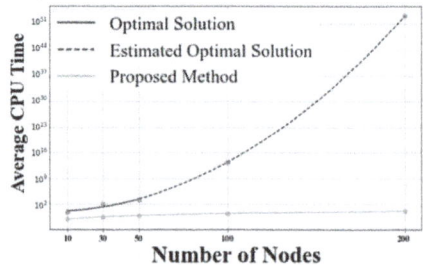

Fig. 3. Average CPU Time

In all cases for $|V| \in \{10, 30, 50\}$, both the proposed algorithm and the exact method successfully identified three valid relay nodes. The average travel costs and CPU times across instances are shown in Fig. 2 and Fig. 3, respectively. These results demonstrate that the proposed method achieves high-quality solutions with significantly lower computation time, even for larger graph sizes.

5   Conclusion

In this paper, we proposed a POI placement algorithm based on betweenness centrality, inspired by the practical problem of placing trash cans in a park. The appropriate selection of relay points is a problem of significant practical relevance. As future work, we plan to evaluate our method using real-world spatial data and user mobility patterns.

Acknowledgements. This work was partially supported by JSPS KAKENHI 24K01112, JST CREST JPMJCR21D1, and JST ERATO JPMJER2301.

References

1. Brandes, U.: A faster algorithm for betweenness centrality. J. Math. Sociol. **25**(2), 163–177 (2001)
2. Erdős, D., Ishakian, V., Bestavros, A., Terzi, E.: A divide-and-conquer algorithm for betweenness centrality. In: Proceedings of the 15th SIAM International Conference on Data Mining (SDM-2015), pp. 433–441 (2015)
3. Freeman, L.: A set of measures of centrality based on betweenness. Sociometry **40**(1), 35–41 (1977)
4. Han, P., Shang, S., Sun, A., Zhao, P., Zheng, K., Zhang, X.: Point-of-interest recommendation with global and local context. IEEE Trans. Knowl. Data Eng. **34**(11), 5484–5495 (2021)
5. Islam, M.A., Mohammad, M.M., Das, S.S.S., Ali, M.E.: A survey on deep learning based point-of-interest (POI) recommendations. Neurocomputing **472**, 306–325 (2022)
6. Li, R., Shen, Y., Zhu, Y.: Next point-of-interest recommendation with temporal and multi-level context attention. In: Proceedings of the 18th IEEE International Conference on Data Mining (ICDM-2018), pp. 1110–1115 (2018)
7. Vinayaraj, P., Najjar, A., Mede, K.: SATPOI-net: deep learning model to predict the point of interest (poi) from satellite imagery. In: Proceedings of the 43rd IEEE International Geoscience and Remote Sensing Symposium (IGARSS-2023), pp. 5257–5260 (2023)
8. Wu, X., et al.: A spatial interaction incorporated betweenness centrality measure. PLoS One **17**(5), e0268203 (2022)
9. Yin, F., Liu, Y., Shen, Z., Chen, L., Shang, S., Han, P.: Next poi recommendation with dynamic graph and explicit dependency. In: Proceedings of the 37th AAAI Conference on Artificial Intelligence, (AAAI-2023), pp. 4827–4834 (2023)

Analytics

Analytics Modelling over Multiple Datasets Using Vector Embeddings

Andreas Loizou[✉] and Dimitrios Tsoumakos

Database and Knowledge Systems Lab, School of ECE, National Technical University
of Athens, Athens, Greece
{antreasloizou,dtsouma}@mail.ntua.gr

Abstract. The massive increase in the data volume and dataset availability for analysts compels researchers to focus on data content and
select high-quality datasets to enhance the performance of analytics
operators. While selecting high-quality data significantly boosts analytical accuracy and efficiency, the exact process is very challenging given
large-scale dataset availability. To address this issue, we propose a novel
methodology that infers the outcome of analytics operators by creating a
model from the available datasets. Each dataset is transformed to a vector embedding representation generated by our proposed deep learning
model *NumTabData2Vec*, where similarity search are employed. Through
experimental evaluation, we compare the prediction performance and the
execution time of our framework to another state-of-the-art modelling
operator framework, illustrating that our approach predicts analytics
outcomes accurately, and increases speedup. Furthermore, our vectorization model can project different real-world scenarios to a lower vector
embedding representation accurately and distinguish them.

Keywords: Data Quality · Analytics Modelling · Vector embeddings ·
Vector Similarity

1 Introduction

Big data technologies daily face the rapid evolution in volume as well as variety
and velocity of processed data [12]. Such big data characteristics routinely force
analytics pipelines to underperform, requiring continuous maintenance and optimization. One major reason for this is bad data quality[1]. Poor data quality leads
to low data utilisation efficiency and even brings forth serious decision-making
errors [6].

Data quality can be improved when focusing on the actual content of the
data. Data-centric Artificial Intelligence (AI) [30] emphasises on the quality,
context, and structure of the data to improve its quality, as well as the analytical or machine learning (ML) algorithmic performance. Understanding the data
context properties, such as data features, origins, relevance, and potential biases,

---

[1] https://tinyurl.com/de62sf48.

© The Author(s), under exclusive license to Springer Nature Switzerland AG 2026
R. Wrembel et al. (Eds.): DEXA 2025, LNCS 16047, pp. 237–253, 2026.
https://doi.org/10.1007/978-3-032-02088-8_17

plays a critical role in modelling more accurate and reliable models. Data-centric AI prioritises the process of refining and enriching datasets to make them more suitable for real-world applications. Similarly, many researchers argue that prioritizing content-focused data quality is essential for achieving superior results [30].

Yet, the plethora of available data sources and datasets in an organisation data repository poses a significant challenge: Deciding the most suitable datasets for analytics workflows to ensure accurate results/predictions. While modern analytics workflows incorporate diverse operators, optimising dataset selection using data-centric AI methods remains an active research area [15]. When dataset selection is left to human experts, prediction performance drops, and it consumes more time. Equally costly and inefficient is the evaluation of all available datasets to identify high-quality inputs.

In previous work [9], predicting the output of an analytics operator assuming a plethora of available input datasets was tackled via the creation of an all-pair similarity matrix, which, relative to the similarity function used, reflected the distance between datasets over a single data quality metric (e.g., data distribution). Data or vector embeddings have been proposed to enhance big data analysis and modern AI systems. Data embedding vectorization [22,24] aims at projecting data from a high-dimensional representation space into a more compact, lower-dimensional space. Extracting meaningful information through data features using deep learning, data is projected to a lower representation space.

To improve the accuracy of a modelled analytic operator (i.e., predict the outcome of a ML algorithm without actually executing it due to its cost), we propose a framework that uses vector embeddings for dataset selection from a large data lake repository. Our method predicts an operator's output for an "unseen" query dataset, by selecting *qualitatively similar* datasets through similarity search over the vector embeddings. The selection of similar datasets reduces the prediction error, as well as the cost to model the operator, under the assumption that realistic analytical operators perform similarly under similar inputs. The embeddings are generated using our deep learning method, *NumTabData2Vec*, which processes entire tabular datasets rather than chunks or metadata, enabling efficient distinction between datasets and flexible modelling of multiple operators. Compared to similar previous work [4,9], our work uses state-of-the-art data representation (vector embeddings) which are able to capture multiple data properties that can be used in order to assess similarity, namely record order, dataset size, data distribution, etc.

The main contributions of our work can be summarised as follows:

- We introduce a framework for operator modelling (open-source prototype[2]) in order to predict its outcome on an unseen tabular input dataset from a plethora of available ones. Our method uses dataset vector embedding representations to improve the prediction performance via selecting the most relevant datasets to base its prediction upon.

---

[2] GitHub Repository.

- We develop a deep learning model architecture that transforms an entire tabular dataset of numerical values to a vector embedding representation.
- We provide an experimental evaluation of our proposed methodology using multiple real-world scenarios and compare it directly to the Apollo system [4,9].

Our evaluation illustrates that our methodology produces low prediction error by adaptively selecting similar quality datasets, achieving significant amortized speed-ups. *NumTabData2Vec* evaluation shows that it effectively projects datasets into vector embeddings while accurately capturing diverse dataset properties within the representation space.

2   Related Work

Prior efforts have focused on boosting algorithm performance by increasing data input (record number) rather than assessing quality. Consequently, we review works that identify optimal data features for analytic operator optimization. Vectorising data to lower embedding representation is a modern method that helps in identifying significant features across data types and datasets. As vector embeddings extract important features from data, we discuss studies that used the feature representation of data tuples to improve ML model prediction.

2.1   Data Quality

Big data applications aim to improve data quality by addressing various challenges. Dagger [26] enhances data quality by detecting pipeline errors using an SQL-like language, while ReClean [3] automates tabular data cleaning via reinforcement learning. IterClean [25] employs a large language model (LLM) to iteratively clean data by labelling initial tuples and using error detection, verification, and repair. In [8], data tuple quality is measured using Shapley values from game theory, with Truncated Monte Carlo Shapley and Gradient Shapley methods estimating a tuple's value to a learning algorithm. Apollo [4,9] is a content-based method predicts analytic operator outcomes by leveraging dataset similarity through three steps: creating a similarity matrix, projecting datasets to a lower-dimensional space, and modelling the operator using a small random subset of datasets. Unlike Apollo, our approach selects the most relevant, high-quality datasets to model analytic operators, aiming to improve prediction performance. Additionally, our vector embeddings incorporate all dataset properties, whereas Apollo's [9] similarity functions target only a single property.

2.2   Dataset Selection Inference

SOALA [11] selects optimal data features through online pairwise comparisons to maintain ML models over time. Its extension, Group-SOALA, introduces group maintenance to identify high-quality feature sets. In [23], the tf.data API framework enables the creation of ML pipelines focused on selecting relevant datasets and features to improve data quality. Similarly, our framework uses dataset vector embeddings to select the most suitable datasets for modelling analytic operators or ML models, enhancing prediction accuracy.

2.3   Data Vectorization and Embeddings

The goal of data vectorization is to project high-dimensional data into a lower-dimensional vector space. Word2Vec [22] (using Continuous Bag of Words and Skip-Gram) [22] leverages word context to generate embeddings. Graph2Vec [24] creates graph embeddings by dividing graphs into sub-graphs with a skip-gram model and aggregating their embeddings. ImageDataset2Vec [7] extracts meta-features from image datasets to generate embeddings, helping to select the most suitable classification algorithm. Dataset2Vec [16] uses meta-features and the DeepSet model to project datasets into embeddings and measure dataset similarity. Table2Vec [31] generates table embeddings by incorporating data features, metadata, and structural elements like captions and column headings. Mix2Vec [32], is an unsupervised deep neural network that projects mixed data into vector embeddings. In a clustering experiment like in their work on the common Adult dataset, our model outperformed Mix2Vec (recent method without publicly available code) by nearly 10%, demonstrating superior performance. Inspired by these methods, we designed a model that generates vector representations of tabular datasets over their record data values, not their metadata.

Vector embeddings, which capture valuable information from data tuples, are widely used in classification tasks. TransTab [28] encodes features with transformer layers to predict classes, leveraging supervised and self-supervised pretraining. FT-Transformer [10] and Res-Net architectures similarly use embeddings of categorical and numerical features, processed through transformer layers for class prediction, while Tab-Transformer [14] combines embedded categorical and normalized continuous features in an MLP for class prediction. Unlike these tuple-level approaches, our framework uses dataset-level embeddings to identify relevant datasets, enhancing analytic operator performance.

3   Methodology

In this section, we describe our proposed framework for modelling analytic operators over a large number of available input datasets. We also describe our approach for vectorizing tabular datasets, *NumTabData2Vec*.

3.1   Framework Architecture

Consider a data lake repository that contains a (possibly large) number n of structured tabular datasets $D = (d_1, d_2, d_3, \ldots, d_n)$. Also, let us consider an analytics operator (e.g., a ML algorithm) Φ and an "unseen" dataset D_o (from the same domain). We assume that each $D_i, 1 \leq i \leq n$ as well as D_o consist of records with numerical values only. Each dataset can, naturally, consist of

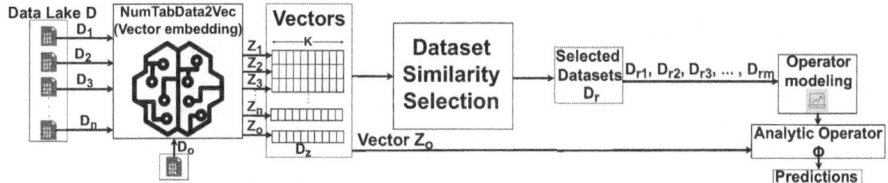

Fig. 1. Pipeline framework architecture

different number of records. Operator Φ consumes a single such dataset as input to produce a single numerical output: $\Phi : D_i \rightarrow \mathbb{R}$. Our goal is to predict $\Phi(D_o)$ with minimal cost and error by modelling the operator's output for D_o using a small subset of similar datasets $D_r \subseteq D$.

Datasets in D_r closely match D_o in their properties (e.g., order, distribution, and size to name a few). Previous work [4,9] had to use separate similarity functions for each such property. In contrast, we leverage the embedding vectorization (D_z) to efficiently identify the most similar datasets using all dataset properties.

Figure 1 depicts the pipeline of our proposed framework. Datasets in D are transformed into k-dimensional vectors using our *NumTabData2Vec* scheme, and these embeddings are stored for reuse. Each time a D_o needs to be inferred relative to an analytics operator Φ, its vector embedding is created. The datasets used for the creation of the model are selected via similarity search to produce a small subset of relevant datasets. These chosen datasets are then used to model and predict $\Phi(D_o)$, ensuring that only high-quality, pertinent data is processed. With this approach, our framework is utilizing "right quality" data in its inference mechanism, with irrelevant and extraneous datasets being excluded from the modelling process.

The datasets are embedded by our *NumTabData2Vec* method, which transforms each entire dataset in D into a k-dimensional vector z that captures all its characteristics. Our framework operates seamlessly across diverse real-world scenarios without modification, requiring only the specification of a repository containing distinct numerical tabular datasets. The Vector embedding z is a lower-dimension representation of the dataset with the entire characteristics of the dataset being encoded. Dataset D_o is similarly embedded as z_o. Using the embedding representation z_o and applying different similarity functions over the vector representations D_z, we may choose the most similar subset of D. The final step of the pipeline involves the operator modelling with any relevant method (e.g., Linear Regression, SVM, Multi-Layer Perceptron, etc.). This model is then used in order to infer the value of $\Phi(D_o)$.

3.2   NumTabData2Vec

This method transforms a dataset D_i into a lower-dimensional vector z using only its numerical values while excluding metadata like column names and file-

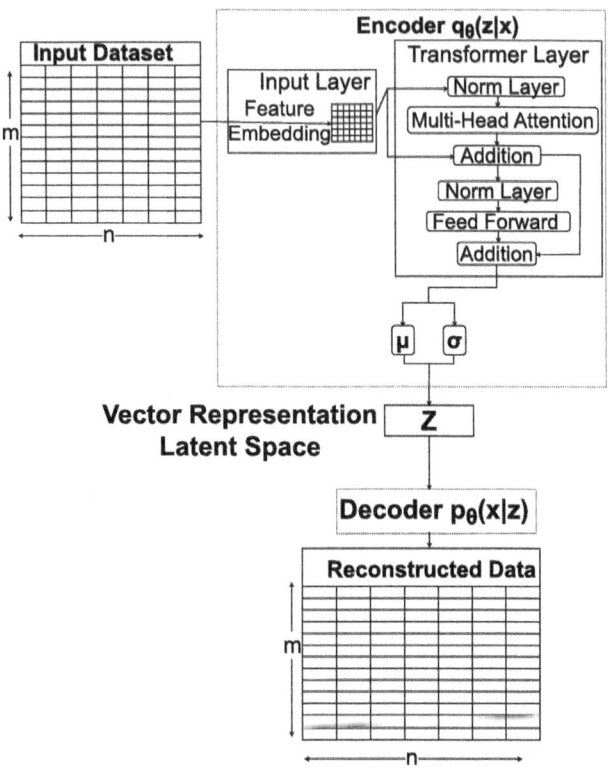

Fig. 2. NumTabData2Vec deep learning model architecture

names. We present a deep learning model architecture based on the variational autoencoder (VAE) [17] concept, that projects high-dimensional data into a $1 \times k$ vector embedding dimension. The proposed method is thus defined as:

$$NumTabData2Vec\,(D_i) \rightarrow z, \tag{1}$$

where the model takes an $m \times n$ dimensional numerical dataset and projects it to a lower k-dimension space z $(k > 1)$. We desire our method to be generally applicable to any dataset by learning to project a vector embedding during training. We also expect this method to learn vector embeddings from diverse data and operate without additional training or fine-tuning. Finally, we ought our scheme to be able to quickly and precisely extract vectors from every input while handling varying dataset dimensions without modifications.

The deep learning model architecture is depicted in Fig. 2, where a dataset D_i with dimensions $m \times n$ passes through the encoder and is projected into a vector embedding representation z, then reconstructed by a decoder that mirrors the encode. The vector representation z is learned using a probabilistic encoder $q_\phi\,(z|x)$, and decoder $p_\theta\,(x|z)$ that learns the distribution utilising the Kullback-

Leibler (KL) divergence [18]. To achieve that, the following condition:

$$minD_{KL}\left(q_\phi\left(z|x\right)\|p_\theta\left(x|z\right)\right) \tag{2}$$

of Kullback-Leibler (KL) divergence must be minimised. To learn the new vector representation z, the dataset must be reconstructed back from z to its input format using the decoder, to verify that the vector representation is compact. This reconstruction loss is part of the overall loss function, defined as:

$$\mathcal{L}_{\theta,\phi}\left(x\right) = \mathbb{E}_{q_\phi(z|x)}\left(\log\left(p_\theta\left(x|z\right)\right)\right) - D_{KL}\left(q_\phi\left(z|x\right)\|p_\theta\left(x|z\right)\right) \tag{3}$$

This loss function is called evidences lower bound (ELBO). While the KL divergence is minimised to learn the vector embedding representation z, the ELBO is maximized so the condition,

$$argmax\mathcal{L}_{\theta,\phi}\left(x\right) \tag{4}$$

must be satisfied. The Decoder $p_\theta\left(x|z\right)$ is only used during the training phase to teach the encoder how to project the vector embedding representation z.

The decoder extracts feature embeddings from the input dataset D_i and processes them through Transformer layers. For the transformer layer we are using the pre-LN Transformer layer [29] instead of the traditional post-LN Transformer layer where the normalisation layer is employed inside the residual connection and before the prediction of the feed-forward layer. Following that, the transformed embedding space is projected into a probabilistic vector space z using the mean (μ) and standard deviation (σ). This lower-dimensional space retains all essential information about D_i, and a higher dimension k in z leads to a more accurate representation by capturing additional features [16].

3.3   Dataset Selection

The selection of the most similar datasets has been implemented using three different approaches. Different similarity functions are easily plugged into our pipeline. The first method uses cosine similarity, which measures the angle between two vectors in the embedding space z independent of their magnitudes, with a higher value indicating greater similarity. The alternative method calculates the Euclidean distance between two vectors to capture their geometric closeness and determine dataset similarity. This approach aids in selecting the most relevant datasets according to organizational requirements. The smaller the distance value then more similar are the datasets. Dataset selection using cosine similarity or euclidean distance selects a fraction of $\lambda, \lambda > 0$ of the closest datasets to dataset D_o. The third approach involves utilising the K-Means [19,21] clustering technique to choose the most relevant datasets. The datasets from D are divided into s ($s > 1$) separate clusters, where datasets with similar features are grouped together in the same cluster based on similarity equations. We determine the optimal number of clusters using silhouette scores [27]. This is done by the following equation:

$$SilhouetteScore(z_i) = \frac{b(z_i) - a(z_i)}{\max(a(z_i), b(z_i))}, \tag{5}$$

Algorithm 1. K-Means Clustering Algorithm for dataset selection

---

Require: Vectors $\mathbf{Z} = \{\mathbf{z}_1, \mathbf{z}_2, \ldots, \mathbf{z}_n\}$, D_o vector z_o, range of clusters $S = (2, \ldots, p)$,
  Maximum size of cluster max_s, Minimum size of cluster min_s
1: **Initialize** the number of cluster s using Silhouette score: $s = SilhouetteScore(Z, S)$
2: **Initialize** the s cluster centroids $\mathbf{C} = \{\mathbf{c}_1, \mathbf{c}_2, \ldots, \mathbf{c}_s\}$ randomly from the vectors $\mathbf{Z}$.
3: **repeat**
4:     **Assignment Step:**
5:     **for** each vector $\mathbf{z}_i \in \mathbf{Z}$ **do**
6:         Assign $\mathbf{z}_i$ to the nearest centroid based on Euclidean distance:
$$\text{Assign } \mathbf{z}_i \text{ to cluster } j = \arg\min_j \|\mathbf{z}_i - \mathbf{z}_j\|^2$$
7:     **end for**
8:     **Update Step:**
9:     **for** each centroid $\mathbf{c}_j$ **do**
10:        Update $\mathbf{c}_j$ as the mean of all vectors assigned to cluster j:
$$\mathbf{c}_j = \frac{1}{|\{\mathbf{z}_i \in C_j\}|} \sum_{\mathbf{z}_i \in C_j} \mathbf{v}_i$$
11:    **end for**
12: **until** Centroids $\mathbf{C}$ do not change significantly
13: Save the cluster model **K-Means**
14: Find in which cluster the vector z belongs, $c = $ **K-Means**(z)
15: Find which datasets D_r are belongs to cluster c
16: Check the number of datasets in D_r and update it if it does not meet the min_s and max_s.
17: **Return** Datasets D_r

---

where for each vector z_i computes the mean intra-cluster distance $(a(z_i))$ which is the distance with the other vectors in the same cluster, and the mean nearest-cluster distance $(b(z_i))$ is the minimum average distance with the other vectors in a different cluster. Silhouette Score ranges from -1 to 1, and the higher value defines the best s number for clusters.

Algorithm 1, outlines the K-means process for selecting relevant datasets D_r based on the target dataset D_o. Using the optimal s (with the highest Silhouette score), the vector representations z of each dataset are clustered. Next, the algorithm uses the vector z_o of dataset D_o to find the closest cluster centroid. All datasets in that cluster are defined as the relevant datasets D_r, which are then used to model the analytics operator. However, we defined a maximum and minimum size for D_r, and if these conditions are not met, datasets are either removed from the cluster or added based on their distance from the cluster centroid. These small adjustments are only made in cases where the clustering technique does not yield results that satisfy our requirements.

4   Evaluation

We compared our framework with Apollo [4,9], which models analytic operators using data content. Two loss functions to measure prediction accuracy are employed: root-mean-square error (RMSE) and mean absolute error (MAE). RMSE is sensitive to outliers, while MAE is not; conversely, RMSE accounts for error direction, which MAE cannot. We further assess efficiency using *Speedup* and *Amortized Speedup* metrics, where *Speedup* is defined as $\frac{T_{op}^{(i)}}{T_{SimOp}^{(i)}+T_{vec}+T_{sim}+T_{pred}}$, where $T_{op}^{(i)}$ is the time to execute operator i on all datasets, $T_{SimOp}^{(i)}$ is the time to model the operator with datasets from similarity search, T_{vec} is the vector embedding computation time, T_{sim} is the similarity search time, and T_{pred} is the prediction time for D_o. Amortized speedup including one-time vectorization per data lake across multiple operators. Three variants with vector sizes 100, 200, and 300 (each with eight transformer layers) were trained for 100 epochs on four NVIDIA A10 GPUs. More experimental evaluation results can be found in the extended version of this work [20].

4.1   Evaluation Setup

Our framework is deployed over an AWS EC2 virtual machine server running with 48 vCPUs of AMD EPYC 7R32 processors at 2.40 GHz, and four A10s GPUs with 24 GB of memory each, 192 GB of RAM memory, and $2TB$ of storage, running over Ubuntu 24.4 LTS. Our code is written in Python (v.3.9.1) and PyTorch modules (v.2.4.0). Apollo was deployed on the same AWS EC2 virtual machine server, utilizing only the vCPUs and RAM, as it does not require a GPU for execution.

4.2   Datasets

We evaluated our framework using four real-world datasets (see Table 1). The NumTabData2Vec module was trained on separate data (60% training, 40% testing). The Household Power Consumption (HPC) dataset [13] contains 401 datasets with 2051 tuples and seven features recorded at one-minute intervals of electric power usage measurements. The Adult dataset [5], used for binary classification, predicts income levels and includes 100 datasets with 228

Table 1. Dataset properties for experimental evaluation

Dataset Name	# Files	# Tuples	# Columns
Household Power Consumption [13]	401	2051	7
Adult [5]	100	228	14
Stocks [1]	508	$1959 - 13$	7
Weather [2]	49	516	7

individuals and socio-economic features. The Stock Market dataset [1] consists of 508 datasets with 13 to 1959 tuples describing daily NASDAQ stock prices. Weather dataset [2] provides hourly measurements from 36 U.S. cities (2012–2017), split into 49 datasets with 516 tuples and seven features. Any categorical feature column in all datasets is transformed to numerical data by one-hot encoding. These datasets were selected to demonstrate our framework's ability to perform consistently across diverse real-world scenarios.

Our framework was evaluated by predicting the outputs of various ML operators without directly executing them. Datasets were projected into k-dimensional spaces with vector dimensions of 100, 200, and 300. For regression datasets (Household Power Consumption and Stock Market), we modelled Linear Regression (LR) and Multi-Layer Perceptron (MLP), while for classification datasets (Weather and Adult), we modelled Support Vector Machine (SVM) and MLP classifiers. Each experiment has executed 10 times, and we report the average error loss and speedup.

4.3   Evaluation Results

Figures 3, 4, 5, and 6 present the evaluation results of different similarity search methods across vector embedding spaces of sizes 100, 200, and 300 (green, blue, and grey bars, respectively). In each sub-figure, the y-axis represents the error loss value, while the x-axis displays the similarity search method applied over the vector embeddings. Figures 3 and 4 display results for the Stock Market and Household Power Consumption datasets, with MLP regression in the bottom sub-figure and LR in the top. Figures 5 and 6 show results for the Weather and Adult datasets, with SVM (SGD) in the top sub-figure and MLP classifier in the bottom. Left sub-figures use RMSE loss, while right sub-figures use MAE loss.

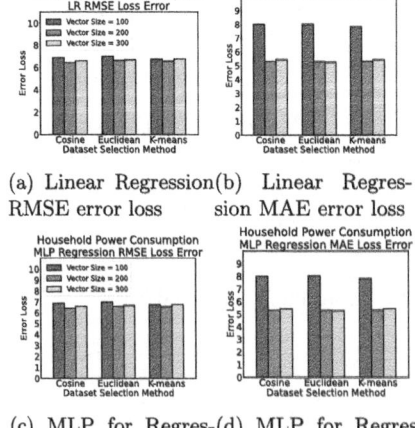

(a) Linear Regression RMSE error loss (b) Linear Regression MAE error loss

(c) MLP for Regression RMSE error loss (d) MLP for Regression MAE error loss

Fig. 3. Household power consumption dataset prediction error loss

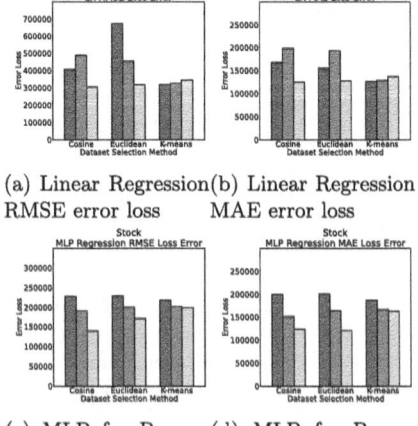

(a) Linear Regression RMSE error loss (b) Linear Regression MAE error loss

(c) MLP for Regression RMSE error loss (d) MLP for Regression MAE error loss

Fig. 4. Stock market dataset prediction error loss

Table 2. Evaluation results of our framework exported analytic operator with lowest prediction error in comparison with Apollo

Dataset Name	Method	Operator	RMSE	MAE	Speedup	Amortized Speedup
Household Power Consumption	300V Cosine	LR	**6.61**	**5.42**	0.0017	**1.99**
	300V SR-0.2	LR	7.77	6.66	0.0018	1.42
	Apollo-SR 0.1	LR	2968.01	2352.55	**0.015**	0.024
	Apollo-SR 0.2	LR	2811.49	2229.50	0.015	0.024
	300V K-Means	MLP Regr.	**6.70**	**3.38**	0.9249	**1.99**
	Apollo-SR 0.1	MLP Regr.	3322.05	2606.99	2.38	1.74
	Apollo-SR 0.2	MLP Regr.	3850.01	2609.36	**2.38**	1.74
Stock	300V Cosine	LR	306382.28	125335.65	0.00085	**1.91**
	300V SR-0.4	LR	21861625.91	5674215.265	0.00087	0.33
	Apollo-SR 0.1	LR	**153665.92**	**118236.48**	**0.00093**	0.00096
	Apollo-SR 0.2	LR	166844.95	133306.68	0.00093	0.00096
	300V Cosine	MLP Regr.	**140236.47**	**123571.12**	0.63	**1.91**
	Apollo-SR 0.1	MLP Regr.	175150.82	145123.09	**0.93**	0.96
	Apollo-SR 0.2	MLP Regr.	174390.81	146338.73	0.93	0.96
Weather	300V Cosine	SVM SGD	**14.13**	**7.63**	1.06	**22.8**
	Apollo-SR 0.1	SVM	69.51	25.52	**2.10**	1.16
	Apollo-SR 0.2	SVM	68.70	22.81	2.10	1.16
	300V Cosine	MLP	**14.29**	**4.03**	1.03	**22.8**
	300V SR-0.4	MLP	15.95	13.31	1.02	1.77
	Apollo-SR 0.1	MLP	69.62	23.10	**1.34**	1.14
	Apollo-SR 0.2	MLP	673.56	84.70	1.32	1.14
Adult	300V Cosine	SVM SGD	**0.36**	**0.2**	0.37	**2.78**
	Apollo-SR 0.1	SVM	68.32	22.95	**0.75**	0.85
	Apollo-SR 0.2	SVM	68.88	22.88	0.74	0.85
	300V K-Means	MLP	**0.36**	**0.19**	0.30	2.78
	300V SR-0.2	MLP	6.01	6.00	0.54	**3.54**
	Apollo-SR 0.1	MLP	71.11	26.51	**1.07**	1.31
	Apollo-SR 0.2	MLP	70.16	25.74	1.05	1.31

Figure 3, for the HPC dataset, shows as increase the vector dimension size there is slightly lower prediction error for all the operator modelling. Different similarity methods do not result in any significant differences in the prediction error loss for all the operator modelling. This suggests that, regardless the similarity selection method, our framework effectively selects the most optimal subset of data to improve model predictions on the unseen input dataset D_o. Additionally, we observe higher error loss with a vector size of 100, which can be attributed to the reduced representation capacity of lower-dimensional vectors. This limitation results in fewer "right" datasets being selected.

For the stock market dataset, Fig. 4 depicts that a vector embedding representation of size 300 models more accurate operators, with cosine similarity performing best in the similarity search and modelling the most optimal operator. However, due to the inherent volatility in Stock market data from different days, all models in the stock market dataset experiments exhibit high loss values.

In the weather dataset, the SVM operator results (Figs. 5a and 5b) show that using 300 sized vectors in the representation space consistently led to more accurate operator models across all similarity methods. Specifically, cosine similarity in combination with the 300-dimensional vector embedding reduced the error rate in operator predictions, demonstrating that projecting datasets into this representation space and applying cosine similarity improves the prediction accuracy on the modelled operator. For the MLP classifier (Figs. 5c and 5d), the results illustrate that using vector embeddings of size 300 and Cosine similarity-produced the most accurate MLP classifier operators.

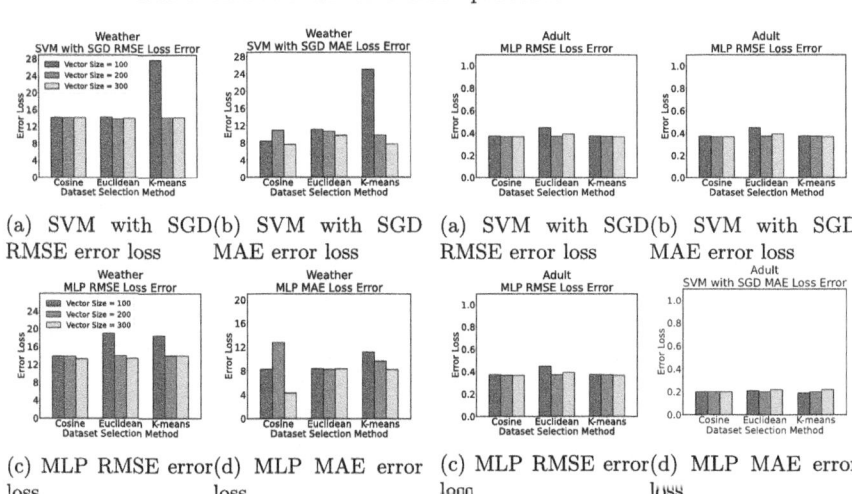

(a) SVM with SGD(b) SVM with SGD (a) SVM with SGD(b) SVM with SGD
RMSE error loss   MAE error loss   RMSE error loss   MAE error loss

(c) MLP RMSE error(d) MLP MAE error (c) MLP RMSE error(d) MLP MAE error
loss   loss   loss   loss

Fig. 5. Weather dataset prediction   **Fig. 6.** Adult dataset prediction error
error loss   loss

On the other hand, the Adult dataset shows the lowest error rates, with error loss values consistently below 0.5 across all vector embedding dimensions and similarity search methods (see Fig. 6). The Adult dataset, besides exhibiting a high number of rows, also has a higher number of columns, which demonstrates that our framework performs consistently well even with larger datasets. Additionally, we observe that the lowest prediction error across all datasets occurs when using higher-dimensional vector embeddings. With a trade-off between accuracy and execution time as the difference to generate all data lake available datasets vector embedding representation between 100 and 300 size dimension in the vector representation space to be less than 60 s. This confirms that a higher number of vector dimensions leads to more accurate predictions, consistent with findings in previous research [22].

We conducted an experimental evaluation using the Sampling Ratio (SR) approach, similar to Apollo [9], but employed neural networks built from the vector embeddings of each dataset. The SR approach involves a unified random selection of $l\%$ datasets from the vector representation space, using this subset to construct a neural network for predicting operator outputs. We tested SR values of 0.1, 0.2, and 0.4, as well as vector embedding dimensions of 100, 200, and 300, across all datasets.

For the HPC and Weather datasets, the SR approach was approximately 15% less accurate in operator prediction compared to all similarity search methods, even as vector embedding dimensions increased. In contrast, the Stock dataset exhibited a significantly larger discrepancy, with the SR approach performing about 70% worse in prediction accuracy across all vector embedding dimensions. Similarly, in the Adult dataset, the SR approach recorded the poorest performance, delivering nearly 90% worse prediction accuracy compared to the similarity search methods.

Table 2 compares model operators, loss functions, and speedup metrics for our framework and Apollo at SR values of 0.1 and 0.2. Methods 100V, 200V, and 300V denote vector embedding dimensions. The lowest prediction errors align with our pipeline's similarity search method. Apollo outperforms our framework on the Stock dataset for the LR analytic operator with the smallest amount of SR. However, our framework excels with the MLP regression operator, improving RMSE and MAE by 20% and 17%, respectively. The LR operator's performance gap on the Stock dataset is minor. For other datasets, our framework consistently surpasses Apollo across different SR values. This demonstrates the effectiveness of our similarity search approach, which enhances data quality and reduces Φ prediction errors by identifying relevant datasets D_r from the data lake directory D. The Adult dataset also highlights our framework's advantage with increasing feature dimensions. Although Apollo achieves better raw speedup due to the higher complexity of our framework's vectorization step, our framework outperforms it in amortized speedup. By excluding the reusable vectorization process, it achieves speed gains of 10% to 60% for most operators. The SR approach, leveraging vector embedding representations, enhances operator prediction compared to Apollo and achieves greater amortized speedup. However, the similarity search method outperforms both Apollo and the SR approach in prediction accuracy and amortized speedup, establishing its clear superiority across most datasets and operator scenarios.

4.4   NumTabData2Vec Evaluation Results

Table 3. Similarity between vectors of different datasets scenarios

NumTabData2Vec Vector Size	Similarity
100	0.54
200	0.18
300	0.16

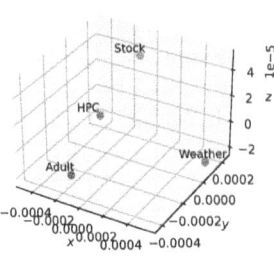

Fig. 7. Vector representation for each dataset from NumTabData2Vec

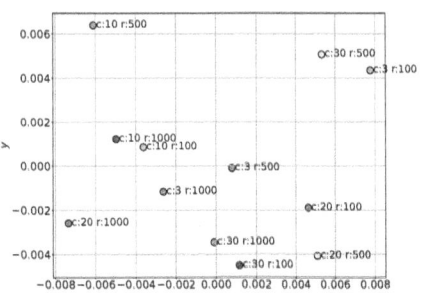

Fig. 8. Synthetic data vector embedding representation

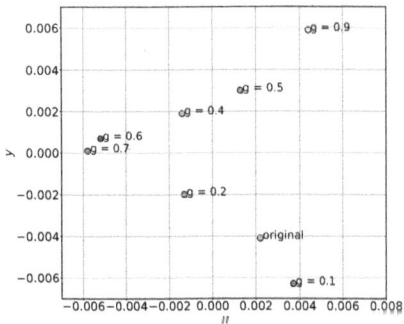

Fig. 9. HPC Dataset vector embedding representation with addition of Noise

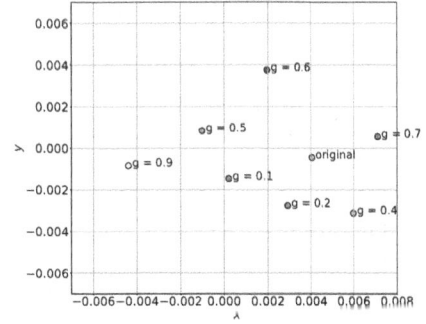

Fig. 10. HPC Dataset vector embedding representation with addition of Noise in the first column

Our proposed model, *NumTabData2Vec*, was evaluated for its ability to distinguish dataset scenarios based on qualitative differences. For each scenario, n random datasets were reduced to 3D vector embeddings using PCA, as shown in Fig. 7, which demonstrates *NumTabData2Vec*'s ability to distinguish datasets with minimal overlap across contexts. Unlike prior methods [22,24], which focus on text or graphs, *NumTabData2Vec* applies to entire datasets. Table 3 further highlights the average cosine similarity between dataset embeddings, showing greater dissimilarity as vector dimensions increase. However, results suggest that dimensions between 100 and 300 are sufficient for accurate distinction, avoiding the need for larger vector sizes.

To evaluate *NumTabData2Vec*'s ability to distinguish datasets with varying row and column counts, we generated synthetic numerical tabular datasets of different dimensions and vectorized them. Figure 8 shows datasets with columns ranging from three to thirty and rows from ten to one thousand, projected from a 200-dimensional space to 2D using PCA. Each bullet caption c and r corresponds to the columns and rows of the dataset, respectively. Datasets with the same number of columns cluster closely in the representation space, and a similar pattern is observed for datasets with the same number of rows. These results indicate that our method effectively distinguishes datasets based on size during vectorization.

To evaluate *NumTabData2Vec*'s ability to distinguish datasets by distribution and order, Gaussian noise was added to *l*% of tuples in an HPC dataset. Figure 9 shows the original and noise-modified datasets projected to 2D using PCA, with greater noise causing larger shifts in the representation space. This demonstrates the model's effectiveness in capturing distribution differences and distinguishing datasets based on ordering.

To assess fine-grained distinctions, we repeated the experiment by adding Gaussian noise exclusively to the first column of the dataset. Figure 10 shows the 2D vector space, where g in the bullet caption indicates the noise level. As noise increases, the representation shifts further from the original dataset, though it remains closer than in Fig. 9, with points more tightly grouped in 2D space.

5   Limitations

There exist a number of limitations in our work as we described it. In this section we briefly highlight them. Firstly, our input datasets comprise records of specific size and type (numerical). This currently excludes data with textual and categorical attributes, or tables with varying number of features inside a set of datasets. Secondly, we currently consider single-input and single-output operator modelling. Finally, our proposed NumTabData2Vec model for data vectorization has a performance limitation, as it cannot deal with datasets bigger than about 3000 tuples. This is mostly a hardware limitation of off-the-shelf GPUs (with at most 24GB of memory available for a budget GPU).

6   Conclusion

In this paper, we presented a novel framework for the modelling of an analytic operator (such as a ML algorithm) when a large number of input data is available and thus no brute-force execution can be performed. We propose a deep learning model, *NumTabData2Vec*, which transforms a dataset to a lower k-dimensional representation space z. Our framework produces vector embeddings for the input datasets using *NumTabData2Vec* and performs a similarity search to identify the most relevant subset of datasets for any unseen input. By modelling the analytic operator based on this selected subset, we are able to accurately predict its output on any given input dataset. In practice, we demonstrated that our framework can accurately model various common algorithms and compared favourably against a similar recent framework [9], in both accuracy and speedup. Furthermore, we showed that *NumTabData2Vec* can create different vector representations for datasets from different scenarios. We also demonstrated that *NumTabData2Vec* can effectively detect when noise is introduced into a dataset.

Acknowledgments. This work is partially supported by the project RELAX-DN, funded by the European Union under Horizon Europe 2021–2027 Framework Programme Grant Agreement number 101072456.

References

1. Stock Market Dataset. Kaggle (2020). https://www.kaggle.com/datasets/jacksoncrow/stock-market-dataset/data
2. Weather Dataset. Kaggle (2020). https://www.kaggle.com/datasets/selfishgene/historical-hourly-weather-data
3. Abdelaal, M., Yayak, A.B., Klede, K., Schöning, H.: Reclean: reinforcement learning for automated data cleaning in ml pipelines. In: 2024 IEEE 40th International Conference on Data Engineering Workshops (ICDEW), pp. 324–330. IEEE (2024)
4. Bakogiannis, T., Giannakopoulos, I., Tsoumakos, D., Koziris, N.: Apollo: a dataset profiling and operator modeling system. In: Proceedings of the 2019 International Conference on Management of Data, pp. 1869–1872 (2019)
5. Becker, B., Kohavi, R.: Adult. UCI Machine Learning Repository (1996). https://doi.org/10.24432/C5XW20
6. Cai, L., Zhu, Y.: The challenges of data quality and data quality assessment in the big data era. Data Sci. J. **14**, 2–2 (2015)
7. Dias, L.V., Miranda, P.B., Nascimento, A.C., Cordeiro, F.R., Mello, R.F., Prudêncio, R.B.: Imagedataset2vec: an image dataset embedding for algorithm selection. Expert Syst. Appl. **180**, 115053 (2021)
8. Ghorbani, A., Zou, J.: Data shapley: equitable valuation of data for machine learning. In: International Conference on Machine Learning, pp. 2242–2251. PMLR (2019)
9. Giannakopoulos, I., Tsoumakos, D., Koziris, N.: A content-based approach for modeling analytics operators. In: Proceedings of the 27th ACM International Conference on Information and Knowledge Management, pp. 227–236 (2018)
10. Gorishniy, Y., Rubachev, I., Khrulkov, V., Babenko, A.: Revisiting deep learning models for tabular data. Adv. Neural. Inf. Process. Syst. **34**, 18932–18943 (2021)
11. Gupta, N., et al.: Data quality toolkit: automatic assessment of data quality and remediation for machine learning datasets. arXiv preprint arXiv:2108.05935 (2021)
12. Hazen, B.T., Boone, C.A., Ezell, J.D., Jones-Farmer, L.A.: Data quality for data science, predictive analytics, and big data in supply chain management: an introduction to the problem and suggestions for research and applications. Int. J. Prod. Econ. **154**, 72–80 (2014)
13. Hebrail, G., Berard, A.: Individual Household Electric Power Consumption. UCI Machine Learning Repository (2006). https://doi.org/10.24432/C58K54
14. Huang, X., Khetan, A., Cvitkovic, M., Karnin, Z.: Tabtransformer: tabular data modeling using contextual embeddings. arXiv preprint arXiv:2012.06678 (2020)
15. Jakubik, J., Vössing, M., Kühl, N., Walk, J., Satzger, G.: Data-centric artificial intelligence. Bus. Inf. Syst. Eng. 1–9 (2024)
16. Jomaa, H.S., Schmidt-Thieme, L., Grabocka, J.: Dataset2vec: learning dataset meta-features. Data Min. Knowl. Disc. **35**(3), 964–985 (2021)
17. Kingma, D.P., Welling, M., et al.: An introduction to variational autoencoders. Found. Trends® Mach. Learn. **12**(4), 307–392 (2019)
18. Kullback, S., Leibler, R.A.: On information and sufficiency. Ann. Math. Stat. **22**(1), 79–86 (1951)

19. Lloyd, S.: Least squares quantization in PCM. IEEE Trans. Inf. Theory **28**(2), 129–137 (1982)
20. Loizou, A., Tsoumakos, D.: Analytics modelling over multiple datasets using vector embeddings. arXiv preprint arXiv:2502.17060 (2025)
21. MacQueen, J., et al.: Some methods for classification and analysis of multivariate observations. In: Proceedings of the Fifth Berkeley Symposium on Mathematical Statistics and Probability, Oakland, CA, USA, vol. 1, pp. 281–297 (1967)
22. Mikolov, T.: Efficient estimation of word representations in vector space. arXiv preprint arXiv:1301.3781 **3781** (2013)
23. Murray, D.G., Simsa, J., Klimovic, A., Indyk, I.: tf. data: a machine learning data processing framework. arXiv preprint arXiv:2101.12127 (2021)
24. Narayanan, A., Chandramohan, M., Venkatesan, R., Chen, L., Liu, Y., Jaiswal, S.: graph2vec: learning distributed representations of graphs. arXiv preprint arXiv:1707.05005 (2017)
25. Ni, W., Zhang, K., Miao, X., Zhao, X., Wu, Y., Yin, J.: Iterclean: an iterative data cleaning framework with large language models. In: Proceedings of the ACM Turing Award Celebration Conference-China 2024, pp. 100–105 (2024)
26. Rezig, E.K., et al.: Dagger: a data (not code) debugger. In: CIDR 2020, 10th Conference on Innovative Data Systems Research, Amsterdam, The Netherlands, January 12-15, 2020, Online Proceedings (2020)
27. Rousseeuw, P.J.: Silhouettes: a graphical aid to the interpretation and validation of cluster analysis. J. Comput. Appl. Math. **20**, 53–65 (1987)
28. Wang, Z., Sun, J.: Transtab: learning transferable tabular transformers across tables. Adv. Neural. Inf. Process. Syst. **35**, 2902–2915 (2022)
29. Xiong, R., et al.: On layer normalization in the transformer architecture. In: International Conference on Machine Learning, pp. 10524–10533. PMLR (2020)
30. Zha, D., et al.: Data-centric artificial intelligence: a survey. ACM Comput. Surv. **57**(5), 1–42 (2025)
31. Zhang, L., Zhang, S., Balog, K.: Table2vec: neural word and entity embeddings for table population and retrieval. In: Proceedings of the 42nd International ACM SIGIR Conference on Research and Development in Information Retrieval, pp. 1029–1032 (2019)
32. Zhu, C., Zhang, Q., Cao, L., Abrahamyan, A.: Mix2vec: unsupervised mixed data representation. In: 2020 IEEE 7th International Conference on Data Science and Advanced Analytics (DSAA), pp. 118–127 (2020). https://doi.org/10.1109/DSAA49011.2020.00024

Towards IoT-Based Smart Mobility Framework for Proactive Road Stress Detection in Individuals with ASD

Barry Amadou Djoulde[1]($\boxtimes$), Nawal Guermouche[1], Viviane Kostrubiec[2], and Pierre Vincent Paubel[3]

[1] LAAS-CNRS, University of Toulouse, INSA, Toulouse, France
{adbarry,nguermou}@laas.fr
[2] CERPPS, University of Toulouse, Toulouse, France
viviane.kostrubiec@univ-tlse2.fr
[3] CLLE, University of Toulouse, Toulouse, France
paubel@univ-tlse2.fr

Abstract. Autism Spectrum Disorder (ASD) is a neurodevelopmental disability that significantly increases the difficulties and risks associated with driving. Individuals with ASD often face a variety of challenges, such as increased sensory sensitivities, difficulty adapting to changing environments, and struggles with unexpected situations on the road. These difficulties can lead to sensory overload, panic attacks, and impaired decision-making, all of which increase the risk of accidents and make driving an especially overwhelming task. In this paper, we propose a novel IoT-based smart mobility framework for predictive stress detection in drivers with ASD, enabling the early identification of potential stressors before they encounter them on the road. This approach leverages AI-based models, including LSTM and CNN-based architectures. Unlike existing methods that focus on reactive stress detection, which may be too late, our approach predicts stress triggers in advance, enabling timely and preventive support.

Keywords: Proactive stress detection · smart mobility · ASD · IoT · deep learning

1 Introduction

"*Since obtaining my driver's license, I have not touched a steering wheel again. I am too afraid of either killing myself or someone else*". This testimony comes from an individual with Autism Spectrum Disorder (ASD), a disability with an estimated global prevalence of 1 in 100[1]. For adults with ASD, driving remains a top priority [14]. Many actively avoid public transportation due to sensory sensitivities and social challenges, making the use of personal vehicle not just a preference but a necessity for daily mobility.

---

[1] https://www.who.int/news-room/fact-sheets/detail/autism-spectrum-disorders.

R. Wrembel et al. (Eds.): DEXA 2025, LNCS 16047, pp. 254–269, 2026.
https://doi.org/10.1007/978-3-032-02088-8_18

However, modern cars and road environments are primarily designed for individuals with typical neurodevelopment, often overlooking the unique needs of people with ASD. They frequently experience sensory hypersensitivity, an intense focus on details, and executive function difficulties, including challenges with attention, planning, decision-making, and cognitive flexibility.

Driving is an intricate task that requires a seamless integration of cognitive and motor skills, the ability to perceive and process the surrounding environment, filter and prioritize relevant information, regulate emotional and motor responses, and manage potentially overwhelming sensory stimuli, such as loud noises or flashing lights. For individuals with ASD, these demands pose significant challenges, amplifying cognitive and emotional strain and heightening the risk of stress, which can lead to dangerous situations on the road.

Stress is an adaptive response to perceived threats, activating the body's fight, flight, or freeze reactions [16]. In individuals with ASD, these stress responses are often heightened, resulting in impaired decision-making, reduced situational awareness, and compromised motor control, all of which are critical for safe driving [12]. Accurate and timely stress detection is essential to mitigating these risks. By providing early alerts, such systems can help alleviate driver stress, reduce sensory overload, and enhance decision-making, ultimately improving both driver well-being and overall road safety.

Stress detection has attracted significant research attention in recent years [1,9,17,21]. Mainly, existing approaches identify stress only after it has already manifested, typically through reactive real-time monitoring and analysis of physiological parameters such as heart rate. While these methods offer valuable insights, reactive stress detection alone is insufficient to prevent the cascading effects of stress on ASD driving performance. These limitations highlight the need for context-aware, proactive stress prediction systems that can anticipate rising stress and enable timely interventions to enhance both driving safety and comfort.

The Internet of Things (IoT) presents a promising avenue for addressing these challenges in the smart mobility domain. By enabling seamless data collection and real-time processing through interconnected sensors and intelligent services, IoT coupled with Artificial Intelligence (AI) can transform the way we approach stress detection and intervention in driving contexts. In this paper, we propose a novel solution that goes beyond traditional reactive stress detection by introducing a proactive IoT-based framework capable of predicting stress before it fully manifests. Our approach leverages advanced predictive models and real-time contextual analysis to assess surrounding environmental events that may impact the driver, even before he reaches the event location. By forecasting how these events could affect the driver's stress in advance, our system can proactively send timely alerts, allowing the driver to take preventive actions before stress is triggered. The proposed framework is built upon a Deep Neural Network (DNN)-based model that harnesses the power of Long Short-Term Memory (LSTM) [10], and Convolutional Neural Networks (CNN) [13]. LSTM networks, a specialized type of Recurrent Neural Network (RNN), are highly effective in capturing long-term dependencies, making them well-suited for handling sequential data. In this work, LSTM is used to predict heart rate variations in response to upcoming road events. Once the heart rate variations are predicted, they are processed by the CNN, which classifies them as either stressful or non-stressful based on identified patterns. This combined approach enables accurate prediction and real-time classification, significantly improving the model's ability to proactively detect and mitigate stress in dynamic driving environments.

The main contributions of this paper can be summarized as follows: 1) We propose a novel IoT-based smart mobility framework dedicated to individuals with ASD, 2) We introduce a deep-learning-based approach for proactive stress detection, tailored to the context of road events, 3) We implemented and evaluated the proposed solution, demonstrating its practical effectiveness in real-world driving scenarios.

The remainder of the paper is structured as follows. Section 2 introduces a real-life scenario. Section 3 provides a review of related works. Section 4 presents a comprehensive overview of the proposed framework. Section 5 details the proposed approach for proactive stress detection. Section 6 discusses the experimental results, and Sect. 7 concludes the paper.

2 Motivating Example: ASD Driving Challenge

Let us consider a real-life case shared by a collaborating psychologist. It concerns a woman with ASD, referred to here as Anna. Anna had a scheduled appointment with her psychologist and, as part of her usual routine, meticulously planned her journey in advance to minimize any potential disruptions. To ensure a timely arrival, she left early and carefully followed a familiar, well-rehearsed route, anticipating a smooth and predictable experience. However, as Anna approached a roundabout, she unexpectedly encountered significant traffic congestion. The abrupt change in traffic conditions disrupted her sense of control and routine. The combination of auditory stimuli, such as honking from impatient drivers and sudden lane changes, contributed to a rapid escalation in Anna's stress levels. Unable to manage this situation, she abruptly stopped her vehicle in the middle of the roundabout, experiencing a state of emotional overload. As panic took over, she exited the vehicle and walked away, leaving the car stationary and obstructing the traffic flow. This situation poses a significant safety risk, not only for Anna but for other road users as it increases the likelihood of traffic accidents and contributes to broader roadway disruptions.

This scenario highlights the critical need for a proactive stress prediction system within intelligent transportation frameworks. If Anna's vehicle had been equipped with an IoT-enabled, context-aware stress anticipation mechanism, it could have detected the risk of increased stress before she encountered the congested roundabout. By analyzing real-time road conditions and estimating their potential impact on the driver's stress levels, such a system could have provided timely alerts. This would have enabled Anna to navigate the situation with increased confidence, reduced stress, and improved safety.

3   Related Works

Several initiatives have been defined to contribute to ongoing development of inclusive mobility solutions that address the diverse needs of ASD users, promoting safer and more accessible environments. A substantial body of research has focused on designing solutions tailored to the unique cognitive, sensory, and behavioral characteristics of individuals with ASD across various domains. For example, Cecchini et al. [5] proposed a comprehensive framework of microscale interventions, including sensory-friendly pedestrian pathways, and traffic-calming designs, aimed at enhancing the navigability and inclusivity of mobility spaces for individuals with Autism. Padmanaban et al. [15], through a design-thinking approach with caregiver input, developed a personalized system that includes customizable interiors, emotional regulation tools, and an intuitive companion mobile app. This solution has demonstrated a significant enhancement in trust and autonomy for neurodiverse passengers, offering a tailored experience that supports their unique needs and promotes greater independence during mobility.

Other works have studied the stress experienced by individuals with ASD during driving. Cox et al. [6] demonstrated that newly licensed drivers with ASD exhibit greater physiological stress responses and poorer performance in driving simulations compared to their neurotypical peers. Fok et al. [8] conducted a study showing that emerging drivers with ASD report elevated emotional stress and greater difficulty managing challenging driving situations, which these difficulties strongly correlating with their perceived struggles in completing driving tasks. Other smart mobility systems dedicated to individuals with ASD have also been proposed. In this context, wearable technologies have been increasingly employed to monitor and support individuals with ASD. Benssassi et al. [3] reviewed assistive devices such as smart glasses and proximity sensors, highlighting their potential to enhance real-time social interaction and sensory regulation while pointing out limitations related to usability, feedback design, and ethical implementation. D. Bian et al. [4] introduced a novel adaptive Virtual Reality (VR) driving simulator for individuals with ASD, which infers user engagement from physiological signals and dynamically adjusts task difficulty in real-time to foster emotionally safe and personalized learning experiences. All these efforts aim to create an ecosystem of mobility solutions that empower individuals with ASD, enhancing their autonomy and inclusivity. However, they fall short in proposing proactive solutions for detecting and anticipating stress states.

Stress detection through physiological signals has been extensively studied. In particular, deep learning techniques have significantly advanced the performance of stress detection models [1,9,18,19,22]. Rastgoo et al. [17] developed a method to process electrocardiogram (ECG) signals for real-time driver stress classification. Similarly, Amin et al. [1] used ECG signals alongside deep transfer learning with fuzzy logic to classify driver stress. Ke Wang et al. [21] presented an ensemble classification model using autoencoders and the AdaBoost framework to classify driving stress levels based on various physiological signals.

While these works have made significant contributions to detecting stress reactively by analyzing observed physiological data, they primarily rely on real-time signals such as ECG to classify stress levels after the stress response has already been triggered. This reactive approach, though effective in many scenarios, presents limitations, particularly within the context of mobility, where it is crucial to anticipate and prevent stress induced by road events and environmental factors before they arise.

4   Global Overview of the IoT-Based Smart Mobility System

Figure 1 illustrates the IoT-based smart mobility framework designed to enable proactive stress detection for drivers with ASD. The system leverages Multi-Access Edge Computing (MEC) servers, which facilitate the acquisition of real-time road event data from geographical areas that the driver is likely to traverse, supporting timely and context-aware decision-making. In this framework, two main categories of data are utilized:

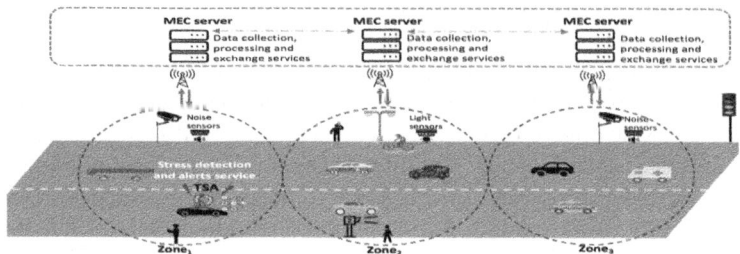

Fig. 1. IoT-based framework for proactive stress detection service

- *Road event data*: Through IoT-enabled services, external environmental and traffic-related conditions, such as noise levels, lighting conditions, weather changes (e.g., fog, rain), and traffic density, are continuously monitored and transmitted to the ASD vehicle analysis system. Moreover, contextual data are also monitored. They refer to information related to the driver's situation and journey progression, such as acceleration patterns, distance traveled, and time elapsed since the start of the trip.
- *Physiological responses*: Stress levels are commonly assessed through various physiological signals, with electrocardiogram (ECG) data being the most widely used due to its reliability in capturing autonomic nervous system responses [23]. In this paper, we leverage two key ECG-derived metrics to assess physiological stress responses: heart rate (HR) and heart rate variability (HRV). HR provides a basic measure of cardiac activity and is expressed in beats per minute (bpm): HR $= \frac{60}{\overline{RR}}$, where $\overline{RR}$ is the mean of the RR

intervals in seconds (i.e., the time between successive R-wave peaks in the ECG signal).

To gain a deeper understanding of autonomic nervous system regulation, we also consider two well-established HRV metrics, each offering complementary insights into stress dynamics [11]:

– *Root Mean Square of Successive Differences (RMSSD)*: It quantifies short-term fluctuations in heart rate, primarily driven by parasympathetic activity, making it a highly sensitive marker for detecting acute stress responses. $\text{RMSSD} = \sqrt{\frac{1}{N-1}\sum_{i=2}^{N}(RR_i - RR_{i-1})^2}$

– *Standard Deviation of NN Intervals (SDNN)*: measures overall HRV and is considered a robust indicator of long-term stress adaptation : $\text{SDNN} = \sqrt{\frac{1}{N}\sum_{i=1}^{N}(RR_i - \overline{RR})^2}$, where RR_i represents the ith RR interval and N is the total number of RR intervals.

Under stress conditions, HR tends to increase, while RMSSD and SDNN typically decrease [20].

Each ASD vehicle is equipped with an onboard AI-driven predictive stress detection model designed to assess the potential impact of context-aware road events on drivers with ASD. The primary objective is to proactively anticipate stress responses triggered by these events, thereby enabling timely interventions and promoting safer, more adaptive driving experiences.

5   Predictive Stress Detection Driven by Event Roads

Figure 2 presents the architecture of the proposed model for proactively assessing stress responses in ASD drivers in the presence of upcoming road events. The framework includes two core components: (1) an LSTM-based prediction module to predict physiological biomarkers (HR, RMSSD, SDNN) in response to road and contextual conditions, and (2) a CNN-based classification module to infer the driver's stress state from these predicted biomarkers. Each component is described in the subsequent sections.

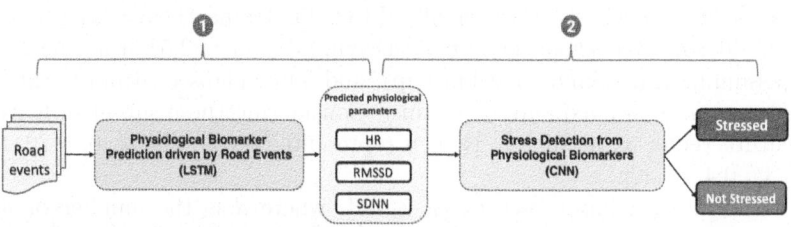

Fig. 2. Global architecture of the proposed stress prediction model

5.1 LSTM-Based Physiological Biomarkers Prediction Driven by Road Events

The proposed LSTM-based model is illustrated in Fig. 3. It captures temporal dependencies and contextual patterns within the input stream of road events and contextual data. A notable advantage of using LSTM in this setting lies in its capacity to model both immediate and delayed physiological responses. For instance, acute stimuli such as sudden loud noises may trigger instantaneous stress reactions, whereas prolonged exposure to traffic congestion may lead to gradual physiological changes. The model comprises two main components: the *LSTM network* and *a dense layer*, presented below.

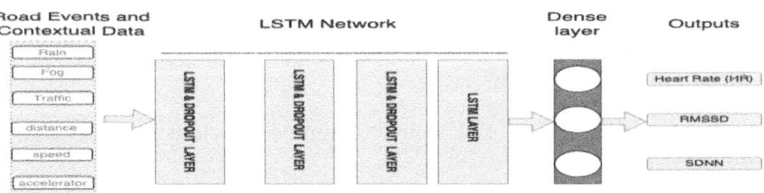

Fig. 3. Architecture of the LSTM-based model

LSTM Network-Based Context and Road Events Encoding: The model takes as input a sequential stream of road events, each characterized by contextual and environmental factors such as traffic dynamics (e.g., congestion levels), environmental conditions (e.g., noise levels, lighting variations, weather changes), and emergency events (e.g., roadblocks, sirens, and accidents). These features are encoded as normalized numerical vectors and processed sequentially to reflect their temporal relationships.

Example: Let us consider an ASD driver on a 15-min urban journey. At a specific moment in time, denoted as t, the vehicle collects and transmits a set of contextual features: Fog (500 m visibility) normalized to 0.92 (assuming a range of 20 m to 6000 m), Rain intensity: 0 (no rain), Noise level (70 dB) normalized to 0.70, Lighting level (40 lux) normalized to 0.45, Traffic: "heavy" as 1.0, Acceleration: -0.4 m/s^2 normalized to 0.35, Driving distance: 2.5 km normalized to 0.42 (assuming a maximum of 6 km trip), and Time elapsed: 6 min normalized to 0.40 (assuming a maximum of 15 min). Considering these subset of features, the feature vector x_i at time t_i is then represented as $x_i = [0.92, 0, 0.70, 0.45, 1.0, 0.35, 0.42, 0.40]$.

The normalized input vectors $x_{t_i} \in \mathbb{R}^d$, where d is the number of input features, are fed into an LSTM-based model, capturing temporal dependencies to forecast physiological responses. The LSTM architecture consists of stacked layers that progressively capture temporal patterns. A Dropout layer is used to reduce overfitting by randomly deactivating a subset of neurons during training.

The output of the final LSTM layer consists of hidden states $\{h_1, \ldots, h_n\}$, where each $h_n \in \mathbb{R}^d$ encodes the learned temporal and contextual representation of the input at time step t_i.

Final Mapping of Physiological Biomarkers Output Through a Dense Layer: The model incorporates a fully connected *Dense* layer that maps the last hidden state output vector $h \in \mathbf{R}^{d_h}$, where d_h is the dimensionality of the final LSTM hidden state, to the final output space corresponding to the predicted physiological biomarkers: HR, RMSSD, and SDNN.

This transformation is defined as: $\hat{y} = W \cdot h + b$, where W is the weight matrix connecting the last hidden state vector h to the output layer, b is the bias vector. The output vector $\hat{y}$ represents the predicted values of the three physiological biomarkers HR, RMSSD, and SDNN.

5.2 Stress Classification Based on Predicted Physiological Biomarkers

The predicted physiological biomarkers HR, RMSSD, and SDNN are used to infer stress states through a structured pipeline that includes directional alignment to standardize biomarker interpretations, statistical feature extraction to capture relevant physiological patterns, personalized labeling based on individual stress score distributions, and deep learning-based classification to distinguish between stressed and non-stressed conditions. The different steps are presented below.

Directional Alignment: Given that HR is positively correlated with stress, whereas RMSSD and SDNN are negatively correlated, we applied a transformation to invert the negatively correlated biomarkers: $1 - \hat{X}_i(t)$, where $\hat{X}_i(t)$ denotes the normalized value of the biomarker X_i at time t. This transformation ensures that all biomarkers contribute consistently to the overall stress estimation, with higher values uniformly indicating higher levels of stress.

Example: Consider a sample where the normalized physiological parameters at time t are : $\hat{HR}(t) = 0.80$, $\hat{RMSSD}(t) = 0.75$, $\hat{SDNN}(t) = 0.60$. Since HR is positively correlated with stress, its normalized value is used directly. However, RMSSD and SDNN are negatively correlated with stress-higher values typically indicate lower stress. To ensure directional consistency across features, we apply the following transformation: $\tilde{RMSSD}(t) = 1 - \hat{RMSSD}(t) = 1 - 0.75 = 0.25$ and $\tilde{SDNN}(t) = 1 - \hat{SDNN}(t) = 1 - 0.60 = 0.40$.

The directionally aligned values are : $\hat{HR}(t) \approx 0.80$, $\tilde{RMSSD}(t) = 0.25$, $\tilde{SDNN}(t) = 0.40$. These values can now be interpreted consistently: higher values uniformly indicate a higher level of physiological stress activation.

Stress Indicators and Stress Scores: To quantify stress, we computed four descriptive statistics: *mean, first quartile, median,* and *third quartile,* for each of the three biomarkers HR, RMSSD, and SDNN within each sample of the output dataset. These statistics, derived from the distribution of each biomarker within the sample, are referred to as *stress indicators.*

So, for each sample, a set of 12 stress indicators $I_i = \{I_i^{(1)}, I_i^{(2)}, \ldots, I_i^{(N)}\}$ was generated. Here, $N = 12$ represents the total number of stress indicators, with each $I_i^{(n)}$ denoting the n-th stress indicator corresponding to the i-th sample. These indicators capture essential characteristics of the physiological responses across the three biomarkers, providing a multi-dimensional representation of the individual's stress profile.

Subsequently, an overall *stress score* S_i for each sample was computed by averaging these 12 stress indicators. This aggregation yields a single scalar value that summarizes the stress level for the sample, representing the combined impact of the various biomarkers' behaviors within the observed time window. The stress score S_i is calculated using the following formula: $S_i = \frac{1}{N} \sum_{n=1}^{N=12} I_i^{(n)}$. A higher value of S_i reflects a higher stress level, as all output biomarkers were aligned to be positively correlated with driver stress.

Stress Labeling: To account for individual variability in physiological stress responses, a personalized, distribution-based approach was used to label stress levels for each driver. Specifically, labels were assigned by analyzing the statistical distribution of stress scores S_i within each driver's dataset. This method avoids reliance on fixed thresholds and instead uses intra-subject variability as a reference as follows [22]:

1. Samples with stress scores greater than the third quartile (Q_3) but below the upper outlier boundary ($Q_3 + 1.5 \times IQR$) were labeled as *stressed.*
2. Samples with stress scores lower than or equal to the third quartile but above the lower outlier boundary ($Q_1 - 1.5 \times IQR$) were labeled as *not stressed.*

Where Q_1 and Q_3 represent the first and third quartiles of the stress score distribution for a given driver, and $IQR = Q_3 - Q_1$ is the interquartile range.

Example: Let us assume that, based on the descriptive statistics computed for each normalized biomarker within the sample window, we obtain the following values:

Statistic	HR	RMSSD	SDNN
Mean	0.82	0.30	0.40
First Quartile (Q_1)	0.78	0.25	0.36
Median	0.81	0.29	0.39
Third Quartile (Q_3)	0.85	0.34	0.43

From these statistics, the 12 stress indicators $I_i^{(n)}$ for this sample are: $I_i = \{0.82, 0.78, 0.81, 0.85, 0.30, 0.25, 0.29, 0.34, 0.40, 0.36, 0.39, 0.43\}$. The overall stress score S_i is then computed as the average of these 12 indicators: $S_i = \frac{1}{12} \sum_{n=1}^{12} I_i^{(n)} = \frac{0.82 + 0.78 + \dots + 0.43}{12} = 0.487$.

To determine the stress label for this sample, we compare the computed score to the statistical distribution of stress scores for the corresponding driver. Assume the following quartile values based on the driver's stress score history: First Quartile $(Q_1) = 0.420$ and Third Quartile $(Q_3) = 0.530$. The interquartile range is: $IQR = Q_3 - Q_1 = 0.530 - 0.420 = 0.110$. From this, we compute the outlier boundaries:

- Lower Bound $= Q_1 - 1.5 \times IQR = 0.420 - 1.5 \times 0.110 = 0.255$
- Upper Bound $= Q_3 + 1.5 \times IQR = 0.530 + 1.5 \times 0.110 = 0.695$

Now we compare the current sample's stress score $S_i = 0.487$ to the distribution. Since $S_i = 0.487 \leq Q_3 = 0.530$ and $S_i \geq Q_1 = 0.420$, the sample is labeled as *not stressed*.

CNN-Based Stress Classification: Figure 4 illustrates the architecture of the proposed CNN-based model, designed to classify an individual's stress state based on predicted physiological biomarkers HR, RMSSD, and SDNN. The model employs a series of one-dimensional convolutional layers to extract hierarchical temporal features from the input sequences, followed by pooling and flattening operations to reduce dimensionality and prepare the data for classification. Fully connected layers then integrate the learned features to produce a binary stress classification.

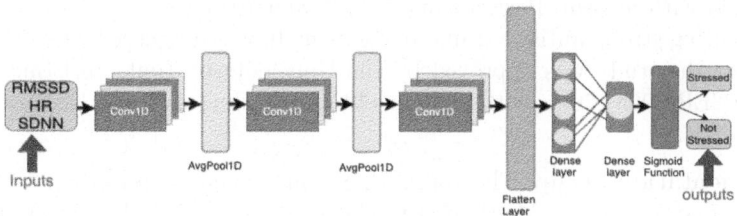

Fig. 4. Architecture of the classification CNN-based model

Once stress is proactively detected, the system can be extended to propose adaptive responses, such as suggesting an alternative route. If rerouting is not possible, it instead describes the road-related threat to help the driver anticipate and prepare for it, offer step-by-step coping strategies to assist in navigating the situation, or initiate a return home or a stop at a nearby rest area. In cases of extreme stress, the system may also trigger an alert to designated caregivers or family members.

6   Experiments and Evaluation

The proposed approach has been implemented and thoroughly evaluated. In the following sections, we present dataset acquisition, experimental setup, and the results obtained across different evaluation metrics.

6.1   Dataset Acquisition and Implementation Setup

Driving Simulator: Data were collected using the *SimulAuto*[2] driving simulator at the University of Toulouse 2, depicted in Fig. 5, featuring a Renault Zoe car and a 360° visual setup. Several scenarios and simulations were built and executed using SCANeR^{TM}studio 9.8[3], capturing data 500 Hz.

Fig. 5. The car simulator *SimulAuto*

Participants: Ten participants (7 women, 3 men; aged 22–28), all licensed drivers with no heart issues, took part. All participants passed the RAADS-14 [7], a screening test designed to identify adults presenting behavioral traits associated with autism. Participants 1, 2, 3, and 10 scored above the cutoff of 14, indicating strong autistic traits; participant 10 was diagnosed with ASD. We note that the study was approved by the University of Toulouse Ethics Committee (CER-2023_709), with consent obtained in line with the Declaration of Helsinki [2].

Implementation Setup: The approach was implemented in Python 3.11 using PyTorch and tested on a MacBook Pro (M3 Max, 36 GB RAM). Data were split into 75% for training and 25% for testing. The Adam optimizer was used for training.

Evaluation Metrics: For the LSTM model, we used Mean Absolute Error (MAE = $\frac{1}{n}\sum_{i=1}^{n}|y_i - \hat{y}_i|$) and Mean Squared Error (MSE = $\frac{1}{n}\sum_{i=1}^{n}(y_i - \hat{y}_i)^2$). For CNN-based classification, metrics included accuracy, recall, and AUC. Recall (TPR) and AUC (based on ROC curve) were computed based on true positive (TP), false positive (FP), true negative (TN), and false negative (FN)

---

[2] https://miroir.univ-tlse2.fr/2019/05/27/com-simulauto-le-simulateur-de-conduite-de-la-plateforme-ccu/.

[3] https://www.avsimulation.com/en/simulators/full-cab/simreal/.

rates as follows: Accuracy $= \frac{TP+TN}{TP+TN+FP+FN}$, TPR $= \frac{TP}{TP+FN}$, FPR $= \frac{FP}{FP+TN}$, AUC $= \int_0^1 TPR \, d(FPR)$. These metrics help evaluate model performance, especially under class imbalance.

6.2   Experimental Evaluations

LSTM Model: Figure 6 and Fig. 7 show the substantial reduction in MAE and MSE, respectively, along with rapid convergence during training for each one of the ten users.

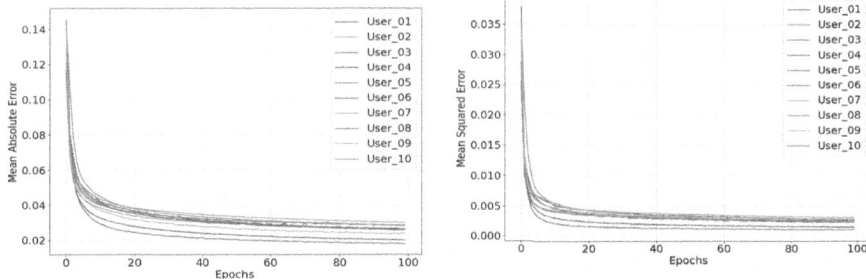

Fig. 6. Training MAE loss of the LSTM model

Fig. 7. Training MSE loss of the LSTM Model

Table 1 presents the performance of the LSTM model for each of the ten users, showing both training and validation losses. Overall, the model achieves low error values across all users, with average MAE and MSE of 0.0249 and 0.0023 for training, and 0.021 and 0.00247 for validation, indicating good generalization and consistent predictive accuracy.

Table 1. The MSE and MAE losses

Users	MAE	MSE	MAE_val	MSE_val
user_01	0.026	0.0028	0.026	0.0031
user_02	0.025	0.0026	0.022	0.020
user_03	0.018	0.001	0.014	0.007
user_04	0.026	0.0023	0.020	0.0015
user_05	0.028	0.0025	0.023	0.0017
user_06	0.019	0.0014	0.016	0.001
user_07	0.03	0.003	0.025	0.0023
user_08	0.026	0.0025	0.023	0.0021
user_09	0.024	0.0024	0.02	0.0022
user_10	0.028	0.0028	0.023	0.0022
Average	0.0249	0.0023	0.021	0.00247

A Mann-Whitney U test comparing participants with low versus high ASD features across all performance metrics (MAE, MSE, MAE_{val}, MSE_{val}) revealed a single statistically significant difference: MSE_{val} was significantly higher in the high-ASD-feature group compared to the low-ASD-feature group ($U = 1.50$, $p < 0.032$).

CNN Classification Model: Figure 8 shows the progressive decrease in binary cross-entropy loss across epochs, reflecting effective learning and convergence of the CNN model. Additionally, Fig. 9 presents the AUC scores for each driver, demonstrating the model's strong discriminative capability in accurately classifying stress versus non-stress conditions.

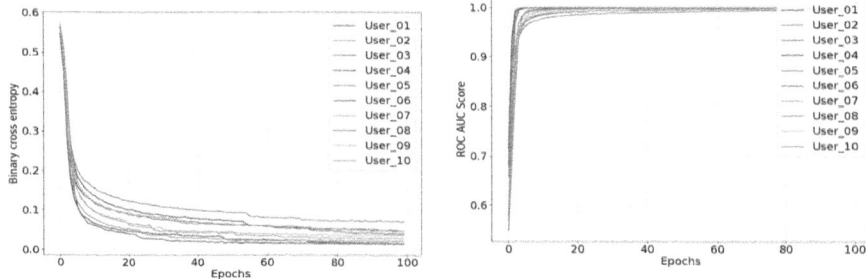

Fig. 8. CNN training binary cross-entropy

Fig. 9. CNN training AUC score

Table 2 summarizes the performance of the CNN classification model for each user during the validation phase. It reports accuracy, recall, and AUC scores, showing consistently high performance across all users. On average, the model achieved 0.987% accuracy, 98.3% recall, and an AUC of 99.8%, highlighting its robustness and reliability in stress classification.

Table 2. The Performance metrics of each user during the validation phase

Users	Accuracy	Recalls	AUC
user_01	0.994	0.994	0.998
user_02	0.990	0.989	0.999
user_03	0.986	0.979	0.998
user_04	0.994	0.990	0.999
user_05	0.982	0.970	0.998
user_06	0.997	0.995	0.999
user_07	0.983	0.975	0.997
user_08	0.975	0.959	0.994
user_09	0.987	0.983	0.998
user_10	0.992	0.987	0.999
Average	0.987	0.983	0.998

The evaluation results demonstrate the effectiveness of the proposed approach in predicting physiological biomarkers HR, RMSSD, and SDNN and detecting stress accurately, while improving stress detection reliability, particularly for ASD drivers.

7   Conclusion

This paper introduces a novel IoT-based smart framework specifically designed to address the unique needs of individuals with ASD in driving contexts. By integrating LSTM and CNN, our approach goes beyond conventional reactive models, enabling the proactive detection of stress responses before they are physiologically observed. The system leverages real-time environmental data to forecast the potential impact of road events. Experimental results demonstrate the model's robustness and accuracy across a diverse range of users and scenarios.

In our future work, we plan to extend the proposed framework by advanced intelligent models capable of further personalizing stress prediction and intervention strategies. Specifically, we aim to refine the system to dynamically adjust preventive actions based on the ASD driver's unique profile, the specific road context, and individual driver constraints. To support this, we will define and incorporate a wider spectrum of driving scenarios and environmental conditions. Furthermore, expanding the participant pool will be crucial to improving the generalizability and robustness of our findings. We also intend to incorporate additional features, such as jerk (the rate of change of acceleration), pedal pressure, steering variability, and headway distance, to enhance both the accuracy of stress detection and the system's sensitivity to contextual nuances.

Acknowledgment. This work has been supported by the Occitanie region, the European Regional Development Fund (ERDF), and the French government, through the France 2030 project managed by the National Research Agency (ANR) with the reference number "ANR-22-EXES-0015."

References

1. Amin, M., et al.: ECG-based driver's stress detection using deep transfer learning and fuzzy logic approaches. IEEE Access **10**, 29788–29809 (2022)
2. Association, W.M.: World medical association declaration of Helsinki: ethical principles for medical research involving human participants. JAMA **332**(15), 1575–1583 (2024)
3. Benssassi, S., Aitken, J.M., Waraich, A.: Wearable assistive technologies for autism: opportunities and challenges. IEEE Pulse **9**(6), 39–42 (2018)
4. Bian, D., Wade, J.W., Swanson, A., Weitlauf, A.S., Warren, Z., Sarkar, N.: Design of a physiology-based adaptive virtual reality driving platform for individuals with ASD. ACM Trans. Accessible Comput. (TACCESS) **12**, 1–24 (2019). https://doi.org/10.1145/3301498

5. Cecchini, A., Corti, S., Ranci, E.: Mobility policies and extra-small projects for improving mobility of people with autism spectrum disorder. Sustainability **10**(5), 1461 (2018)

6. Cox, D., et al.: A pilot study comparing newly licensed drivers with and without autism and experienced drivers in simulated and on-road driving. J. Autism Dev. Disord. **50**, 1258–1268 (2020)

7. Eriksson, J.M., Andersen, L.M., Bejerot, S.: RAADS-14 screen: validity of a screening tool for autism spectrum disorder in an adult psychiatric population. Mol. Autism **4**, 49 (2013). https://doi.org/10.1186/2040-2392-4-49

8. Fok, M., Owens, J., Ollendick, T., Scarpa, A.: Perceived driving difficulty, negative affect, and emotion dysregulation in self-identified autistic emerging drivers. Front. Psychol. **13** (2022)

9. Healey, J., Picard, R.W.: Detecting stress during real-world driving tasks using physiological sensors. IEEE Trans. Intell. Transp. Syst. **6**, 156–166 (2005)

10. Hochreiter, S., Schmidhuber, J.: Long short-term memory. Neural Comput. **9**(8), 1735–1780 (1997)

11. Immanuel, S., Teferra, M.N., Baumert, M., Bidargaddi, N.: Heart rate variability for evaluating psychological stress changes in healthy adults: A scoping review. Neuropsychobiology **82**, 187–202 (2023)

12. Jabon, M., Bailenson, J., Pontikakis, E., Takayama, L., Nass, C.: Facial expression analysis for predicting unsafe driving behavior. IEEE Pervasive Comput. **10**(4), 84–95 (2011)

13. LeCun, Y., Bottou, L., Bengio, Y., Haffner, P.: Gradient-based learning applied to document recognition. Proc. IEEE **86**(11), 2278–2324 (1998)

14. Cheak-Zamora, N., Tait, A., Coleman, A.: Assessing and promoting independence in young adults with autism spectrum disorder. J. Dev. Behav. Pediatr. **43**(3), 130–139 (2022)

15. Padmanaban, V., Sudhakaran, A., Sabnis, S., Geetha, S.: An autonomous driving system—dedicated vehicle for people with asd and their caregivers. arXiv preprint arXiv:2108.04367 (2021)

16. Porges, S.W.: Polyvagal theory: a science of safety. Front. Integr. Neurosci. **16**, 871227 (2022)

17. Rastgoo, M., Nakisa, B., Maire, F., Rakotonirainy, A., Chandran, V.: Automatic driver stress level classification using multimodal deep learning. Expert Syst. Appl. **138** (2019)

18. Seo, W., Kim, N., Kim, S., Lee, C., min Park, S.: Deep ECG-respiration network (deeper net) for recognizing mental stress. Sensors (Basel, Switzerland) **19** (2019)

19. Seo, W., Kim, N., Park, C., Park, S.M.: Deep learning approach for detecting work-related stress using multimodal signals. IEEE Sens. J. **22**, 11892–11902 (2022)

20. Shaffer, F., McCraty, R., Zerr, C.L.: A healthy heart is not a metronome: an integrative review of the heart's anatomy and heart rate variability. Front. Psychol. **5** (2014)

21. Wang, K., Guo, P.: An ensemble classification model with unsupervised representation learning for driving stress recognition using physiological signals. IEEE Trans. Intell. Transp. Syst. **22**, 3303–3315 (2020)

22. Wang, K., Murphey, Y., Zhou, Y., Hu, X., Zhang, X.: Detection of driver stress in real-world driving environment using physiological signals. 2019 IEEE 17th International Conference on Industrial Informatics (INDIN), vol. 1, pp. 1807–1814 (2019)

23. Zepf, S., Hernandez, J., Schmitt, A., Minker, W., Picard, R.W.: Driver emotion recognition for intelligent vehicles: A survey. ACM Comput. Surv. **53**(3), 1–30 (2021)

A Divisive Unsupervised Feature Selection Approach for Explainable Remaining Useful Life Prediction

Mouhamadou Lamine Ndao[1,2(✉)] , Genane Youness[1,2] , Ndèye Niang[2] ,
and Gilbert Saporta[2]

[1] Laboratoire LINEACT CESI, Nanterre, France
{mlndao,gyouness}@cesi.fr
[2] Laboratoire CEDRIC MSDMA, Paris, France
{n-deye.niang-keita,gilbert.saporta}@cnam.fr

Abstract. Predicting the Remaining Useful Life (RUL) in maintenance often encounters challenges such as high dimensionality, feature redundancy, and limited explainability. This paper presents a novel approach that combines Interpretable Divisive Feature Clustering (IDFC) with Long Short-Term Memory (LSTM) networks. The IDFC algorithm leverages the strengths of variable clustering methods (VARCLUS) and the Clustering of Variables around Latent Components (CLV) to identify significant features and non-orthogonal latent components. This method enables effective dimensionality reduction by selecting key features rather than combining them. Integrating IDFC with a single-layer LSTM and Shapley Additive Explanations (SHAP) results in a robust and interpretable framework for RUL prediction, achieving a balance between accuracy and transparency. Experimental results on a bearing dataset show that the IDFC + LSTM model outperforms traditional methods while enhancing interpretability through the identification of key energy-related features which influence the RUL prediction more.

Keywords: Dimensionality reduction · XAI · LSTM · variable classification · Predictive Maintenance

1 Introduction and Related Works

1.1 Introduction

One key challenge in predictive maintenance is dimension reduction due to data availability in the industry, particularly with the rise of the Internet of Things (IoT) connected objects [16] and the effectiveness of Deep Learning models such as Long Short-Term Memory (LSTM, [6]) for Remaining Useful life (RUL) prediction. This is due to the ability of models such as LSTM to tackle temporal dependencies and handle complex datasets, such as multivariate time series signals, as shown in the literature [23,25]. Additionally, LSTM models address the

© The Author(s), under exclusive license to Springer Nature Switzerland AG 2026
R. Wrembel et al. (Eds.): DEXA 2025, LNCS 16047, pp. 270–286, 2026.
https://doi.org/10.1007/978-3-032-02088-8_19

issue of vanishing gradients, which occurs when the hidden layer is unrolled multiple times [6]. However, the complexity of these models does not allow one to understand the results they provide. They are often called black boxes because their outputs are uninterpretable. This lack of transparency hinders stakeholders' ability to assess their reliability.

Regarding the dimension reduction challenge, integrating multiple data sources in industrial settings often leads to the inclusion of irrelevant features for RUL prediction and the presence of highly correlated or redundant features. This complicates the modeling process and hinders the interpretation of results by introducing bias through multicollinearity and the presence of irrelevant features [14, 19]. This challenge is critical because high-dimensional data can lead to problems such as overfitting, increased computational complexity, and reduced interpretability. To address this challenge, numerous studies have explored various techniques to reduce the number of features while preserving critical information, which combines original features into synthetic features through techniques such as PCA (e.g. Principal Component Analysis [1]), Autoencoder (EA) [24]. Even though these approaches perfectly correct feature redundancy, they are not suitable when the aim is to use XAI method to explain the result of models. These methods create a latent space represented as a combination of features, making them inherently uninterpretable. As a result, even with XAI methods, it is difficult to explain which feature most directly influences degradation, as it lacks direct correspondence to the original features.

To address the lack of interpretability associated with synthetic features generated through dimension reduction, one solution is to employ a feature clustering method followed by data fusion within each feature cluster [4]. This approach reduces feature redundancy by generating sparse synthetic features, which aggregate only features within relevant clusters. However, this method has several limitations. Many approaches in the literature require predefined numbers of clusters, and it is often noted that some clusters are atypical and require special consideration. Furthermore, despite generating sparser synthetic features, the literature highlights persistent challenges in interpreting results derived from feature clustering methods, as the relationships between synthetic features and original features often remain uninterpretable.

This study addresses two key challenges in predicting the RUL of mechanical components, such as bearing elements: first, dimensionality reduction, which addresses feature redundancy and high dimensionality, and mitigates bias caused by multicollinearity [14, 19] and second, the lack of explainability inherent in RUL prediction models. We propose an Interpretable Divisive Feature Clustering (IDFC) approach to bridge the gap between predictive performance and interpretability in industrial prognostics.

The paper is organized as follows: Sect. 1 presents an overview of the research problem and highlights key challenges in RUL prediction, particularly those related to dimensionality reduction and explainability. Section 2 presents the materials and methods, including the theoretical background (Subsect. 2.1), the proposed methodology (Subsect. 2.2), and the experimental setup (Subsect. 2.7), which covers the dataset description, preprocessing steps, and hyperparameter configuration. Section 3 presents the results and discussion, comparing the proposed model with traditional approaches, analyzing its performance on benchmark datasets, and explaining its prediction using the SHAP method.

1.2   Related Works

Explainability and dimensionality reduction are two significant challenges in predicting Remaining Useful Life (RUL). Utilizing Explainable Artificial Intelligence (XAI) methods is essential for enhancing transparency in RUL prediction [7]. In the context of RUL prediction, recent studies have employed dimensionality reduction (DR) techniques to manage the complexity of high-dimensional data. Methods such as Principal Component Analysis (PCA) effectively reduce data dimensions and improve model efficiency; however, they often compromise interpretability by transforming features into latent spaces. Some researchers, such as [20], have combined PCA with models like Random Forest (RF) or Support Vector Machine (SVM) to achieve reduced dimensionality while accurately predicting bearing RUL. Even when using less complex models, the dimensionality reduction approach does not always yield interpretable RUL predictions. Hybrid models that incorporate unsupervised dimensionality reduction techniques, such as EA-DNN (Deep Neural Networks) [17], PCA-CNN (Convolutional Neural Network)-LSTM [9], self-organizing map (SOM)-PCA-BiLSTM [10], and PCA-GCN (Graph Convolutional Network) [8], show promise in predicting RUL. However, these methods also face similar limitations in transparency. Advanced techniques like fractal dimensions and spatiotemporal GCNs [5] enhance accuracy by extracting spatial characteristics from signals but still lack transparency. Recent studies have integrated XAI methods to address this gap. For example, [3,25] utilize feature selection layers to rank contributions, while [26] combines feature clustering with SHAP values. However, this approach requires a predefined number of clusters and produces indirect explanations based on latent features (components). This underscores the critical need for explainability in dimensionality reduction, emphasizing the importance of providing synthetic features that align with domain knowledge and avoid opaque latent spaces.

To overcome these limitations and accurately predict bearing RUL using an unsupervised dimensionality reduction approach, we propose an LSTM-based model integrated with our Interpretable Divisive Feature Clustering (IDFC) method. Unlike traditional approaches, this novel framework automatically identifies the optimal number of features to retain, ensuring that predictions are grounded in domain-relevant features while enhancing both model efficiency and explainability.

2   Materials and Methods

We denote $X = \{x^1, x^2, \ldots, x^p\}$ a set of p features x_j describing N instances; $\tilde{X} = \{c_1, c_2, \ldots, c_K\}$ a projection of X on a smaller latent space of dimension

$K < p$, $\mathcal{P}_{\mathcal{K}} = \{C_1, \ldots C_K\}$ is a partition of the features into K clusters where each cluster C_k is represented by the component c_k.

2.1 Background

There are several ways to reduce the dimension of X: by projection onto a low-dimensional $K < p$ space $(\tilde{X})$, resulting in a combination of features into components (PCA), by feature selection, or a combination of both: clustering and latent feature. The third approach identifies clusters of features and defines their latent components. Each cluster C_k is represented by a latent feature (synthetic feature c_k) that summarizes, in the best way, its corresponding features, resulting in sparser and more interpretable components [4,22]. In the literature, two main algorithms are often used to perform feature clustering: The VARCLUS procedure of SAS software, which has never been the subject of scientific articles and clustering of variables around latent components (CLV) [21,22].

VARCLUS is a divisive, top-down hierarchical algorithm. It starts by dividing the features into two clusters based on their correlation with the first two principal components obtained from a PCA of the entire set of features. The algorithm then iteratively refines these clusters until each cluster contains only one eigenvalue that is larger than one $(\lambda_1 > 1)$ when performing PCA on that specific feature cluster. The condition on the eigenvalues gives VARCLUS two important advantages: the number of clusters K is naturally obtained, and the resulting clusters, which are associated with a first large eigenvalue, are somewhat unidimensional. Clustering features into unidimensional blocks is a simple and efficient way to search for the so-called factor. Each block can then be represented by a single component combining only its features, which is necessarily sparse relative to the total number of features. When the cluster size is still too large, further simplification is needed. VARCLUS does not force orthogonality between cluster components and does not require the number of clusters K to be specified in advance. However, the method can lead to many small clusters.

CLV algorithm [22] is an alternative to the VARCLUS approach. It is a k-means-like algorithm that clusters features around latent components by both identifying clusters of features and a latent component associated with each cluster. For a given number of clusters, K, CLV algorithm seeks a partition $\mathcal{P}_{\mathcal{K}}$ of the features and $\tilde{X} = \{c_1, \ldots, c_K\}$ of K latent features, each associated with one cluster k, to maximize the internal coherence within the clusters. When examining the relationship between features using their covariance or correlation coefficients, the clusters found are called directional groups, regardless of the sign of the coefficients. In this case, the aim is to define the partition $\mathcal{P}_K$ and the group's latent features $\tilde{X}$ to obtain the optimal value T^* of the CLV criterion, so that:

$$T^* = \max_{(\delta_{k,j}, c_K)} \sum_{k=1}^{K} \sum_{j=1}^{p} \delta_{kj} \operatorname{Cov}(x_j, c_k)^2, \tag{1}$$

under the constraint $\|c_k\| = 1$ with $\delta_{kj} = 1$ if the j-th feature belongs to the cluster C_k, and $\delta_{kj} = 0$ otherwise. For each cluster k, c_k is the first principal component of its features PCA. A comparison of the two described algorithms is given by Table 1. Unlike the VARCLUS algorithm, CLV reduces the number of clusters and does not produce a unidimensional feature cluster.

Both algorithms provide $\tilde{X}$ with latent features. Therefore, when used in conjunction with an XAI method, it may not be possible to directly explain the results provided by XAI method, even if all potential biases due to multicollinearity and redundancy are potentially reduced. If the explanation is unnecessary, it is generally also advantageous to concentrate on the most significant components, as it captures the essential patterns in the data. To correct all these issues, we propose the feature clustering approach described in the section below.

2.2   Interpretable Divisive Feature Clustering (IDFC)

Our proposed method integrates the strengths of both VARCLUS and CLV algorithms. IDFC approach involves three steps: an initialization step based on the VARCLUS algorithm, a clustering step based on the CLV algorithm, and a final step to extract interpretable features instead of latent features.

Initialization Step This step involves determining an initial partition $\mathcal{P}_K$ using VARCLUS. However, VARCLUS may produce a large number of small clusters. To address this, clusters with only one feature are considered noise clusters and are merged into a single cluster. This initial partition is defined in the Algorithm 1 and illustrated in Fig. 1.

Algorithm 1 Initial Partition

Input: $X = \{x^1, x^2, \ldots, x^p\}$.   ▷ Set of all features
Output: $\mathcal{P}_{K'}$ ▷ $K' = K + 1$ cluster merging all small clusters.
1: Perform PCA on X:
2: **if** $\lambda_2 > 1$ **then** :
   Divide the x^j of X based on the first two components, denoted c_1 and c_2, obtained from a PCA on X.
3: **end if**
4: **for** $C_k \in \{C_1, C_2\}$ **do**
   Repeat 1 and 2 on $X_{1_{[\delta_{k,j}, j \in [1:p]]}}$.   ▷ Splitting
5:     **for** $x^j \in X$ **do**:
   Reassign it to the C_k such that

$$\max_{x^j \in C_k} r^2 \left(x^j, c_k \right)$$

   Stop the division when $|C_k| = 1$ or if ($\lambda_2^k < 1$).   ▷ Stopping Criterion
6:     **end for**
7: **end for**
8: **Obtain** $\mathcal{P}_K$ with a potentially high number of single-feature clusters.
9: **Return** $\mathcal{P}_{K'}$ with $K' = K + 1$ cluster merging all small clusters.

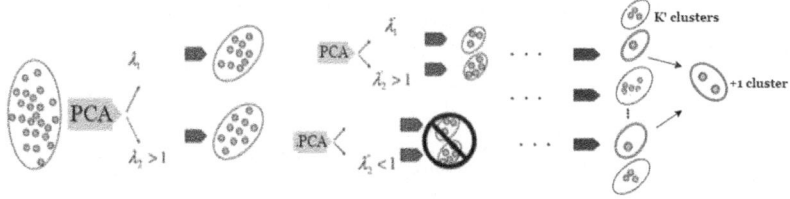

Fig. 1. Initialization of the partition using an extension of VARCLUS

This algorithm provides an initial partition. The next step consists of using an approach based on the CLV algorithm to consolidate the initial partition and consider the positive and negative correlations between features.

Refinement of the Partition Based on CLV Algorithm with a $K + 1$ partition: In this step, the partition $\mathcal{P}_K$ identified in the first step serves as the initialization partition. Then, a CLV algorithm is fitted using the 'K+1' approach proposed by [21], as illustrated in Fig. 2. It consists of finding T_{new} given by:

$$T_{\text{new}} = \max_{\delta_{jk}, c_k} \sum_{k=1}^{K} \sum_{j=1}^{p} \delta_{kj} \operatorname{cov}^2 (x_j, c_k) + \sum_{j=1}^{p} \left(1 - \sum_{k=1}^{K} \delta_{kj}\right) \rho^2 \operatorname{var}(x_j) \qquad (2)$$

For each cluster C_k, c_k is its first principal component. In Eq. 2, an additional noise cluster can be represented by a prototypal feature that is expected to have the same correlation with all the observed features. This noise cluster is controlled with a tuning parameter ρ to be chosen between 0 and 1. If $\rho = 0$, Eq. 1 will be considered with no noise cluster. To find the refined partition, Algorithm 2 is considered.

Algorithm 2 Refinement using CLV

Input: $X = \{x^1, x^2, \ldots, x^p\}$; $\mathcal{P}_{K'+1}$.

Output: the refined partition $\mathcal{P}_{K''+1}$ and the components of each cluster.

1: **for** each C_k **do**
2:     Compute the first principal component c_k.
3: **end for**
4: **for** each x^j **do**
5:     Assign x^j to the cluster that maximizes the internal homogeneity criterion T_{new}.
6: **end for**
7: Repeat until convergence.
8: **Return** $\mathcal{P}_{K''+1}$ and $c_1, \ldots, c_{K''+1}$. ▷ The refined partition and the components of each cluster

---

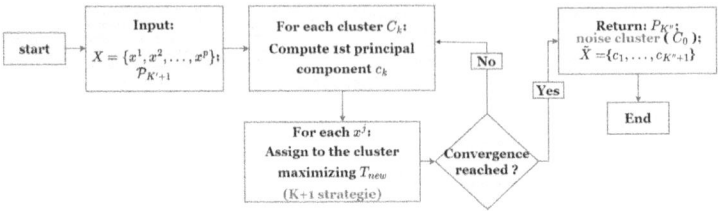

Fig. 2. Partition refinement using CLV algorithm with a $K + 1$ strategy

Feature Extraction Step: This step involves selecting features according to Eq. 3. To do this, the first principal component is computed for each cluster. Then, in each cluster, the most correlated feature is chosen as a prototype for each latent feature, given by:

$$x^k = \arg \max_{x_j \in C_k} r^2 (x_j, c_k) \tag{3}$$

Table 1 compares the proposed interpretable Feature Clustering with two of the most used feature clustering algorithms (CLV, VARCLUS). Unlike CLV, which requires the number of clusters to be specified beforehand, VARCLUS and the proposed method determine clusters without prior constraints. Additionally, while only CLV and the proposed method can distinguish positive and negative correlations, VARCLUS and the proposed method ensure unidimensional clusters, a feature absent in CLV. The proposed method also offers optional sparse components and is not sensitive to initialisation, similar to VARCLUS. Most importantly, it is the only method, compared to the other two, that produces interpretable results, making it a significant improvement over CLV and VAR-CLUS in terms of flexibility, handling of correlations, and overall interpretability.

Table 1. Comparison between CLV, VARCLUS and proposed method (IDFC)

	CLV	VARCLUS	IDFC
Choice of Number of clusters K	✗	✓	✓
Unidimensionality	✗	✓	✓
Differentiate positive and negative correlations.	✓	✗	✓
Sparse components	✓	✗	✓
Initialization-sensitive	✗	✓	✓
Isolate noise features	✓	✗	✓
Interpretable result	✗	✗	✓

2.3   The RUL Prediction Problem

The main objective of this analysis is to provide a parsimonious model to accurately predict the RUL and allow the use of XAI method to understand the model's result. Therefore, the first step will entail reducing the dataset dimension. Then, an optimal predictive model, denoted as f, will be identified to achieve the highest accuracy in RUL predictions. This is an optimization problem given by:

$$\arg\min_{f \in \mathcal{F}}(\|y - f(x)\|) \tag{4}$$

$\mathcal{F}$ is the space of all possible models, and y the response feature (RUL) to predict. The $RMSE$ (Root-Mean-Square Error) metric is used to choose the optimal prediction model f on $\mathcal{F}$.

After predicting the RUL, an optimal local post-hoc XAI method, based on a surrogate model g, is identified. This method provides the optimal $\xi = g(x, f)$ to explain the feature importance corresponding to $\text{RUL}'_i = f(x_i)$.

2.4   Explainable Artificial Intelligence (XAI) in RUL Prediction

In this study, our aim is to identify which part of the bearing impacts the RUL more. So, we use one of the most used, reliable and robust XAI post-hoc methods, the SHAP method [11,12,18]. SHAP is a local post-hoc XAI method that identifies the most relevant features according to their feature importance.

For the SHAP method, the prediction $f(x)$ is considered as the final payoff in a game where the players are the features j. The Shapley value ϕ_j corresponds to the marginal contribution of feature j in the prediction of $f(x)$, considering all possible combinations. According to theoretical considerations, Shapley values are the only values that meet three essential properties: Local Accuracy, which requires the explanation model to accurately reflect the behavior of the complex model in specific contexts; Missingness, which asserts that features designated as missing should not impact predictions; and Consistency, which states that if a feature's contribution increases, its assigned importance must not decrease.

2.5   Proposed Feature Clustering-LSTM Network Model: Experimental Setting

The proposed pipeline for explainable RUL prediction in the context of predictive maintenance is illustrated in Fig. 3. The process begins with data acquisition, during which raw sensor measurements are collected. In the feature extraction step, these measurements are transformed into meaningful features. A preprocessing phase follows, which includes normalization and data smoothing to reduce noise. A fixed-length sliding time window (TW) is then applied to emphasize the degradation phase during model training. The proposed IDFC algorithm is then applied for dimensionality reduction, grouping features into homogeneous clusters and selecting interpretable representatives from each cluster to minimize redundancy. The preprocessed data are used to train a single-layer LSTM model.

This architecture was chosen based on previous works [11, 26]. Finally, the SHAP method provides clear insights into the impact of features on the model's predictions, ensuring a transparent interpretation of the results.

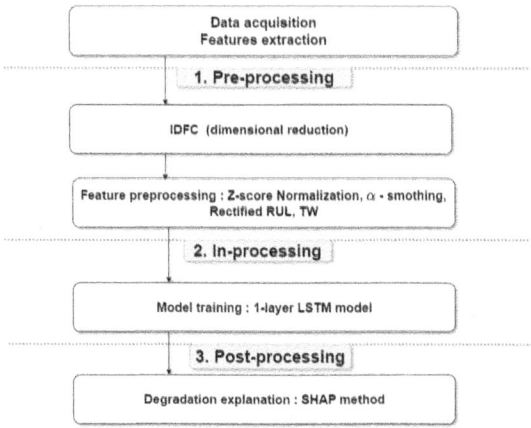

Fig. 3. Flowchart of the proposed method

2.6 FEMTO Bearing Dataset

The dataset used is the bearing operation data collected by FEMTO-ST on the PRONOSTIA platform [13]. Rolling bearings are crucial components in rotating machinery, and their failure can result in the malfunction of adjacent components and even the entire machine.

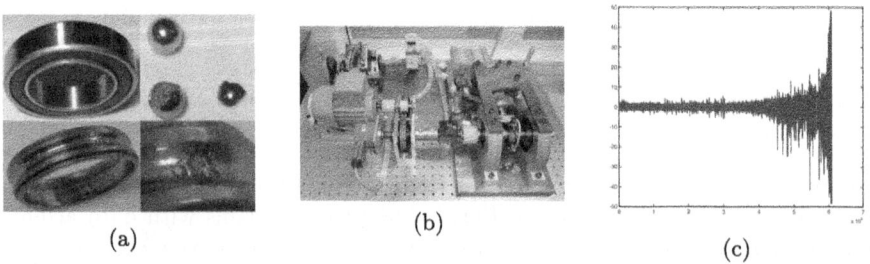

Fig. 4. (a) Normal and degraded bearing; (b) Ovrview of PRONOSTIA platform; (c) A vibration raw signal

The FEMTO bearing dataset was created by accelerating the degradation of bearings on the PRONOSTIA experimental platform (cf. Fig. 4)(a). This dataset includes acceleration signals recorded in both horizontal and vertical directions. The sampling frequency is 25.6 kHz, with each sampling lasting 0.1 s, resulting in 2,560 data points recorded per sampling. The sampling interval is set at 10 s. Bearing RUL prediction consists of two stages. The first one involves extracting the degradation features from the measured vibration data of the target bearing and the second stage involves predicting the RUL according to the degradation features. According to [2], a total of 111 features were manually extracted from each of the signals in two orientations: horizontal and vertical. This study utilizes bearing data under three different operating conditions. The training and test sets used for this study are presented in Table 2. The target feature is the Time (or cycles) until bearing failure, calculated as follows: $RUL_{i,t} = T_i - t$, where: $RUL_{i,t}$ is the Remaining Useful Life of bearing i at time step t, T_i represents the failure time (i.e., the last cycle) of bearing i, t is the current time step.

Table 2. Description of the PRONOSTIA Dataset: Bearing (sample size); Training and Test Sets under Different Conditions

	Condition 1	Condition 2	Condition 3
Training sets	B 1-1 (2803)	B 2-1 (911)	B 3-1 (515)
	B 1-2 (871)	B 2-2 (797)	B 3-2 (1637)
Test sets	B 1-3 (1802)	B 2-3 (1202)	B 3-3 (352)
	B 1-4 (1139)	B 2-4 (612)	
	B 1-5 (2302)	B 2-5 (2002)	
	B 1-6 (2302)	B 2-6 (572)	
	B 1-7 (1502)	B 2-7 (172)	

2.7 Experimental Setting

Standardization is used to scale feature data to have a zero mean and unit variance, which helps facilitate model convergence. Additionally, data smoothing is performed using exponential smoothing, where the parameter α adjusts the weight assigned to recent versus historical observations. This effectively reduces noise in the sensor signals. A fixed-length sliding time window (TW) is then applied to transform the multivariate time series data. Table 3 summarizes the preprocessing parameters, as well as the hyperparameters, for the LSTM-based model applied to the bearing dataset.

2.8 Evaluation Metrics: RMSE and MAE

To evaluate the performance of our predictive models, we use two standard error metrics: the Root Mean Squared Error (RMSE) and the Mean Absolute Error (MAE). The RMSE is defined as follows: $RMSE = \sqrt{\frac{1}{n}\sum_{i=1}^{n}(\hat{y}_i - y_i)^2}$; The

Table 3. Hyperparameters used for the LSTM model

Hyperarameters	Condition 1	Condition 2	Condition 3
TW	45	20	50
Alpha	0.01	0.05	0.01
Nb layers	1	1	1
Nodes	256	64	128
Dropout	0.3	0.3	0.3
Activation	tanh	relu	relu
Batch size	64	32	64
Learning rate	0.001	0.0001	0.001
Optimizer	Adam	Adam	Adam
Nb features	33	33	31
Epochs	50	50	50
Standardisation		Z-score	
Data smoothing		exponential smoothing	

MAE is defined as: $\text{MAE} = \frac{1}{n} \sum_{i=1}^{n} |\hat{y}_i - y_i|$. In these formulas, $\hat{y}_i$ represents the predicted values, and y_i represents the true values for a set of n observations. Both RMSE and MAE quantify the difference between predicted and actual values. Lower values for both metrics indicate better predictive performance.

3   Results and Discussion

3.1   Interpretable Divisive Feature Clustering Results

Table 4 presents the clusters of features obtained, along with the number of features in each cluster ($|C_k|$). Cluster C_0 corresponds to the noise cluster. In

Table 4. Distribution of features by clusters: C_k represents cluster k and $|C_k|$ represents the number of features in cluster C_k

Condition 1	C_k	C_0	C_1	C_2	C_3	C_4	C_5	C_6	C_7	C_8	C_9	C_{10}	C_{11}	C_{12}	C_{13}	C_{14}	C_{15}	C_{16}
	$|C_k|$	17	3	7	15	6	6	5	9	4	2	6	2	13	4	3	6	3
	C_k	C_{17}	C_{18}	C_{19}	C_{20}	C_{21}	C_{22}	C_{23}	C_{24}	C_{25}	C_{26}	C_{27}	C_{28}	C_{29}	C_{30}	C_{31}	C_{32}	C_{33}
	$|C_k|$	9	4	6	10	11	10	11	12	5	6	9	5	3	2	2	2	4
Condition 2	C_k	C_0	C_1	C_2	C_3	C_4	C_5	C_6	C_7	C_8	C_9	C_{10}	C_{11}	C_{12}	C_{13}	C_{14}	C_{15}	C_{16}
	$|C_k|$	22	11	7	7	7	6	4	21	5	1	10	4	4	2	7	2	7
	C_k	C_{17}	C_{18}	C_{19}	C_{20}	C_{21}	C_{22}	C_{23}	C_{24}	C_{25}	C_{26}	C_{27}	C_{28}	C_{29}	C_{30}	C_{31}	C_{32}	C_{33}
	$|C_k|$	9	3	5	1	4	13	4	7	2	6	6	5	2	12	1	5	10
Condition 3	C_k	C_0	C_1	C_2	C_3	C_4	C_5	C_6	C_7	C_8	C_9	C_{10}	C_{11}	C_{12}	C_{13}	C_{14}	C_{15}	C_{16}
	$|C_k|$	20	5	3	13	10	3	10	6	4	7	4	16	5	7	4	6	12
	C_k	C_{17}	C_{18}	C_{19}	C_{20}	C_{21}	C_{22}	C_{23}	C_{24}	C_{25}	C_{26}	C_{27}	C_{28}	C_{29}	C_{30}	C_{31}		
	$|C_k|$	18	4	10	8	6	6	3	2	2	2	5	3	1	15	2		

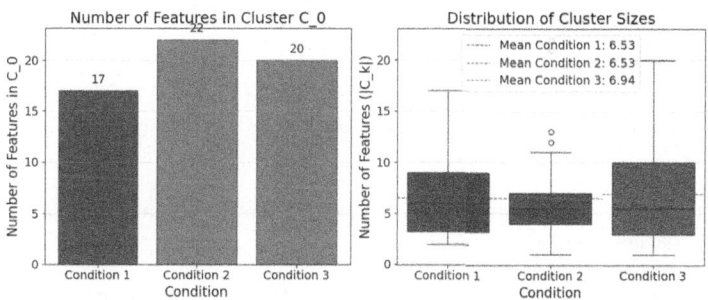

Fig. 5. Distribution of the Number of features per Cluster across Conditions

Table 5. Remaining features grouped by their type on the Condition 1

Type	Features
Energy	energy_band_13_freq_h, energy_band_22_freq_h
	energy_band_6_freq_v, energy_band_13_freq_v,
	energy_band_16_freq_v, energy_band_20_freq_v, energy_h, energy_v
Statistical Moments	moment_7_h, moment_8_v, derivative_moment_9_v
Derivative Features	derivative_peak_to_peak_v, derivative_std_h,
	derivative_margin_factor_h, derivative_margin_factor_v, derivative_moment_9_v
Integral Features	integral_std_h, integral_margin_factor_h, integral_moment_6_h,
	integral_max_abs_v, integral_impulse_factor_v,
	integral_moment_7_v, integral_moment_11_v, integral_moment_6_v
RMS & IMF	imf_entropy_3_tf_h, imf_rms_1_tf_h, imf_rms_0_tf_v,
	imf_rms_1_tf_v, imf_rms_2_tf_v, rms_freq_freq_h
Skewness	form_factor_freq_freq_h, skewness_freq_freq_v
Mean and frequency	mean_freq_freq_v, mean_v

Condition 1, 17 features are identified as noise, while there are 22 noise features for Condition 2 and 20 for Condition 3. The mean number of features is shown in Fig. 5.

The results from the IDFC analysis differ depending on the experiment. For instance, both Condition 1 and Condition 2 yield 33 feature clusters in addition to one noise cluster, whereas Condition 3 produces 31 feature clusters along with a single noise cluster.

When analyzing the size of the obtained clusters, the minimum cluster size is 2 (specifically C_3, C_{30}, C_{31}, and C_{32}). Notably, some clusters contain a greater number of features. For example, in Condition 1, cluster C_3 includes 15 features, while cluster C_{24} contains 12 features. This indicates that 15 features are represented by a single feature within cluster C_3, and 12 features are represented by one feature in cluster C_{24}.

Table 5 shows the 33 features extracted and grouped according to the type of feature selected on the subset Condition 1.

3.2   Ablation Study

To evaluate the effectiveness of the proposed feature clustering approach for RUL prediction, we conducted an ablation study. This involved comparing our model (IDFC+LSTM) with a classical alternative, PCA+LSTM, using K components,

Table 6. Performance by bearing and method, with averages including other state-of-the-art approaches. mean(std)

Cond	Bearing	Methods	RMSE	MAE
1	1-3	IDFC+LSTM	**0,063**	**0,051**
		PCA+LSTM	0,075	0,064
	1-4	IDFC+LSTM	0,071	0,056
		PCA+LSTM	**0,059**	**0,050**
	1-5	IDFC+LSTM	**0,151**	**0,121**
		PCA+LSTM	0,165	0,124
	1-6	IDFC+LSTM	**0,005**	**0,004**
		PCA+LSTM	0,012	0,010
	1-7	IDFC+LSTM	**0,133**	**0,088**
		PCA+LSTM	0,178	0,123
	Moyenne	IDFC+LSTM	**0.084 (0.052)**	**0.064 (0.039)**
		PCA+LSTM	0,098 (0.064)	0.074 (0.044)
		LSTM [15]	0,152	0,120
		GRU [10]	0,181	0,147
		HDCN [27]	0,232	0,198
		SOM+PCA+BiLSTM [10]	0,128	0,105
2	2-3	IDFC+LSTM	**0,039**	**0,030**
		PCA+LSTM	**0,039**	**0,030**
	2-4	IDFC+LSTM	0,037	0,031
		PCA+LSTM	**0,035**	**0,030**
	2-5	IDFC+LSTM	**0,243**	**0,195**
		PCA+LSTM	0,247	0,199
	2-6	IDFC+LSTM	**0,110**	**0,088**
		PCA+LSTM	0,113	0,090
	2-7	IDFC+LSTM	0,200	0,172
		PCA+LSTM	**0,198**	**0,169**
	Moyenne	IDFC+LSTM	**0.126 (0.084)**	**0.103 (0.069)**
		PCA+LSTM	0.127 (0.085)	0.104 (0.070)
		LSTM [15]	0,270	0,219
		GRU [10]	0,226	0,171
		HDCN [27]	0,283	0,244
		SOM+PCA+BiLSTM [10]	0,164	0,121
3	3-3	IDFC+LSTM	0,029	**0,001**
		PCA+LSTM	**0,028**	0,022

where K represents the number of features selected by IDFC for each subset. Table 6 presents the results of our model in terms of accuracy, utilizing the Root Mean Square Error (RMSE) and Mean Absolute Error (MAE) metrics. Analyzing the results shown in the table, we observe that our proposed model outperforms the classical PCA+LSTM method. For instance, in Condition 1, our model predicts RUL with an RMSE of 0.084 and an MAE of 0.064, while the PCA+LSTM model yields an RMSE of 0.098 and an MAE of 0.074. In Condition 2, the proposed algorithm achieves even more precise predictions, with an RMSE of 0.126 and an MAE of 0.103, compared to the PCA+LSTM model's RMSE of 0.127 and MAE of 0.104.

These results demonstrate that our model outperforms the classical method combining PCA and LSTM across both conditions. The effectiveness of the proposed model is demonstrated in Fig. 6, which compares the predicted RUL of bearing 1_3 to its actual RUL. It is evident that the predicted RUL aligns closely with the actual RUL, with both trends reaching zero at the end of the RUL period. To enhance transparency, it is essential to identify the features that most significantly influence the predicted RUL. To accomplish this, we employed the SHAP method.

3.3   Prediction Explanation Using SHAP XAI Method

The proposed dimensionality reduction method allows for the application of Explainable Artificial Intelligence (XAI) techniques to clarify the RUL predictions made by an LSTM model. Unlike traditional methods, this new approach produces interpretable features that yield meaningful results, regardless of the components involved. By using SHAP, we can evaluate which features have the most significant impact on RUL predictions.

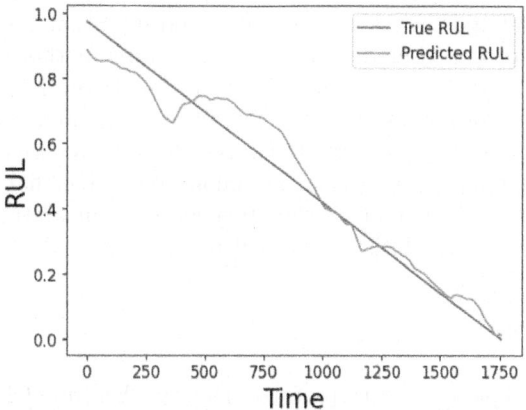

Fig. 6. Proposed Method: Prediction results on Bearing 1_3

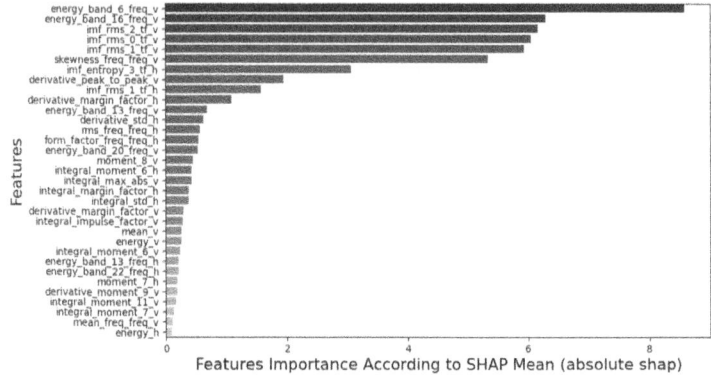

Fig. 7. Feature importance provided by the SHAP method for bearing 1_3.

Figure 7 illustrates the feature importance determined by the SHAP method for the RUL prediction of bearing 1_3. It shows that the most relevant features for RUL prediction are energy_ban_6_freq_v and energy_ban_16_freq_v, which relate to the energy contained in the 6th and 16th frequency bands of the vertical signal component. Among the other features, the least relevant ones are moment_8 and energy_v. In contrast, classical methods like PCA and t SNE do not provide such explanations, as they generate synthetic features that are not interpretable in terms of original features.

3.4   Comparison with Other Studies

We have compared the proposed model with several studies that used the same dataset for dimensionality reduction. Table 6 presents a performance comparison between our model and these studies, focusing on RMSE and MAE. The results indicate that the proposed model outperforms existing methods, achieving lower RMSE and MAE values. Specifically, for condition 1, the RMSE is 0.084, and the MAE is 0.064; for condition 3, the RMSE is 0.126, and the MAE is 0.103. These findings demonstrate the effectiveness of the proposed model using IDFC. Furthermore, our model is the only one among the three that permits the use of XAI methods to offer genuine explanatory features, in contrast to the studies against which we compared our results with.

4   Conclusion

This study introduces the Interpretable Divisive Feature Clustering (IDFC) method for predicting Remaining Useful Life (RUL), emphasizing dimensionality reduction and interpretability. IDFC effectively reduces features and provides transparent insights in contrast to PCA-based methods.

Experimental results from the FEMTO bearing dataset demonstrate that IDFC, when combined with Long Short-Term Memory (LSTM) networks, outperforms the PCA + LSTM approach. For Bearing 1, the Root Mean Square Error (RMSE) decreases from 0.98 to 0.084, and the Mean Absolute Error (MAE) drops from 0.74 to 0.064. For Bearing 2, the MAE significantly improves from 0.022 to 0.001. Furthermore, SHAP analysis reveals that specific energy-related features have a strong influence on RUL prediction, confirming the interpretability of the method.

As part of future work, additional empirical studies will be conducted using other datasets to compare the clustering methods of the IDFC algorithm, rigorously evaluating clustering quality and assessing the robustness of our approach. We will also explore alternative dimensionality reduction techniques, such as sparse PCA, kernel PCA, and autoencoder-based methods, to analyze their effects on both the quality and interpretability of the resulting clusters. These extensions will contribute to a deeper understanding of the methodological choices behind IDFC and refine its applicability to real-world scenarios.

References

1. Abdi, H., Williams, L.J.: Principal component analysis. Wiley Interdisciplinary Rev. Comput. Stat. **2**(4), 433–459 (2010)
2. Ayman, A., Onsy, A., Attallah, O., Brooks, H., Morsi, I.: Feature learning for bearing prognostics: a comprehensive review of machine/deep learning methods, challenges, and opportunities. Measurement, p. 116589 (2024)
3. Barraza, J.F., Droguett, E.L., Martins, M.R.: Towards interpretable deep learning: a feature selection framework for prognostics and health management. Case Stud. Health Diagnostics RUL Prediction **12**, 45–67 (2021)
4. Chavent, M., Kuentz-Simonet, V., Liquet, B., Saracco, J.: Clustofvar: an r package for the clustering of variables. J. Stat. Softw. **50**, 1–16 (2012)
5. Ding, G., Wang, W., Zhao, J.: Prediction of remaining useful life of rolling bearing based on fractal dimension and cnn. J. Prognostics **32**, 112–125 (2022)
6. Hochreiter, S., Schmidhuber, J.: Long short-term memory. Neural Comput. **9**(8), 1735–1780 (1997)
7. Kobayashi, K., Alam, S.B.: Explainable, interpretable, and trustworthy ai for an intelligent digital twin: a case study on remaining useful life. Eng. Appl. Artif. Intell. **129**, 107620 (2024)
8. Kumar, A., Parkash, C., Kundu, P., Tang, H., Xiang, J.: Enhanced deep learning framework for accurate near-failure rul prediction of bearings in varying operating conditions. Adv. Eng. Inform. **65**, 103231 (2025)
9. Li, H., Li, Y., Wang, Z., Li, Z.: Remaining useful life prediction of aero-engine based on pca-lstm. In: 2021 7th International Conference on Condition Monitoring of Machinery in Non-Stationary Operations (CMMNO), pp. 63–66. IEEE (2021)
10. Liu, J., Yang, Z., Xie, J., Wang, R., Liu, S., Xi, D.: A feature fusion-based method for remaining useful life prediction of rolling bearings. IEEE Trans. Instrumentation Measur. (2023)
11. Ndao, M.L., Youness, G., Niang, N., Saporta, G.: Enhancing explainability in predictive maintenance: investigating the impact of data preprocessing techniques on xai effectiveness. In: The 37th International Conference of the Florida Artificial Intelligence Research Society-FLAIRS-37, vol. 37. FLAIRS (2024)

12. Ndao, M.L., Youness, G., Niang, N., Saporta, G.: Improving predictive maintenance: evaluating the impact of preprocessing and model complexity on the effectiveness of explainable artificial intelligence methods. Eng. Appl. Artif. Intell. **144**,
110144 (2025). https://doi.org/10.1016/j.engappai.2025.110144

13. Nectoux, P., et al.: Pronostia: an experimental platform for bearings accelerated
degradation tests. In: IEEE International Conference on Prognostics and Health
Management, PHM'12, pp. 1–8. IEEE Catalog Number: CPF12PHM-CDR (2012)

14. Nicodemus, K.K., Malley, J.D.: Predictor correlation impacts machine learning
algorithms: implications for genomic studies. Bioinformatics **25**(15), 1884–1890
(2009)

15. Qiu, H., Lee, J., Lin, J., Yu, G.: Wavelet filter-based weak signature detection
method and its application on rolling element bearing prognostics. J. Sound Vib.
289(4–5), 1066–1090 (2006)

16. Ran, Y., Zhou, X., Lin, P., Wen, Y., Deng, R.: A survey of predictive maintenance:
Systems, purposes and approaches. arXiv preprint arXiv:1912.07383 (2019)

17. Ren, L., Sun, Y., Cui, J., Zhang, L.: Bearing remaining useful life prediction based
on deep autoencoder and deep neural networks. J. Manuf. Syst. **48**, 71–77 (2018)

18. Ribeiro, M.T., Singh, S., Guestrin, C.: "Why should i trust you?" explaining the
predictions of any classifier. In: Proceedings of the 22nd ACM SIGKDD International Conference on Knowledge Discovery and Data Mining, pp. 1135–1144 (2016)

19. Strobl, C., Boulesteix, A.L., Kneib, T., Augustin, T., Zeileis, A.: Conditional variable importance for random forests. BMC Bioinform. **9**, 1–11 (2008)

20. Tayade, A., Patil, S., Phalle, V., Kazi, F., Powar, S.: Remaining useful life (rul)
prediction of bearing by using regression model and principal component analysis
(pca) technique. Vibroengineering Procedia **23**, 30–36 (2019)

21. Vigneau, E.: Dimensionality reduction by clustering of variables while setting aside
atypical variables. Electron. J. Appl. Stat. Anal. **9**(1), 134–153 (2016)

22. Vigneau, E., Qannari, E.: Clustering of variables around latent components. Commun. Stat.-Simul. Comput. **32**(4), 1131–1150 (2003)

23. Wahid, A., Breslin, J.G., Intizar, M.A.: Tcrscanet: harnessing temporal convolutions and recurrent skip component for enhanced rul estimation in mechanical
systems. Human-Centric Intelligent Systems, pp. 1–24 (2024)

24. Wang, Y., Yao, H., Zhao, S.: Auto-encoder based dimensionality reduction. Neurocomputing **184**, 232–242 (2016)

25. Wang, Z., Liu, N., Chen, C., Guo, Y.: Adaptive self-attention lstm for rul prediction
of lithium-ion batteries. Inf. Sci. **635**, 398–413 (2023)

26. Youness, G., Aalah, A.: An explainable artificial intelligence approach for
remaining useful life prediction. Aerospace **10**(5) (2023). https://doi.org/10.3390/
aerospace10050474

27. Zheng, L., He, Y., Chen, X., Pu, X.: Optimization of dilated convolution networks
with application in remaining useful life prediction of induction motors. Measurement **200**, 111588 (2022)

Data Storytelling to Unlock the Communicative Power of Digital Twins

Faten El Outa[(✉)] [iD], Hugo Breuillard[iD], and Guillaume Dechambenoit[iD]

French Geological Survey (BRGM), Orléans, France
{f.elouta,h.breuillard,g.dechambenoit}@brgm.fr
https://www.brgm.fr/fr

Abstract. Digital Twin (DT) systems are increasingly applied in environmental contexts to support real-time monitoring, forecasting, and decision-making. While technically advanced, many DT implementations fall short in communicating insights in ways that are accessible, interpretable, and actionable for diverse audiences. This paper introduces a user-centered workflow to integrate data storytelling into environmental DTs, enhancing their communicative impact. The workflow is illustrated through environmental prediction use case, demonstrating its applicability and its potential to effectively communicate digital twin results to a specific audience.

Keywords: Environmental Digital Twins · Data Storytelling · Narrative Workflow

1 Introduction

Digital Twin (DT) systems fuse real-time data, simulation models, and AI to *monitor* and *forecast* complex phenomena in environmental, industrial, and urban domains. However, despite advances in technical accuracy and system architectures [2,6], their outputs still rely on generic dashboards and static charts that fall short of **user-centered, context-aware** storytelling and thus limit decision-making impact. In parallel, narrative visualization research has shown how structured data stories—even driven by high-level "intentional operators" [7] and enriched with design patterns [1,5] can transform raw analytics into engaging, actionable narratives. Building on our foundational process [4], this paper (i) operationalizes a four-phase storytelling workflow within a three-level formalization framework for environmental DTs and (ii) validates it through a full groundwater-forecast case study.

2 Integrating Data Storytelling into DT Pipelines

To enhance the communicative power of Digital Twins, we propose a user-centered workflow built around a four-phase data storytelling cycle [3,4]: (i)

R. Wrembel et al. (Eds.): DEXA 2025, LNCS 16047, pp. 287–292, 2026.
https://doi.org/10.1007/978-3-032-02088-8_20

Explore, which collects, integrates and preprocesses data to uncover and validate insights; (ii) **Answer Questions**, which aligns analytical goals with user intent and distills findings into clear messages; (iii) **Structure Answers**, which arranges those messages into narrative units—acts and episodes—guided by a storytelling plot; and (iv) **Present**, which generates interactive visual reports and dashboards to convey the narrative effectively. Figure 1 illustrates how these phases, color-coded pink, purple, yellow and blue, interlock with three concentric formalization levels (Personal, Interpersonal, Community) to progressively transform raw analysis into audience-tailored communication. In more detail:

- **Individual Circle – Preliminary Knowledge Exploration:** In this initial circle, knowledge is largely unstructured and meant for individual interpretation. Guided by user-specified analytical objectives and parameters such as complement of goal and type of audience, the system selects relevant datasets and algorithms, trials them, and visualizes preliminary results such as trends, distributions, or risk zones, without applying predefined narrative structures. At this exploratory stage, a finding emerges from the exploration phase, understood as a combination of the dataset used, the algorithm applied, and the resulting output, providing the user a basis for reflection and informal sense-making.
- **Interpersonal Circle – Shared Knowledge Structuring:** At this level, knowledge begins to take on a clearer form for collaborative interpretation. The system develop potential messages interpreting the finding retrieved, selecting appropriate visualizations and applying narrative design patterns (e.g. *Call-to-Action*) to support shared review. Messages are framed to support collaborative review and dialogue such as experts, journalists, or analysts.
- **Community Circle – Public Knowledge Communication:** In this final circle, knowledge is fully formalized and prepared for wide dissemination. The system selects an appropriate narrative structure—such as a plot type—followed by the development of storytelling acts, which are then translated into visual formats using communicative templates (e.g., dashboards, episodic narratives, or structured reports). These outputs are tailored to the intended audience (e.g., general public, policymakers), ensuring that the final communication is accessible, context-aware, and actionable across diverse communities.

This workflow constitutes a core contribution of our work, proposing a novel integration of storytelling into DT communication pipelines. By operationalizing the four phases of data storytelling within each circle of formalization, we enable DT systems to generate outputs that are not only technically sound, but also semantically structured, intentional, and accessible across varied audiences.

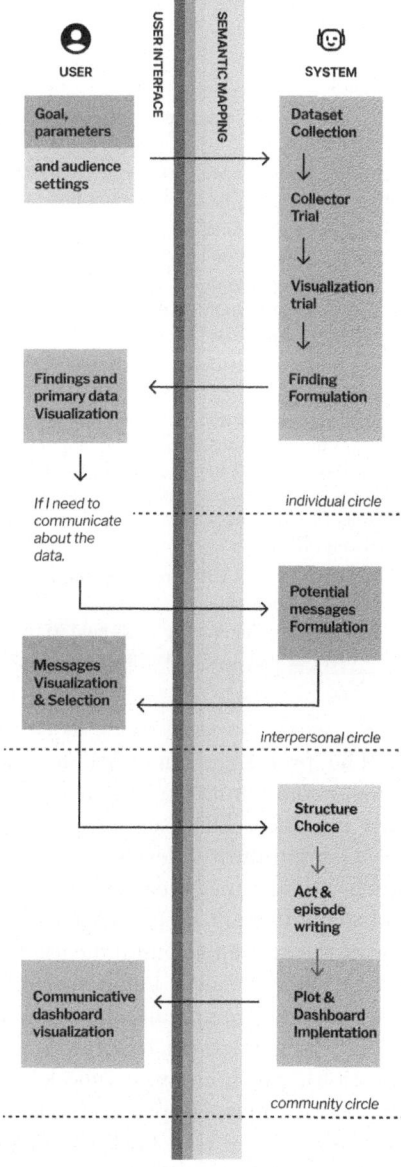

Fig. 1. User-System Communication Workflow for Digital Twins (Color figure online)

3   Case Study: Predicting Groundwater Levels

This scenario illustrates how the model functions through user-system interaction across the three communication circles. A Figma-based dashboard prototype was also developed but is not shown due to space limitations (Table 1).

Table 1. Groundwater level forecast scenario in Digital Twin for Piezometer 03266X0009/P for the year 2019

Element	Description
Intentional Operator	*Predict* — Used to anticipate environmental conditions and guide preventive actions.
Parameters	Groundwater Level, Location (Piezometer 03266X0009/P), Date (2019-01-01), Forecast Horizon (12 months).
Dataset	Historical time series: weekly groundwater levels, local rainfall, and evapotranspiration from past years.
Algorithm	Random Forest Regression — Applied to model complex, non-linear interactions between inputs (e.g., rainfall, evapotranspiration) and outputs (e.g., groundwater levels).
Type of Result	Numerical values — Predicted weekly groundwater levels for 2019 (in meters).
Narrative Pattern	Gradual Reveal + Call-to-Action — Forecast results are progressively unveiled to build audience engagement, followed by a clear directive. Example: "Week by week, groundwater levels have steadily dropped. By July, projections indicate values falling below 113.5 meters, the drought alert threshold. To prevent water shortages, initiate conservation efforts before the end of spring."
Plot Structure	Water Tower — Begins with an impactful introduction and sequentially delivers accessible insights.
Visualization	Line Chart + Area Chart with Threshold Marker — Tracks changes over time and flags critical levels intuitively.

- **User Component:** The process begins with a non-expert user (e.g., a concerned citizen or local community stakeholder) interacting with the DT through a user friendly interface. The user specifies their analytical goal, to *Predict* future (forecast) groundwater levels—along with relevant environmental parameters such as location (Piezometer 03266X0009/P), temporal scope (year 2019), and thematic context (potential drought risk). This declaration of intent activates the system's communication pipeline.
- **System Component:** When the user invokes a high-level operator (e.g., *Predict*), the system relies on a semantically driven mapping table that specifies, for each operator, which datasets to draw (e.g., groundwater levels, rainfall, evapotranspiration), which analytical model to apply (e.g., Random Forest Regression), and which message patterns, visual formats, and narrative structures to use. This ensures that both the analysis and the resulting communication are perfectly aligned with the user's intent and audience.
- **Individual Circle – Preliminary Knowledge Exploration:** At this stage, the user is presented with raw data outputs such as historical and predicted groundwater trends, allowing for informal, individual sense-making. The system surfaces unstructured visual cues—such as line charts showing rainfall and aquifer level correlations, without imposing a specific narrative. These early visuals help the user explore patterns and reflect on recent environmental changes, laying the groundwork for further interpretation.

- **Interpersonal Circle – Shared Knowledge Structuring:** As engagement deepens, the system translates the explored data into potential communicative messages using predefined narrative patterns. Here, the "Gradual Reveal" pattern is applied to disclose forecast information incrementally, enhancing user curiosity and involvement. This is paired with a "Call-to-Action" to emphasize the urgency of the situation and prompt behavioral response. For example, "Week by week, groundwater levels have steadily dropped. By July, projections indicate values falling below 113.5 m, the drought alert threshold. To prevent water shortages, initiate conservation efforts before the end of spring." This structured narrative supports interpersonal discussion among stakeholders (e.g., local experts, journalists, or environmental communicators), guiding collaborative understanding of the forecasted risk.
- **Community Circle – Public Knowledge Communication:** In the final stage, the knowledge is formalized into a cohesive story tailored for public dissemination. The system employs the "Water Tower" plot structure, organizing the narrative into distinct acts and episodes: Act 1 opens with a powerful contextual headline on aquifer depletion, Episode 1 then unveils the first supporting insight; Act 2 follows with an "Episode of Trend Analysis" that highlights critical seasonal fluctuations; and the final act, "Call-to-Action," presents conservation recommendations. Each episode delivers a focused message in sequence, preserving clarity and guiding the audience through the story without overload. Visually, this is implemented through a combination of a line chart (showing weekly level trends) and an area chart (highlighting cumulative decline), overlaid with a threshold marker at 113.5 m to signify the drought risk boundary. This layered design ensures that the message remains accessible and visually intuitive to non-expert audiences. The final output ends with a clear directive, encouraging water conservation measures during spring, transforming data into action and promoting community preparedness.

This example demonstrates how the proposed workflow increases interpretability of complex forecasts, sustains user engagement through progressive storytelling, and ultimately supports timely and informed decision-making in environmental risk contexts.

4   Conclusion

We have proposed a user-centered workflow that demonstrates how data storytelling can enhance the communicative power of environmental Digital Twins. By integrating a four-phase narrative cycle with multi-level formalization, our approach bridges the gap between complex analytic and actionable insights. The groundwater-forecast case study illustrates its practical applicability and potential to guide users from raw data exploration to community-ready narratives.

Moving forward, we plan to develop a complete prototype, implementing the mapping-table engine, narrative planner, and interactive dashboard, to validate the workflow end-to-end. This will be followed by empirical testing with environmental practitioners to evaluate usability, interpretability, and decision-support effectiveness. Finally, we will explore the integration of large language models for dynamic, personalized narrative generation, targeting an initial LLM-powered prototype.

Acknowledgments. This work has been realized within the JUNON Program of Environmental Digital Twin creation and has received financial support under the Ambition Recherche et Développement JUNON Program, funded by the Région Centre-Val de Loire (France).

Disclosure of Interests. The authors declare that they have no competing interests relevant to the content of this article.

References

1. Bach, B., Kerracher, N., Hall, K., Kennedy, J., Riche, N.H., Carpendale, S.: Narrative design patterns for data-driven storytelling. In: Proceedings of the 2018 CHI Conference on Human Factors in Computing Systems. ACM (2018)
2. Osama, H.: From photogrammetry to real-time decision: the digital twin pipeline in smart environments. In: Elsevier Smart Cities (2024)
3. El Outa, F., Francia, M., Marcel, P., Peralta, V., Vassiliadis, P.: Towards a conceptual model for data narratives. In: Dobbie, G., Frank, U., Kappel, G., Liddle, S.W., Mayr, H.C. (eds.) ER 2020. LNCS, vol. 12400, pp. 261–270. Springer, Cham (2020). https://doi.org/10.1007/978-3-030-62522-1_19
4. Outa, F.E., Marcel, P., Peralta, V., Vassiliadis, P.: Highlighting the importance of intentional aspects in data narrative crafting processes. Inf. Syst. Frontiers **26** (2024)
5. Segel, J., Heer, J.: Narrative visualization: telling stories with data. IEEE Trans. Visual Comput. Graph. **16**(6), 1139–1148 (2010)
6. Skoury, L., Leder, S., Menges, A., Wortmann, T.: Digital twin architecture for the AEC industry: a case study in collective robotic construction (2024)
7. Vassiliadis, P., Marcel, P., Rizzi, S.: Beyond roll-up's and drill-down's: an intentional analytics model to reinvent OLAP. Inf. Syst. (2019)

Queueing Theory for Verifying the Utilization Rate of an Image Processing System

Jaqueline Donin Noleto Noleto[1,2]([✉]) [iD], Thiago Germano do Nascimento[2] [iD], and Pedro Henrique Malheiros Costa Martins[1] [iD]

[1] Universidade Federal da Paraíba - UFPB, João Pessoa, PB 58051-900, Brazil
inovatecjpa@gmail.com
[2] Agência de Inovação Tecnológica de João Pessoa - INOVATEC-JP, João Pessoa, PB 58030-021, Brazil
https://www.inovatecjp.com.br/

Abstract. Highway monitoring is crucial to ensuring the safety, efficiency, and maintenance of road networks. This process allows for the identification of dangerous traffic, rapid detection of accidents, management of vehicle flow, and route optimization, among other benefits. For public agencies, monitoring facilitates the collection of traffic data, which is essential for complying with laws and preventing incidents. However, the large volume and speed of data can make processing difficult, impacting decision-making. In this context, parallel processing stands out as an effective solution, especially in image analysis, through the division of subtasks and the use of GPU. Using virtual machines with different memory and processing resources, the delays in image processing were analyzed, using queueing theory to propose machine configurations that allow processing close to real time to minimize the delays generated during the processing of these images. The analysis showed that the use of GPU significantly reduces processing time; however, the use of machines with CPU obtained satisfactory performance from the orchestration of different machines in parallel. It is concluded that parallel processing, combined with queueing theory, optimizes highway monitoring, balancing performance and resource costs.

Keywords: Queuing theory · parallel processing · highway monitoring · image processing

1 Introduction

Highway monitoring is essential for ensuring safety, efficiency, and road maintenance, encompassing activities such as accident detection, traffic management, identification of structural issues, and enforcement of traffic laws [1]. Through real-time data collection, public agencies can make informed decisions, such as

R. Wrembel et al. (Eds.): DEXA 2025, LNCS 16047, pp. 293–298, 2026.
https://doi.org/10.1007/978-3-032-02088-8_21

identifying cargo fraud or irregular vehicles, thereby improving the transportation system and quality of life. These data support preventive and enforcement actions, contributing to incident reduction and optimized traffic flow [2].

However, the large volume and high speed of data generation in monitoring pose significant challenges. Processing infrastructure must analyze images quickly to support agile decisions, but delays in processing and low-quality images, which hinder license plate recognition, can compromise efficiency and lead to critical information loss [3]. These issues underscore the need for robust systems capable of handling intense demands without sacrificing reliability or speed.

To address these challenges, parallel processing emerges as an effective solution, dividing tasks into subtasks executed simultaneously, thus accelerating image processing. Additionally, the use of GPUs, which outperform CPUs in matrix computations, offers significant performance gains. This approach is particularly advantageous in monitoring systems, where rapid processing is crucial for real-time responses, such as incident detection or traffic management.

This study employs queueing theory to model and optimize image processing in highway monitoring systems, comparing scenarios with CPUs and GPUs and using the YOLOv8 algorithm. The research evaluates the impact of different hardware configurations on latency, aiming to minimize queue waiting time and maximize efficiency. Queueing theory enables the identification of bottlenecks, resource allocation adjustments, and system behavior modeling, especially in resource-constrained contexts, ensuring a balance between performance and cost.

The results indicate that GPUs offer superior performance, with significantly lower latencies, even in high-demand scenarios [4], while CPUs require more resources to achieve stability. Integrating queueing theory with parallel processing proves to be a promising strategy for real-time monitoring systems, enabling fast and accurate decisions. Thus, adopting GPUs and continuous queueing theory modeling is recommended to optimize operational efficiency and highway safety.

2   Queueing Theory

Queueing theory is a critical tool for optimizing resource management, reducing costs, and improving operational efficiency, applied through modeling characteristics such as arrival, service, discipline, waiting capacity, and system models. Arrival refers to the demand for services, which can be deterministic (fixed intervals) or stochastic (random), with the arrival rate denoted by λ and often modeled using a Poisson distribution [5]. Service can also be deterministic or random, with service duration represented by μ, requiring the calculation of average service time and its probabilistic distribution. The service discipline defines the order of service, such as FCFS (first-come, first-served), LCFS (last-come, first-served), priority (preemptive or non-preemptive), or random [6].

Waiting capacity can be finite, limiting the number of customers in the queue, or infinite, allowing new customers regardless of queue size. Kendall's notation

(A/B/c/K/Z) is widely used to describe queues, with the M/M/1 model being the simplest, featuring Poisson arrivals and exponential service times, a single server, and an infinite waiting queue [7]. However, the M/M/1 model may be inadequate in some cases, as shown in simulations, leading to the adoption of the M/M/c model, which uses multiple servers to prevent overload and ensure the utilization rate ($\rho < 1$) allows queue clearance at some point [6].

3   Apache Kafka

Apache Kafka is an open-source, distributed messaging system used for processing large volumes of real-time data. It enables producers to send messages to topics, which are organized data categories, while consumers subscribe to these topics to process the messages. Kafka is highly scalable, fault-tolerant, and supports low latency, storing messages in partitioned and replicated logs across a cluster of servers (brokers) [8]. This ensures durability and message ordering, making it ideal for real-time data pipelines, system integration, event analysis, and event-driven architectures [9].

4   Datasets

This study modeled a queueing system based on the characteristics of the Paraíba State Finance Department (SEFAZ-PB) system, which processes events from monitoring state and federal highways captured by cameras, using Apache Kafka as the data manager, configured to handle up to 1500 events per minute. The arrival rate (λ) was estimated from daily event samples, enabling analysis of processing delays. Traffic images, sourced from a Bangladesh road dataset available on Kaggle[1], with a standardized resolution of 640×359 pixels, were classified using YOLOv8, covering elements such as cars, motorcycles, and pedestrians at various times. The research configured test scenarios for simulations, applying queueing theory to evaluate performance and optimize real-time event processing.

5   Simulations

The CPU processing test environment was set up using virtual machines (VMs) in VirtualBox, with varying hardware configurations, such as core count and RAM, detailed in[2]. Each scenario processed 300 images from the same dataset, run on a notebook with an Intel Core i3-6100U processor (2.3 GHz) and 12 GB of RAM, without a dedicated graphics card. The required dependencies, including YOLOv8 and associated libraries, are available in footnote. For CPU and GPU

---

[1] https://www.kaggle.com/datasets/ashfakyeafi/road-vehicle-images-dataset/data.

[2] https://github.com/Jaquedonin/Article-Monitoring-system-utilization\discretionary-/rate/blob/main/README.md.

simulations, a random arrival rate following a Poisson distribution was adopted, illustrated in Fig. 1 available in the footnote, showing temporal variations in arrivals.

Initial simulations with the M/M/1 model, using VM-1 (1 core, 1 GB RAM), revealed systemic instability, with an arrival rate ($\lambda = 1000$) far exceeding the service rate ($\mu = 110.19$), resulting in $\rho \approx 9.07$. This indicates a single server is insufficient, leading to infinite queues and impractical waiting times, as shown in Table 1 available in the footnote. The image processing time by YOLOv8, presented in Fig. 2 available in the footnote, confirmed slowness in basic configurations. The analysis suggests the M/M/1 model is unsuitable for high-demand scenarios, necessitating a multi-server model to stabilize the system.

To address instability, a model with $c = 10$ servers was adopted, reducing the utilization rate to $\rho = 0.907$, as detailed in Table 2 available in the footnote. Table 3, available in the footnote, presents the updated parameters, including adjusted server counts per scenario, queue waiting times (W_q), system times (W), and system utilization. These adjustments ensured efficient flow, with viable waiting times and no indefinite queue buildup. The results highlight the importance of properly sizing server counts and suggest that, for high-demand scenarios, configurations with more resources or alternative queueing models, such as M/M/c, are essential for optimizing image processing.

The GPU test scenario was built using Docker to create the processing environment, configured with the necessary dependencies. The image used was obtained from Docker Hub[3]. The machine used for testing was a notebook with a Ryzen 5600x processor, an RTX 4060 8GB GPU, and 32 GB of RAM. The dependencies required for the tests are listed in[4]. Unlike CPU tests, the GPU achieved satisfactory results with just one server. The utilization rate ρ was 0.303, meaning only 30.3% of the server's capacity was used on average.

6   Related Work

The analysis of Age of Information (AoI) in queueing systems, critical for real-time applications like 6G communications and remote monitoring, is explored in Reddy and Venkatesh [10], which uses the Proactive Obsolete Packet Management (POPMAN) strategy to optimize data freshness in FCFS single-hop, LCFS multi-hop, and LCFS multi-source systems. The POPMAN approach proactively discards obsolete packets, reducing average AoI by up to 43.75% in two-server systems compared to conventional FCFS. Using Stochastic Hybrid Systems (SHS) and Relative Freshness Markov Chain (RF-SHS-MC) analysis, the study derives analytical expressions for average and peak AoI, showing that multi-hop network output rates are inversely proportional to average AoI, with queueing theory (M/M/1/1) and heterogeneous parallel servers enhancing efficiency.

---

[3] https://hub.docker.com/search?q=GPU.

[4] https://github.com/Jaquedonin/Article-Monitoring-system-utilization-rate/blob/main/README.md.

Selvaraj [11] proposes a cloud-based batch processing system using Docker containers and RabbitMQ queues, overcoming limitations of sequential systems. An Initiator component sends data blocks to a three-node RabbitMQ cluster, with containers processing tasks in parallel and auto-scaling based on queue size. This approach reduces processing time from 650 to 200 min for 11 tasks, minimizing costs via a pay-per-use model and container decommissioning, validated by simulations and analytical modeling, highlighting the efficiency of parallel processing in scalable environments.

Zafarzade and Ataie [12] address capacity planning for an image classification application in microservices on a Function-as-a-Service (FaaS) platform, using Extended Queueing Networks (EQNs) and Java Modeling Tools (JMT). The dynamic strategy adjusts function instances when utilization exceeds 90%, reducing response time from 43 h to 11 s at 5000 requests/hour, ensuring Quality of Service (QoS). Compared to these studies, our work stands out by applying queueing theory to optimize image processing with YOLOv8 in resource-constrained hardware contexts, offering insights for systems with limited computational resources, such as highway monitoring.

7   Discussion

With GPU usage, results were significantly superior, starting with a utilization rate of $\rho = 0.303$, indicating a stable and highly scalable system. The average queue waiting time (W_q) was only 0.000132 min, and the total system time (W) was approximately 0.000435 min. The average number of customers in the queue (L_q) was below 1, indicating low latency and exceptional response capacity.

Analysis of Table 3[5] shows that virtual machines MV-1 to MV-3 operate with utilization rates above 87%, indicating imminent saturation, while MV-4 and MV-5 show better load balancing, with MV-5 being the most efficient among CPUs at 82.51% utilization. In contrast, the GPU exhibits only 30.3% utilization, suggesting underutilization and high scalability potential. Ideal utilization rates, between 70% and 90%, minimize overload risks, making the GPU a robust solution for high demands. Queue waiting times reinforce this disparity: MV-1 and MV-2 exceed 0.007 min, indicating slowness, while MV-4 and MV-5 are below 0.004 min, and the GPU achieves 0.000132 min, practically real-time.

The average number of customers in the queue highlights congestion in MV-1 and MV-2, with over seven customers waiting, compared to 3.26 for MV-5, the best among CPUs. The GPU, with fewer than one customer waiting, demonstrates exceptional response capacity. While MV-1 and MV-2 require 10 servers for stability, MV-4 to MV-7 operate with 5 to 6, and the GPU requires only one, reducing costs and complexity. Although CPUs with more resources improve performance, gains diminish with increased cores and memory. The GPU, however, offers superior efficiency, with near-instantaneous response and scalability, justifying its use in critical applications like real-time monitoring, as evidenced by performance and cost graphs.

---

[5] https://hub.docker.com/search?q=GPU.

8   Conclusion

This study demonstrated that queueing theory is an effective tool for optimizing resources in image processing, highlighting the superiority of GPUs over CPUs in reducing response times and maintaining performance under high demand. Tests with virtual machines of varying configurations revealed that increasing cores and RAM reduces processing time, but with diminishing returns, with Scenario 6 (4 cores, 4 GB RAM) being an optimal cost-performance point. CPU-based systems require horizontal scalability for stability, while GPUs deliver superior results with a single server. Queueing theory modeling, combined with parallel processing, is recommended for real-time monitoring systems, such as highways, promoting operational efficiency and road safety.

References

1. Kul, S., Tashiev, I., Şentaş, A., Sayar, A.: Event-based microservices with apache Kafka streams: a real-time vehicle detection system based on type, color, and speed attributes. IEEE Access **9**, 83 137-83 148 (2021)
2. Di Renzo, A.B., et al.: Desenvolvimento de metodologia para monitoramento remoto de rodovias: Vantrod," Master's thesis, Universidade Tecnológica Federal do Paraná (2017)
3. Ribeiro, M.V.: Tecnicas de processamento de sinais aplicadas a transmissão de dados via rede eletrica e ao monitoramento da qualidade de energia." Ph.D. dissertation, University of Campinas, Brazil (2005)
4. Souza, M.L.D.: Evolução dos processadores e seu futuro. Monografia apresentada ao Curso de Engenharia Elétrica da Universidade São Francisco (2012)
5. Prado, D.: Teoria das Filas e da Simulação. Falconi Editora, vol. 2 (2022)
6. Costa, L.C.: Teoria das filas, acessado em (2000). http://www.decom.ufop.br/prof/rduarte/cic271.TeoriaFilas_Cajado.pdf
7. Magalhaes, M.N.: Introdução a rede de filas. Abe (1996)
8. Le Noac'h, P., Costan, A., Bougé, L.: A performance evaluation of apache Kafka in support of big data streaming applications. In: 2017 IEEE International Conference on Big Data (Big Data), pp. 4803–4806. IEEE (2017)
9. Kim, Y.-K., Jeong, C.-S.: Large scale image processing in real-time environments with Kafka. In: Proceedings of the 6th AIRCC International Conference on Parallel, Distributed Computing Technologies and Applications (PDCTA), pp. 207–215 (2017)
10. Reddy, Y.A.K., Venkatesh, T.: Age of information analysis for queueing systems with proactive obsolete packet management: multi-hop, multi-source, and priority mechanisms. IEEE Trans. Network Sci. Eng. (2025)
11. Selvaraj, A.: High performance cloud ready parallel batch processing using auto scaling containers and queue technique. In: 2024 4th International Conference on Sustainable Expert Systems (ICSES), pp. 630–639. IEEE (2024)
12. Zafarzade, Z., Ataie, E.: Capacity planning of a microservices-based image classification application using analytic modeling. Computing **107**(5), 1–31 (2025)

Effect of Frequency Features of ELA Maps on the Detection Performance of Image Manipulation Based on DCT and FFT Basis Features

Jarosław Kobiela[✉] [ID] and Piotr M. Dzierwa [ID]

Institute of Computer Science, University of Opole, Opole, Poland
{jaroslaw.kobiela,piotr.dzierwa}@uni.opole.pl

Abstract. Detecting manipulations in digital images is critical for ensuring their authenticity and integrity. This study evaluated the impact of frequency descriptors (DCT/FFT) derived from Error Level Analysis (ELA) maps on manipulation detection performance. Using Random Forest, SVM, XGBoost, and Light-GBM on IMD2020, we compared models trained with baseline original image features against those augmented with ELA map features. The results show that ELA-derived features significantly improved the tree-based models. XGBoost and LightGBM yielded the best performance (F1-score ≈ 0.81 validation; XGBoost 0.815 test), demonstrating strong generalization. This study highlights the informative value of combining ELA maps and original image frequency analysis for effective classical machine learning-based detection.

Keywords: Image manipulation detection · Image forensic analysis · Error Level Analysis (ELA) · Discrete Cosine Transform (DCT) · Fast Fourier Transform (FFT) · Machine learning · Feature engineering

1 The Growing Challenge of Image Forgery

Rapid advancements in digital technologies have made the verification of image authenticity increasingly critical. Manipulation techniques, from basic to advanced AI-driven methods, create false visual content that is often indistinguishable from the original [1]. Such forgeries spread disinformation and undermine public trust, necessitating effective and robust detection methods [2]. Challenges in detecting manipulated images and videos, particularly deepfakes [3], continue to evolve, necessitating ongoing methodological and database development.

2 Analysis of Frequency and Compression Artifacts

Traditional Image Manipulation Detection (IMD) leverages frequency-domain analysis, as modifications often reveal clearer artifacts in this domain. Key transforms include the Discrete Cosine Transform (DCT) and Fast Fourier Transform (FFT). The DCT [4]

© The Author(s), under exclusive license to Springer Nature Switzerland AG 2026
R. Wrembel et al. (Eds.): DEXA 2025, LNCS 16047, pp. 299–304, 2026.
https://doi.org/10.1007/978-3-032-02088-8_22

is foundational for lossy compression, such as JPEG, with its coefficient distribution revealing anomalies such as double compression [5]. The FFT [6] efficiently computes the Discrete Fourier Transform (DFT), offering a full frequency-domain representation (amplitude and phase). The analysis of both amplitude and phase enables the detection of manipulations such as Copy-Move Forgery Detection (CMFD) [7] or texture modifications [8]. Both DCT and FFT are widely used in forensic image analysis, often combined with other data analysis or machine learning techniques to identify various manipulations, including double compression, deepfakes, and hybrid approaches [9, 10]. Additionally, Error Level Analysis (ELA) [11] detects inconsistencies in JPEG compression errors by re-saving images at various quality levels, generating error maps that highlight manipulated regions [12]. These fundamental analyses underpin the feature engineering approach proposed in this study.

3   Related Work

Image manipulation detection research has evolved from analyzing specific spatial/frequency artifacts to advanced deep learning. Early work used hand-projected features such as LBP and DCT [13]. Deep learning networks (CNNs, Transformers) now dominate, for example, in detecting computer-generated imagery (CGI) [14]. Despite their success, generalization and interpretability remain challenging [15]. Complementary approaches include robust hashing [16] and multimodal analyses [17]. This work explores the combination of "classical" frequency features (DCT, FFT) with compres sion artifact analysis (ELA) [11] using modern machine learning algorithms (Random Forest, SVM, XGBoost, LightGBM) on the IMD2020 dataset [18].

4   Dataset

The IMD2020 dataset [18], a large-scale, diverse, and annotated collection for visual manipulation detection, was used in this study. It comprises 35,000 authentic images from 2322 camera models and 35,000 synthetically manipulated images (classical techniques and GANs), with binary masks indicating the modified areas. The dataset used in this study is publicly available.

5   Methodology

The research process involved data preparation, feature extraction, and classification model optimization and evaluation. Input data from IMD2020 [18] were labeled (original '0', manipulated '1') and split into training (70%), validation (15%), and testing (15%) subsets with class balance. The images were converted into numerical features. For the original images, after grayscaling, the first 100 low-frequency coefficients of 2D DCT and 2D FFT (amplitude and phase) were computed. Additionally, Error Level Analysis (ELA) was applied by re-saving images at JPEG quality levels 75, 85, and 95, generating three error maps. From each ELA map, 100 DCT and 100 FFT coefficients (amplitude and phase) were obtained. This yielded two feature vectors: a base vector (original image DCT/FFT) and an extended vector (base features plus DCT/FFT from all three ELA maps).

Four binary classifiers were compared: Random Forest (RF), Support Vector Machine (SVM) with RBF kernel, XGBoost, and LightGBM. Feature vectors were standardized using StandardScaler and trained solely on the training set. Hyperparameter optimization (HPO) employed RandomizedSearchCV with 3-fold cross-validation on the training set, maximizing the F1-score (fewer iterations for SVM due to the computational cost). The optimal HPO configuration, based on the validation F1-score, was then used to train the final model on the combined training and validation data. Performance was evaluated using Accuracy, F1-Score, Precision, and Recall.

6 Experimental Results

Table 1 presents the model evaluation results for the validation set. Adding ELA features significantly improved the performance of tree-based models (LightGBM, XGBoost, and Random Forest) across all key metrics. For instance, the F1-score of LightGBM increased from 0.626 (without ELA) to 0.816 (with ELA), and that of XGBoost increased from 0.690 to 0.810. The AUC values also increased substantially (e.g., 0.900 for Light-GBM with ELA). Models without ELA tended to classify most samples as manipulated (high recall, low precision). The SVM model was ineffective, regardless of the feature set. LightGBM (with ELA) and XGBoost (with ELA) achieved the highest overall efficiencies in the validation set.

Table 1. Model results on the validation set for scenarios with and without ELA.

Model	Variant Cech	Accuracy	AUC	F1	Precision	Recall	CV_F1_Score
LightGBM	w/o ELA	0.503	0.441	0.626	0.502	0.831	0.591
LightGBM	**w/ ELA**	**0.806**	**0.900**	**0.816**	**0.776**	0.860	**0.811**
XGBoost	w/o ELA	0.560	0.551	0.690	0.532	**0.982**	0.691
XGBoost	w/ ELA	0.796	0.887	0.810	0.758	0.870	0.806
RandomForest	w/o ELA	0.537	0.443	0.664	0.521	0.914	0.642
RandomForest	w/ ELA	0.708	0.778	0.749	0.657	0.869	0.749
SVM	w/o ELA	0.228	0.864	0.248	0.242	0.255	0.375
SVM	w/ ELA	0.377	0.695	0.336	0.359	0.316	0.398

Note: Bold values indicate the best score in a given column

Feature importance analysis of the XGBoost model (Fig. 1 and Fig. 2) revealed a fundamental shift. In the baseline scenario (without ELA, Fig. 1), the original Fourier transform phase features (fft_phase_orig) dominated. Among the ELA features (Fig. 2), the FFT magnitude from the ELA maps (fft_mag_ela) was the most important, followed by the ELA FFT phase (fft_phase_ela) and ELA DCT (dct_ela). Original image features lost importance, suggesting that the models learned to effectively use compression artifact information from the ELA.

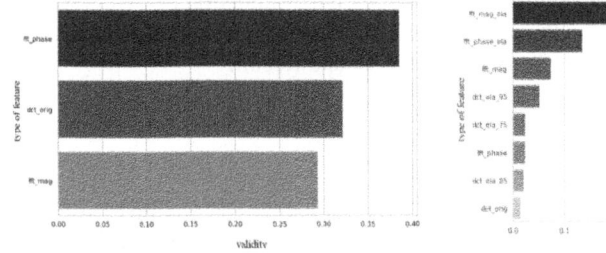

Fig. 1. Validity of Guild Groups- XGBoost (without Ela)

Fig. 2. Validity of Feature Groups-XGBoost (with ELA)

Analyzing the predicted probabilities of misclassified test samples revealed distinct error distributions (Fig. 3). Without ELA, the models (RandomForest, LightGBM, and XGBoost) predominantly produced high-confidence false-negative (FN) errors, with P(Manipulation) near zero. Conversely, the ELA features significantly altered this. The XGBoost model (with ELA) showed a more balanced FP/FN error profile, with misclassifications occurring closer to the decision boundary rather than at probability extremes, suggesting less-confident errors.

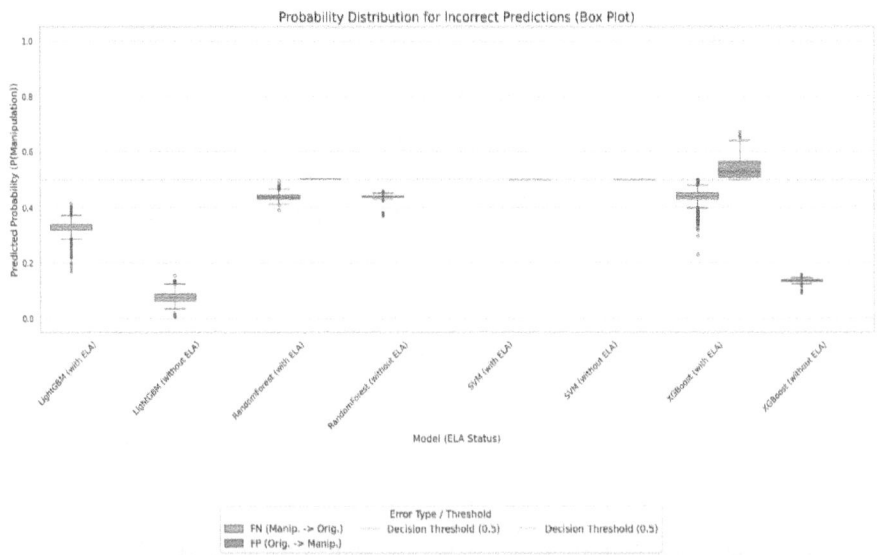

Fig. 3. Probability Distribution for Incorrect Predictions

The XGBoost model using ELA features, selected as the best configuration (validation F1-score 0.810, AUC 0.887), was finally evaluated on an unseen test set. Results: Accuracy: 0.7995; AUC: 0.8949; F1-score: 0.8150; Precision: 0.7565; Recall: 0.8834.

The test set metrics were very close to the validation set metrics, confirming good generalization ability and stable high performance.

7 Discussion and Conclusions

Including DCT/FFT descriptors from ELA error maps significantly improves the manipulation detection performance of tree-based models (RandomForest, XGBoost, and LightGBM). The ELA's ability to capture local JPEG compression inconsistencies, confirmed by feature importance analysis, proved to be more informative. Gradient boosting models performed the best, whereas SVM was ineffective. The ELA also led to more balanced error profiles.

Limitations include reliance on the IMD2020 dataset, which requires further generalization validation across other datasets and manipulation types. The effectiveness of this method against novel falsification techniques, anti-forensics, or extreme compression remains unexplored, limiting its direct external validity.

The implications of this study are both practical and theoretical. Practically, ELA analysis enhances classical ML-based detection pipelines. Theoretically, this study highlights the ongoing importance of feature engineering, which exploits manipulation-specific signals such as compression artifacts, even with deep learning advances.

8 Future Research

Future research should verify the results on other datasets and assess their robustness against advanced falsification and anti-forensic techniques. Comparing this feature-engineering approach with end-to-end deep learning models on the same dataset is also of interest. Further exploration of advanced feature selection/weighting and hybrid digital-physical authenticity methods is required.

References

1. Zhou, P., Han, X., Morariu, V.I., Davis, L.S.: Learning Rich Features for Image Manipulation Detection (2018). arXiv. https://doi.org/10.48550/ARXIV.1805.04953
2. Bayar, B., Stamm, M.C.: Constrained convolutional neural networks: a new approach towards general purpose image manipulation detection. IEEE Trans. Inf. Forensics Secur. **13**(11), 2691–2706 (2018). https://doi.org/10.1109/TIFS.2018.2825953
3. Maras, M.-H., Alexandrou, A.: Determining authenticity of video evidence in the age of artificial intelligence and in the wake of Deepfake videos. Int. J. Evid. Proof **23**(3), 255–262 (2019). https://doi.org/10.1177/1365712718807226
4. Ochoa-Domínguez, H., Rao, K.R.: Discrete Cosine Transform, Second edition. Boca Raton, FL : Taylor & Francis Group, CRC Press, 2019. I Revised edition of: Discrete cosine transform : algorithms, advntages, applications / K. R. Rao, P. Yip. 1990.: CRC Press (2019). https://doi.org/10.1201/9780203729854

5. Bianchi, T., Piva, A.: Image forgery localization via block-grained analysis of JPEG arti-facts. IEEE Trans. Inf. Forensics Secur. **7**(3), 1003–1017 (2012). https://doi.org/10.1109/TIFS.2012.2187516

6. Cooley, J.W., Lewis, P.A.W., Welch, P.D.: The fast fourier transform and its applications. IEEE Trans. Educ. **12**(1), 27–34 (1969). https://doi.org/10.1109/TE.1969.4320436

7. Mishra, M., Adhikary, M.C.: Detection of Clones in Digital Images (2014). arXiv. https://doi.org/10.48550/ARXIV.1407.6879

8. Srivastava, P., Khare, A.: Integration of wavelet transform, local binary patterns and moments for content-based image retrieval. J. Vis. Commun. Image Represent. **42**, 78–103 (2017). https://doi.org/10.1016/j.jvcir.2016.11.008

9. Kim, B., Abuadbba, S., Kim, H.: DeepCapture: Image Spam Detection Using Deep Learning and Data Augmentation (2020). arXiv. https://doi.org/10.48550/ARXIV.2006.08885

10. Qian, Y., Yin, G., Sheng, L., Chen, Z., Shao, J.: Thinking in Frequency: Face Forgery Detection by Mining Frequency-aware Clues' (2020). arXiv. https://doi.org/10.48550/ARXIV.2007.09355

11. Krawetz, N., Solutions, H.F.: A picture's worth. Hacker Factor Solut. **6**(2), 2 (2007)

12. Mahdian, B., Saic, S.: A bibliography on blind methods for identifying image forgery. Signal Process. Image Commun. **25**(6), 389–399 (2010). https://doi.org/10.1016/j.image.2010.05.003

13. Kunbaz, A., Saghir, S., Arar, M., Sonmez, E.B.: Fake image detection using DCT and local binary pattern. In: 2019 Ninth International Conference on Image Processing Theory, Tools and Applications (IPTA), Istanbul, Turkey: IEEE, November 2019, pp. 1–6. https://doi.org/10.1109/IPTA.2019.8936108

14. Peng, J., Liu, C., Pang, H., Gao, X., Cheng, G., Hao, B.: GP-Net: image manipulation detection and localization via long-range modeling and transformers. Appl. Sci. **13**(21), 12053 (2023). https://doi.org/10.3390/app132112053

15. Zhai, Y., Luan, T., Doermann, D., Yuan, J.: Towards generic image manipulation detection with weakly-supervised self-consistency learning. In: 2023 IEEE/CVF International Conference on Computer Vision (ICCV), Paris, France: IEEE, pp. 22333–22343, October 2023. https://doi.org/10.1109/ICCV51070.2023.02046

16. Tanaka, M., Kiya, H.: Fake-image detection with Robust Hashing (2021). arXiv. https://doi.org/10.48550/ARXIV.2102.01313

17. Singh, B., Sharma, D.K.: Predicting image credibility in fake news over social media using multimodal approach. Neural Comput. Appl. **34**(24), 21503–21517 (2022). https://doi.org/10.1007/s00521-021-06086-4

18. Novozamsky, A., Mahdian, B., Saic, S.: IMD2020: a large-scale annotated dataset tailored for detecting manipulated images. In: 2020 IEEE Winter Applications of Computer Vision Workshops (WACVW), Snowmass Village, CO, USA: IEEE, March 2020, pp. 71–80. https://doi.org/10.1109/WACVW50321.2020.9096940

ALPHA: A Multi-Attention Enhanced YOLO Framework for Robust Photovoltaic Defect Detection

Bechir Ben Tekfa[1], Amira Mouakher[2(✉)], and Naeem Ayoub[3]

[1] Department of Computer Science, University of Tunis El Manar, Tunis, Tunisia
[2] Espace-Dev UMR 228, UPVD, IRD, UM, UA, UG, Perpignan, France
amira.mouakher@univ-perp.fr
[3] Department of Technology and Innovation, SDU Technology Entrepreneurship and Innovation, Sønderborg, Denmark

Abstract. Photovoltaic systems are crucial for sustainable energy, yet manufacturing defects and environmental degradations compromise their efficiency. While deep learning approaches show promise, existing methods face limitations in precision, computational efficiency, and generalization across diverse defect categories. We introduce ALPHA, an attention-enhanced YOLO framework integrating Feature Cross-Attention, Channel Attention, and Efficient Multi-scale Attention to improve spatial and channel-wise feature representations. Comprehensive experiments conducted on electroluminescence image datasets demonstrate that ALPHA outperforms state-of-the-art models in terms of accuracy, robustness, and inference speed.

Keywords: Photovoltaic Defect Detection · Attention Mechanisms · YOLO Framework · Deep Learning

1 Introduction

Photovoltaic (PV) systems are essential for sustainable energy, but their efficiency is often reduced by manufacturing defects and operational degradations. Accurate and real-time defect detection is crucial to ensure reliability and reduce energy loss and maintenance costs [1]. Manual inspection and traditional vision methods struggle with the variability in PV defect appearance [2], highlighting the need for automated, robust detection solutions [1]. Recent deep learning approaches, especially CNNs with attention mechanisms, have improved defect detection. However, issues with generalization, precision, and computational overhead limit their real-time applicability. We propose ALPHA, a multi-attention YOLO-based framework integrating Feature Cross-Attention (FCA), Channel Attention (CA), and Efficient Multi-scale Attention (EMA) to enhance feature extraction, localization, and generalization, while keeping the model lightweight for deployment. The remainder of the paper covers related work (Sect. 2), our approach (Sect. 3), experiments (Sect. 4), and conclusions (Sect. 5).

© The Author(s), under exclusive license to Springer Nature Switzerland AG 2026
R. Wrembel et al. (Eds.): DEXA 2025, LNCS 16047, pp. 305–310, 2026.
https://doi.org/10.1007/978-3-032-02088-8_23

2   Related Work

Recent deep learning advances have significantly improved defect detection in PV systems. Attention mechanisms enhance feature extraction by focusing on critical image regions. For instance, Chen et al. [3] used a multi-scale Faster R-CNN with a convolutional block attention module for precise crack detection. Architectural enhancements such as spatial attention and spatial pyramid pooling have been integrated into models like Mask R-CNN [4], improving detection of small defects. Tao et al. [5] incorporated deformable convolutions and attention to refine boundary box regression. Multi-scale feature extraction and fusion have also proven effective. Su et al. [6] proposed a Bidirectional Attention Feature Pyramid Network, while Jiang et al. [7] combined classification and segmentation with multi-scale attention for real-time crack detection. To overcome limited datasets, GANs have been used for augmentation, enhancing model robustness [5]. Lightweight architectures have emerged for real-time applications. Akram et al. [8] introduced a VGG-based CNN for crack detection with low computational cost, and Meng et al. [1] developed YOLO-PV to balance efficiency and performance.

Despite these efforts, challenges remain in real-time and generalizable detection across diverse defect types. Our proposed framework, ALPHA, addresses these gaps by integrating multi-attention mechanisms into a YOLO architecture to enhance localization, feature learning, and robustness.

3   The ALPHA Framework

ALPHA addresses PV defect detection challenges through three integrated attention mechanisms in YOLOv10: Coordinate Attention (CA) for directional sensitivity, Feature Channel Attention (FCA) for channel-wise emphasis, and Efficient Multi-scale Attention (EMA) for local-global context capture.

Algorithm 1 outlines the ALPHA framework execution for 12 defect categories. The framework initializes model components with scale factors $[0.33, 0.50, 1024]$, then trains for 600 epochs with SGD optimization. The INITIALIZE_MODEL function constructs network components using compound scaling, incorporating attention mechanisms (CA, FCA) in the backbone and bidirectional feature pyramid with EMA in the neck. The TRAIN_MODEL function optimizes parameters through structured training with SGD (lr=0.01, momentum=0.937), cosine annealing scheduler, and early stopping (30 epochs patience). Each epoch processes batches of 204 images with augmentation, feature extraction, pyramid processing, and validation. The APPLY_AUGMENTATIONS function enhances generalization through domain-specific transformations: resizing to 1024 pixels (scale=0.7), geometric flips (0.5 probability), HSV adjustments (h=0.015, s=0.5, v=0.4), and compositional techniques including Mosaic (1.0) and Mixup (0.7) augmentations. The EXTRACT_FEATURES function, described by Algorithm 5, implements a hierarchical feature extraction process enhanced with attention mechanisms to improve the model's ability to capture both local and global feature dependencies. This function takes as input an image and a backbone parameter set $\Theta_{backbone}$, and it outputs a set of extracted feature maps at different levels of abstraction. In Line 2, the function initializes feature extraction by computing F_1, which processes the input image through an initial set of convolutional layers (64 and 128 channels), C2f modules, and a Coordinate Attention (CA) module with 128 channels. The CA module is particularly effective in enhancing directional feature sensitivity, which is crucial for detecting linear defects. This is achieved by embedding positional information directly into channel attention through height-wise and width-wise pooling. These operations allow the model to aggregate global contextual information along spatial dimensions while maintaining fine-grained local details. In Line 3, the function extracts F_3 by processing F_1 through additional convolutional layers (256 channels), C2f modules, and a FCA mechanism. The latter improves feature representation by adaptively emphasizing informative channels while suppressing less relevant ones. This mechanism selectively amplifies the most discriminative features while mitigating background noise, thereby refining the feature map for subsequent layers. Lines 45 continue the hierarchical extraction process, computing deeper features F_4 and F_5 through successive transformations. These deeper representations encode high-level semantic information, essential for downstream tasks such as object detection or defect classification. Finally, in Line 6, the function returns the extracted feature maps $\{F_3, F_4, F_5\}$, which serve as the multi-scale feature representations for subsequent processing stages.

Algorithm 1. The Alpha framework process.

Require: Training dataset D_{train}, Validation dataset D_{val}, Classes $nc = 12$, Scale factors s
1: $\Theta_{backbone}, \Theta_{neck}, \Theta_{detect} \leftarrow$ Initialize_Model(nc, s)
2: $model \leftarrow$ Train_Model($\Theta_{backbone}, \Theta_{neck}, \Theta_{detect}, D_{train}, D_{val}, epochs = 600$)
3: **return** $model$

Algorithm 2. The Initialize_Model function.

1: $\Theta_{backbone} \leftarrow$ Scaled_Backbone($s[0], s[1], s[2]$)
2: $\Theta_{neck} \leftarrow$ Scaled_Neck($s[0], s[1], s[2]$)
3: $\Theta_{detect} \leftarrow$ v10Detector(nc)
4: **return** $\{\Theta_{backbone}, \Theta_{neck}, \Theta_{detect}\}$

The Process_Feature_Pyramid function implements a bidirectional feature pyramid network to enhance multi-scale feature representations. This function described in Algorithm 6, refines hierarchical feature maps through a combination of top-down and bottom-up processing, leveraging attention mechanisms to improve feature selectivity and positional awareness. In the top-down

---

Algorithm 3. The TRAIN_MODEL function.

1: $optimizer \leftarrow$ SGD($lr = 0.01, momentum = 0.937, weight\_decay = 0.0005$)
2: $scheduler \leftarrow$ Cosine_AnnealingLR($T_{max} = epochs$)
3: $stopper \leftarrow$ Early_Stopping($patience = 30$)
4: **for** $epoch = 1$ to $epochs$ **do**
5:     **for** each batch $\{images, labels\}$ in DataLoader($D_{train}, batch\_size = 204$) **do**
6:         $features \leftarrow$ EXTRACT_FEATURES(APPLY_AUGMENTATIONS($images$), $\Theta_{backbone}$)
7:         $predictions \leftarrow$ DETECT_OBJECTS(PROCESS_FEATURE_PYRAMID($features, \Theta_{neck}$), Θ_{detect})
8:         Update parameters using COMPUTE_LOSS($predictions, labels$)
9:     Validate and save best model if improved
10: **return** $\{\Theta_{backbone}, \Theta_{neck}, \Theta_{detect}\}$

Algorithm 4. The APPLY_AUGMENTATIONS function.

1: $images_{aug} \leftarrow \emptyset, labels_{aug} \leftarrow \emptyset$
2: **for** each $(image, label)$ in $(images, labels)$ **do**
3:     $images_{aug} \leftarrow images_{aug} \cup \{image\}$
4:     $labels_{aug} \leftarrow labels_{aug} \cup \{label\}$
5: **return** $\{images_{aug}, labels_{aug}\}$

---

pathway, high-level semantic information propagates to lower-resolution features while preserving spatial details. Line 2 initializes P_5 by enhancing F_5 using the ENHANCE_FEATURES function. This step refines the highest-level feature representation, which encodes the most abstract and contextual information. Line 3 constructs P_4 by fusing F_4 with the upsampled P_5 features using CA mechanism. Line 4 creates P_3 by merging F_3 with the upsampled P_4 features using EMA partitions which input features into G groups. This method enables the model to capture fine-grained textures while maintaining contextual consistency. In the bottom-up pathway, previously generated pyramid features are refined to ensure information consistency across scales. Line 5 refines P_4 using CA with a reduction ratio $r = 32$, ensuring computational efficiency while preserving essential spatial dependencies. Line 6 refines P_5 using Feature Channel Attention (FCA), which recalibrates the feature map at the channel level. FCA refines feature importance by computing an adaptive reweighting function. Finally, line 7 returns the refined multi-scale feature maps $\{P_3, P_4', P_5'\}$, which are optimized for detecting various defect sizes. This bidirectional processing strategy ensures that both high-level semantic information and fine-grained spatial details are effectively captured, improving the model's ability to identify defects across a range of scales and textures. The PROCESS_FEATURE_PYRAMID function implements bidirectional feature fusion with attention mechanisms. EMA partitions features into groups $X = [X_1, X_2, \ldots, X_G]$ with attention scores $s_i = \sigma(f_{1 \times 1}(X_i) \odot f_{3 \times 3}(X_i))$, enabling multi-scale context capture. The DETECT_OBJECTS function performs

---

Algorithm 5. The EXTRACT_FEATURES function.

1: $F_1 \leftarrow$ Initial_Features($image, \Theta_{backbone}$)
2: $F_3 \leftarrow$ Mid_Features($F_1, \Theta_{backbone}$)
3: $F_4 \leftarrow$ Deep_Features($F_3, \Theta_{backbone}$)
4: $F_5 \leftarrow$ Context_Features($F_4, \Theta_{backbone}$)
5: **return** $\{F_3, F_4, F_5\}$

multi-scale detection across pyramid levels, classifying regions into 12 defect categories, localizing bounding boxes, and applying non-maximum suppression for final predictions.

Algorithm 6. The PROCESS_FEATURE_PYRAMID function.

1: $P_5 \leftarrow$ ENHANCE_FEATURES(F_5, Θ_{neck})
2: $P_4 \leftarrow$ FEATURE_FUSION(F_4, P_5, CA)
3: $P_3 \leftarrow$ FEATURE_FUSION(F_3, P_4, EMA)
4: $P_4', P_5' \leftarrow$ REFINE_FEATURES(P_3, P_4, P_5)
5: **return** $\{P_3, P_4', P_5'\}$

Algorithm 7. The DETECT_OBJECTS function.

1: $detections \leftarrow \emptyset$
2: **for** each scale P_i in $pyramid$ **do**
3:     $detections \leftarrow detections \cup \{$CLASSIFY($P_i$), LOCALIZE($P_i$), SCORE($P_i$)$\}$
4: **return** NONMAX_SUPPRESSION(COMBINE_PREDICTIONS($detections$))

---

4   Experimental Results

We evaluate ALPHA on the PVEL-AD dataset using stratified 7:3 train/validation split across 12 defect categories. Performance is assessed using mAP@50, mAP@50-95, precision, and recall metrics. ALPHA achieves 85.0% mAP@50, surpassing YOLOv10 (76.5%) and YOLOv11 (77.6%) by 8.5% and 7.4% respectively, demonstrating the effectiveness of integrated attention mechanisms. The model shows strong precision (83.9%) and maintains competitive computational efficiency (23.5 GFLOPs, 9.18M parameters). Most notably, ALPHA achieves 99.5% mAP@50 for horizontal dislocation detection compared to YOLOv10's 0.02%, highlighting the CA module's effectiveness in capturing orientation-specific features. Our ALPHA model achieves superior detection performance with a parameter count of 9.17M, closely matching YOLOv10 (8.93M) while remaining more efficient than YOLOv11 (9.42M). This highlights the effectiveness of our architectural enhancements, which integrate advanced attention mechanisms to improve feature extraction without significantly increasing model complexity. A notable improvement is observed in the detection of horizontal dislocation defects, where ALPHA attains an mAP@50 of 99.5%, a dramatic increase compared to YOLOv10's near-zero performance (0.02%) and YOLOv11's 1.22%. This substantial gain underscores the effectiveness of our CA module in capturing orientation-specific features that are often misclassified due to their subtle appearance and similarity to background patterns. Beyond horizontal dislocations, ALPHA consistently outperforms YOLOv10 and YOLOv11 in several defect categories. For short circuits, it achieves 99.5% mAP@50, surpassing

YOLOv10 (98.7%) and YOLOv11 (99.2%). Similarly, in thick line detection, it registers 89.4%, outperforming YOLOv10 (89.3%) and YOLOv11 (88.4%). Additionally, for crack defects, the ALPHA model improves upon YOLOv10 (70.4%) with an mAP@50 of 71.8%, although it falls short of YOLOv11's 74.5% (Table 1).

Table 1. Overall detection metrics (values in %).

Model	mAP50	mAP50-95	Precision	Recall	Parameters	**GFLOPs**
Alpha	**85.0**	**60.3**	**83.9**	74.8	**9,176,800**	**23.5**
YOLOv10	76.5	55.2	78.2	73.7	8,930,776	25.4
YOLOv11	77.6	56.3	76.5	**75.1**	9,417,444	21.3

5   Conclusion

We present ALPHA, an attention-enhanced YOLO framework integrating FCA, CA, and EMA modules for accurate PV defect detection. ALPHA achieves 85.0% mAP@50, surpassing YOLOv10 and YOLOv11 by 8.5% and 7.4% respectively, while maintaining computational efficiency. The framework demonstrates particular effectiveness for challenging defects like horizontal dislocations (99.5% vs 0.02% for YOLOv10). Future work will focus on multi-scale feature fusion enhancements, domain adaptation, and deployment optimization for real-time industrial applications.

References

1. Meng, Z., et al.: Defect object detection algorithm for electroluminescence image defects of photovoltaic modules based on deep learning. Energy Sci. Eng. **10**(3), 800–813 (2022)
2. Hijjawi, U., Lakshminarayana, S., Xu, T., Fierro, G.P.M., Rahman, M.: A review of automated solar photovoltaic defect detection systems: approaches, challenges, and future orientations. Sol. Energy **266**, 112186 (2023)
3. Chen, H., Pang, Y., Hu, Q., Liu, K.: Solar cell surface defect inspection based on multispectral convolutional neural network. J. Intell. Manuf. **31**, 453–468 (2020)
4. Guo, S., Wang, Z., Lou, Y., Li, X., Lin, H.: Detection method of photovoltaic panel defect based on improved mask R-CNN. J. Internet Technol. **23**(2), 397–406 (2022)
5. Tang, W., Yang, Q., Xiong, K., Yan, W.: Deep learning based automatic defect identification of photovoltaic module using electroluminescence images. Sol. Energy **201**, 453–460 (2020)
6. Su, B., Chen, H., Zhou, Z.: BAF-detector: an efficient CNN-based detector for photovoltaic cell defect detection. IEEE Trans. Ind. Electron. **69**(3), 3161–3171 (2021)
7. Jiang, Y., Zhao, C.: Attention classification-and-segmentation network for micro-crack anomaly detection of photovoltaic module cells. Sol. Energy **238**, 291–304 (2022)
8. Akram, M.W., et al.: CNN based automatic detection of photovoltaic cell defects in electroluminescence images. Energy **189**, 116319 (2019)

Security/Privacy

Secure Approach for Blockchain-Based Anonymous Attribute-Based Searchable Encryption Scheme for Data Sharing

Dhruv Kalambe⬤, Nish Shah⬤, Payal Chaudhari$^{(\boxtimes)}$⬤,
and Priyanshi Manglani⬤

Department of Computer Science and Engineering, Pandit Deendayal Energy
University, Gandhinagar, India
payal.chaudhari@sot.pdpu.ac.in

Abstract. Attribute-Based Encryption (ABE) is one of the most powerful paradigms which enables fine grained access-control mechanism by encrypting sensitive data using the attributes of the user/receiver. Anonymous Attribute-Based Encryption (AABE) is an extension of ABE which provides confidentiality to the attributes and identity of the user during encryption and decryption. However, these schemes often involve complex pairing operations that increase encryption and decryption time, limiting scalability in resource-constrained environments. Searchable encryption a cryptographic mechanism that enables searching on encrypted data helps to retrieve a selected subset of encrypted documents from remote storage without losing the data confidentiality. Anonymous Attribute-Based Searchable Encryption (AABSE) ensures sender and receiver anonymity while allowing search on encrypted data. Zhang *et al.* have proposed a blockchain-based AABSE scheme for secure data sharing. In the cryptanalysis of their scheme, we found a security loophole that compromises receiver anonymity. In this work, we explore this security flaw and demonstrate how an adversary can determine user attributes. We also propose an enhanced scheme to overcome this vulnerability.

Keywords: Attribute Based Encryption · Blockchain · Searchable Encryption · Receiver Anonymity

1 Introduction

As cloud computing progresses steadily, it provides scalable storage solutions and computational resources, enabling users to store and access data from anywhere. However, outsourcing data to third-party servers introduces significant security and privacy concerns. Data owners must ensure that sensitive information remains confidential and is only accessible to authorized users. Traditional encryption techniques fail to provide fine-grained access control and are impractical in dynamic environments where multiple users with different access privileges interact.

© The Author(s), under exclusive license to Springer Nature Switzerland AG 2026
R. Wrembel et al. (Eds.): DEXA 2025, LNCS 16047, pp. 313–328, 2026.
https://doi.org/10.1007/978-3-032-02088-8_24

Attribute-based encryption (ABE) [3], which supports one-to-many encryption, has emerged as a promising solution to address these challenges by allowing data to be encrypted under specific access policies tied to user attributes, allowing a data owner to define access policies that allow multiple users to decrypt data if their attributes satisfy the conditions. It is classified into two main models: Key-Policy ABE (KP-ABE) and Ciphertext-Policy ABE (CP-ABE). In KP-ABE, access policies are embedded in the users private key, and decryption is possible when the ciphertext attributes meet the key policy. However, CP-ABE allows data owners to specify access conditions within the ciphertext, granting access to users whose attributes satisfy these conditions. Although ABE enables precise access control, it inherently reveals attribute information during encryption and decryption. This exposure allows an adversary to infer sensitive details about the recipients identity or the nature of encrypted content, leading to privacy breaches.

To address this issue, Anonymous Attribute-Based Encryption (AABE) [8] extends ABE by concealing both the access policy and the user's attributes during decryption. AABE is an advanced form of ABE that enhances user privacy by ensuring that the identity of the user or their attributes remain hidden during the encryption and decryption process. It is particularly useful in scenarios where confidentiality and user anonymity are critical, such as secure messaging, healthcare, and blockchain-based systems. However, conventional AABE schemes require users to perform full decryption on every ciphertext to determine if they are eligible, which imposes substantial computational overhead.

Searchable encryption (SE) [10] allows the user to perform search operations on the encrypted data. In addition, it also aims to maintain the confidentiality of the data and the queries made. This technique enables the user to delegate searching operations to third parties, without revealing their private key or need for them to decrypt the data. This saves computational overhead caused by AABE (as discussed above). However, most of the schemes existing in the literature have assumed that the identity of the receiver is known by the sender [4].

This creates a necessity for researchers to propose solutions that could allow keyword-based search operations on encrypted data without disclosing the identity of the receiver. This led to the invention of Attribute Based Searchable Encryption (ABSE). An ABSE scheme allows users to store their data on third-party systems, such as cloud servers, by encrypting the data. This encryption takes place according to the underlying policy/ access structure, which determines which recipient is eligible to search and decrypt the data [4,11]. Several ABSE schemes have been presented in the literature [2,4,12]. Although ABSE schemes allow receiver anonymity, in order to achieve total anonymity (both sender and receiver) within the system, Anonymous ABSE(AABSE) schemes were introduced. Several researchers have proposed schemes that use AABSE to improve the security of the system [1,6,7,14].

While reviewing the existing AABSE schemes, we find the scheme proposed by Zhang *et al.* [13]. The scheme proposes a blockchain-based anonymous attribute-based searchable encryption scheme for data sharing (BADS), which combines ABSE with blockchain for achieving integrity, non-repudiation, and tamper proofness. The user encrypts the data and stores them on an Inter-Planetary File System (IPFS). The secure index of the same is stored on the blockchain. The scheme allows a matching algorithm to perform a fixed number of pairing operations before performing the searching algorithm. The matching algorithm is run by the blockchain before performing the search. This reduces the computation and improves the security. However, we find that by using the public parameters along with the trapdoor components and the encrypted indexes, any user or adversary can determine the attributes possessed by the data owner.

1.1   Literature Review

The authors in [14] address the inefficiency caused in the notable works of AABE, where in the trial-and-error approach to decrypt is computationally expensive especially when it involves complex pairing operations. It introduces the concept of match-then-decrypt mechanism which allows users to verify their attribute keys before decrypting without having to incur the full cost of computation.

In [6], the authors have proposed an efficient multi keyword ABSE scheme (EMK-ABSE), which helps in addressing the limitations caused by ABE and ABSE schemes like searching inefficiency and limited query expressiveness due to single keyword search. EMK-ABSE stores its encrypted data on the cloud while the encrypted indices are offloaded on the edge nodes making it much easier and faster for multi keyword search and partial decryption.

Unifying ABE, receiver anonymity and keyword based search all in one whole framework, is the aim of the work presented in [1]. It ensures that only authorized users can generate valid trapdoors, using hidden access policies, without having to reveal their attribute sets or identities.

The authors in [7] propose a refined ABSE framework that addresses the concerns like policy exposure risk and limited attribute universe support that are persistent in ABSE. It introduces a pre-matching stage which allows for secure attribute matching without revealing the identity to the cloud server and therefore preserving the privacy of the policy. Additionally, it decouples the number of public parameters from the size of the universe of attributes, making the scheme scalable and efficient for real-world smart health deployments. All these existing ABSE schemes with search facility have not included the concept of Blockchain, which can support a decentralized framework.

The authors in [5] Hierarchical Searchable Encryption with Blockchain-based Indexing (HSE-BI) scheme, which integrate hierarchical(indexing) key derivation with Directed Acyclic Graph(DAG)-based access policy which helps and allows for a stepwise key derivation, addressing the challenge of handling hierarchical access structures. However, it has not included receiver anonymity in its objectives. One more scheme that uses blockchain to store the search index and perform search operation is described in [9]. But its work is towards symmetric encryption and they do not address the fine-grained access control policy.

1.2   Our Contribution

In this paper we claim to provide the following.

- We explore the security flaw present in the scheme [13] with the mathematical proof.
- We propose a modified construction for this scheme that is able to mitigate the security flow present in [13].
- The security analysis of the proposed customized algorithms is presented with its mathematical correctness.

Organization of the Paper: The remainder of the paper is organized as follows. Section 2 discusses the preliminaries, while Sect. 3 reviews the scheme proposed by Zhang *et al.* [13]. Section 4 highlights the security flaws present in [13]. Section 5 introduces our proposed modified construction in detail, while its security analysis is presented in Sect. 6. The performance analysis of the modified scheme is discussed in Sect. 7, while the conclusion of this work is presented in Sect. 8.

2   Preliminaries

2.1   Complexity Assumptions

The security claim for the modified algorithms proposed in this paper, we take the support of complexity to solve the Decisional Diffie-Hellman problem in polynomial time.

Definition 1. *Decisional Diffie Hellman (DDH) Assumption:* Let $\mathbb{G}$ be a cyclic multiplicative group of prime order q, where g is the generator of $\mathbb{G}$. The DDH assumption states that for three random elements x, y and z from $\mathbb{G}_q$ such that $z \neq xy$, it is computationally infeasible for any probabilistic polynomial-time adversary to distinguish between the tuples $\{g, X = g^x, Y = g^y$ and $Z = g^z\}$ and $\{g, X = g^x, Y = g^y$ and $Z = g^{xy}\}$.

2.2   Access Policy

The access policy is an amalgamation of AND gates that support multi-valued attributes. Let the universe consist of n attributes and let $A = \{att_1, att_2, \ldots, att_n\}$ represent the set of attributes. Let each attribute i be denoted by att_i. For any $att_i \in A$ $(1 \leq i \leq n)$, there exists a set of values $V_i = \{v_{i,n_1}, v_{i,n_2} \ldots, v_{i,n_i}\}$ that represents the set of possible values of att_i. Here, n_i represents the number of possible values att_i can possess. Let $AL = [AL_1, AL_2, \ldots, AL_n]$ be the user attribute list, where $L_i \in V_i$ $(1 \leq i \leq n)$. The access policy is defined as $AP = [AP_1, AP_2, \ldots, AP_n]$, where $AP_i \in V_i$ for all $1 \leq i \leq n$. Furthermore, a function F is defined so that it yields $F(AL, AP) = 1$ iff the attribute list AL fulfills the requirements of AP, else $F(AL, AP) = 0$.

2.3   Notations

For a better understanding of the scheme proposed by [13] and our proposed solution scheme, we provide the details of the notations used in this paper in Table 1.

Table 1. Notations

Notation	Description
SGP	Global Parameter of System
PuK	Public Key of the System
PrK	Master Private Key
$(\mathcal{ID}_u, \mathcal{D}_u)$	Information specific to the user u
AL	Attribute List
SK_L	Secret key of user
PT_F	Plaintext of file
KS	Keyword set of file F
F	File
AP	Access Policy
I	Encrypted Index of keyword set
CT_F	Ciphertext of file F
Sig	Signature
KS'	Keyword set interested by user
$TD_{KS'}$	Search trapdoor for KS'
UL	User List
$\perp$	Invalid result

3   Zhang *et al.*'s Scheme [13]

3.1   Scheme Definition

The Scheme in [13] comprises of four key algorithms Setup, KeyGen, Encrypt, and Decrypt. The following is a definition for each:

- **Setup** $(1^l, u) \longrightarrow (SGP, PuK, PrK, (ID_u, D_u))$: The algorithm takes a security parameter l and a data user u as input and returns the global parameter of the system SGP, the public key of the system (PuK), the Master private key (PrK) along with the identity components (ID_u, D_u); here ID_u depicts the distinctive identity associated with the data user u.
- **KeyGen** $(SGP, PuK, PrK, AL) \longrightarrow SK_L$: In input of attribute list AL, system public key PuK, master private key PrK, and global parameter SGP, the algorithm outputs the secret key SK_L, associated with the attribute set AL.

- **Encrypt** $(SGP, PuK, PT_F, KS, AP) \longrightarrow (I, CT_F, Sig)$: This algorithm takes as input the public key PuK, system global parameter SGP, plaintext of file PT_F, keyword set KS, from the file F and access policy AP, which gives the output as ciphertext CT_F, encrypted indexes I and signature Sig.
- **Trapdoor** $(SGP, PuK, SK_L, KS') \longrightarrow TD_{KS'}$: This algorithm takes as input the global parameter of the system SGP, the public key of the system PuK, the secret key SK_L, along with the set of keywords KS' that output the search Trapdoor $TD_{KS'}$, which works as a search token.
- **Search** $(SGP, I, TD_{KS}, UL) \longrightarrow (CT_F, I_2)$ *or* $\perp$: This algorithm takes into account input global parameter SGP, encrypted index I, user list UL, trapdoor TD_{KS} and outputs (CT_F, I_2), where I_2 is auxiliary information required for decryption with $\perp$ being an invalid result.
- **Decrypt** $(SGP, SK_L, (CT_F, I_2)) \longrightarrow PT_F$: This algorithm is run by DUs with the mentioned inputs to output the plaintext of the file PT_F .

3.2   Detailed Construction

- **Setup** $(1^l, u) \longrightarrow (SGP, PuK, PrK, (ID_u, D_u))$: The setup phase is conducted by the Attribute Center (AC). It selects three cyclic multiplicative groups $\mathbb{G}_0, \mathbb{G}_1, \mathbb{G}_2$ with prime order q. It randomly selects g_0 as the generator of $\mathbb{G}_0$ and two multiple-three linear mapping functions $e_0 : \mathbb{G}_0 \times \mathbb{G}_0 \to \mathbb{G}_1$ and $e_1 : \mathbb{G}_0 \times \mathbb{G}_1 \to \mathbb{G}_2$. In addition, it selects two hash functions $H_1 : \{0,1\}^* \to \mathbb{G}_0, H_2 : \mathbb{G}_0 \times \mathbb{G}_1 \to \mathbb{Z}_q^*$ and a symmetric encryption algorithm $SyE = (SyE.Enc, SyE.Dec)$. The Master private Key PrK is chosen as $\langle \alpha, \beta, r \in_r \mathbb{Z}_q \rangle$. For every data user u, AC randomly selects $x_u \in_r \mathbb{Z}_q$ as the unique identity ID_u of user u. In addition, it computes $Y_1 = g_0^\alpha, Y_2 = e_0(g_0, g_0)^\alpha, Y_3 = g^\beta$ and $Y_u = Y_1^{x_u}$ and the corresponding public key PuK $\langle Y_1, Y_2, Y_3, Y_U \rangle$ is published. It finally computes $D_u = Y_2^{r x_u}$ and sends the tuple (ID_u, D_u) to the blockchain to store it as a content of UL.
- **KeyGen** $(SGP, PuK, PrK, AL) \longrightarrow SK_L$: The generation of keys is carried out by AC with the inputs being the global parameter of the system SGP, the public key PuK, the master private key PrK and a set of attributes AL. For $AL (= (AL_1, \ldots, AL_n))$ possessed by DUs, the corresponding attribute secret key is generated. Each AL_i in AL represents the value $v_{i,j}$ the DU has for an attribute i. The secret key SK_L is generated with addition of randomization. The key components of SK_L are $\langle D_1 = Y_1 g_0^r \prod_{i=1}^n H_1(i \parallel AL_i)^\alpha, D_2 = g_0^{\frac{r}{\beta}} \rangle$.
- **Encrypt** $(SGP, PuK, PT_F, KS, AP) \longrightarrow (I, CT_F, Sig)$: Encryption is performed by the DOs. Let the access policy AP be defined as $[AP_1, \ldots, AP_n]$. Assume that AP_j is the attribute value of $1 \le j \le n$ for an attribute j and the value of each $AP_j \in V_i$. The algorithm outputs the encrypted index of the set of keywords KS given by $I = (I_0, I_1, I_2, I_3, \{I_{kw}\}_{kw \in KS})$. It randomly selects $z, v \in \mathbb{Z}_q$, to calculate the components of I as $\langle$

$I_0 = g_0^z \prod_{j=1}^{n} H_1(j \parallel AP_j)^{zv}$, $I_1 = g_0^{zv}$, $I_2 = Y_3^{zv} = g_0^{zv\beta}$, $I_3 = g_0^{z(v-1)}$, $I_{kw} = e_1(H_1(kw), e_0(Y_1, C_3)e_0(g_0, Y_u)), \rangle$.

A symmetric key k is randomly selected for encrypting the plaintext PT_F of file to $SyE.Enc(PT_F)$. Then it is uploaded to IPFS and returns M of the ciphertext file as $M = URL(SyE.Enc(PT_F))$. These components are used to generate ciphertext elements as follows $\langle CT_k = ke_0(g_0, I_1), CT_M = Me_0(g_0, I_1) \rangle$.

Here CT_k is the symmetric key ciphertext and CT_M is the ciphertext of the file address. The output ciphertext is generated as $CT_F = (CT_k, CT_M)$.

Lastly, the following signature elements are generated by DOs, randomly selecting $r_\sigma \in \mathbb{Z}_q$. $Sig = \langle S_1 = r_\sigma H_2(I), S_2 = g_0^{r_\sigma} \rangle$

BC receives identity information (ID_u, D_u), I, ciphertext CT_F of the symmetric key and file address, and signature Sig. The encrypted index I, and the signature Sig are taken from the data set by the consensus node to calculate $H_2(I)$. Equation $e_0(Y_1, S_2)^{H_2(I)} = Y_2^{S_1}$ is checked to be true, if it is, then it is believed to be authorized, and a validation confirmation message is sent.

- **Trapdoor** $(SGP, PuK, SK_L, KS') \longrightarrow TD_{KS'}$: Upon receiving SGP, PuK, SK_L and the set of keywords KS' in which DU is interested as input, DU selects $s \in \mathbb{Z}_q$ uniformly at random. The trapdoor components of the selected keyword set are generated as follows. $\langle \tilde{D} = D_1^s, \hat{D} = D_2^s, D_0 = Y_1^s, \tilde{D}_{kw'} = H_1(kw')^{\frac{1}{s}}, \hat{D}_{kw'} = H_1(kw')^{\frac{1}{r}} \rangle$

Finally, the trapdoor for the set of keywords interested in DUs is published as $TD_{KS'} = (\tilde{D}, \hat{D}, D_0, \{\tilde{D}_{kw'}, \hat{D}_{kw'}\}_{kw' \in KS'})$. Then it is sent to the BC for performing the search operation.

- **Search** $(SGP, I, TD_{KS'}, UL) \longrightarrow (CT_F, I_2)$ or $\perp$: The inputs provided are encrypted index I, trapdoor $TD_{KS'}$ and UL. The BC runs this algorithm and checks if the unique identity ID_u of the data user u exists on UL of the registered DUs or not. If it does not exist, it produces $\perp$. If it does exist, then the following equation is checked:

$$\frac{e_0(I_1, \tilde{D})}{e_0(I_0, D_0)e_0(I_2, \hat{D})} = e_0(I_3, D_0)$$

If this equation holds, it implies that the set of attributes possessed by DUs matches successfully with the access policy of ciphertext. Consequently, $F(AL, AP) = 1$, which essentially produces $C'' = e_0(I_3, D_0)$ and D_u. Otherwise, it outputs $\perp$.

Furthermore, BC proceeds to calculate $I_{kw'}$ for each $kw' \in KS'$ with the help of components of $TD_{KS'}$ as shown below:

$$I_{kw'} = e_1(\tilde{D}_{kw'}, C'')e_1(\hat{D}_{kw'}, D_u)$$

Lastly, it verifies if $I_{kw'} = I_{kw}$ for each $I_{kw} \in I$. If it holds, then it generates the output as (CT_F, I_2), otherwise $\perp$.

– **Decrypt** $(SGP, SK_L, (CT_F, I_2)) \longrightarrow PT_F$: Utilizing the inputs SGP, SK_L and (CT_F, I_2), the algorithm performs following operations to compute k, receive the address M:

$$D' = e_0 \left(D_2^{\frac{1}{r}}, I_2 \right) = e_0 \left(g_0^{\frac{r}{\beta} \cdot \frac{1}{r}}, g_0^{zv\beta} \right) = e_0 \left(g_0, g_0^{zv} \right)$$

With the help of D', k and M could be calculated as follows:

$$k = \frac{CT_k}{D'} = \frac{ke_0(g_0, I_1)}{e_0(g_0, g_0^{zv})} = \frac{ke_0(g_0, g_0^{zv})}{e_0(g_0, g_0^{zv})}$$

$$M = \frac{CT_M}{D'} = \frac{Me_0(g_0, I_1)}{e_0(g_0, g_0^{zv})} = \frac{Me_0(g_0, g_0^{zv})}{e_0(g_0, g_0^{zv})}$$

Finally, the DUs request with M to the IPFS to get the encrypted file $SyE.Enc(PT_F)$. The plaintext of the file PT_F, is obtained by symmetric key decryption (using k), that is, $PT_F = SyE.Dec(SyE.Enc(PT_F))$.

4   Security Flaws in Zhang *et al.*'s Scheme

In [13], Zhang *et al.* propose a Blockchain-based AABSE Scheme for data sharing (BADS). It facilitates the matching algorithm that performs a fixed number of pairing operations before the searching algorithm is run. The authors in [13], claim that any adversary $\mathscr{A}$ who is aware of the search trapdoor and encrypted indexes cannot determine the keyword set and the access policy'. However, it is found that by using the search trapdoor and public parameters, $\mathscr{A}$ can easily determine the attributes of the Data Owner(DO) in polynomial time. The aim of this section is to show that receiver's anonymity is not being provided by the scheme proposed in [13]. To prove this, consider $\mathscr{A}$ receiving the search trapdoor in the following format:

$$\tilde{D} = D_1^s = g_0^{(r+\alpha)*s} * \prod_{i=1}^{n} H_1(i||AL_i)^{\alpha s}$$

$$= D_1^s = g_0^{\alpha s} * g_0^{rs} * \prod_{i=1}^{n} H_1(i||AL_i)^{\alpha s}$$

$$\hat{D} = D_2^s = g_0^{\frac{rs}{\beta}}$$

$$D_0 = Y_1^s = g_0^{\alpha s}$$

Furthermore, $\mathscr{A}$ is also aware of the public parameters g_0, g_0^{α}, and g_0^{β}. Next, $\mathscr{A}$ chooses a list of attribute values, using which he shall try to find out the actual attribute values possessed by DO. $\mathscr{A}$ proceeds by picking one value AL_i for each attribute i, and computes $\prod_{i=1}^{n} H_1(i||AL_i)$ for each pair (i, AL_i). Further, he performs the following computation to check if the chosen set of attribute values matches that of the owner or not:

$$R_1 = e_0(D_0, g_0) \cdot e_0(\hat{D}, g_0^\beta) \cdot e_0(D_0, \prod_{i=1}^{n} H_1(i||AL_i))$$

$$= e_0(g_0^{\alpha s}, g_0) \cdot e_0(g_0^{\frac{rs}{\beta}}, g_0^\beta) \cdot e_0(g_0^{\alpha s}, \prod_{i=1}^{n} H_1(i||AL_i))$$

$$= e_0(g_0, g_0)^{\alpha s} \cdot e_0(g_0, g_0)^{rs} \cdot e_0(g_0, \prod_{i=1}^{n} H_1(i||AL_i))^{\alpha s}$$

$$R_2 = e_0(\widetilde{D}, g_0)$$

$$= e_0(g_0^{(r+\alpha)*s} * \prod_{i=1}^{n} H_1(i||AL_i)^{\alpha s}, g_0)$$

$$= e_0(g_0, g_0)^{\alpha s} \cdot e_0(g_0, g_0)^{rs} \cdot e_0(g_0, \prod_{i=1}^{n} H_1(i||AL_i))^{\alpha s}$$

It should be noted that the hash calculated by $\mathscr{A}$, is used in R_1, while the hash present in R_2 ($\prod_{i=1}^{n} H_1(i||AL_i)^{\alpha s}$), consists of the actual values of attributes possessed by DO. Hence, if $R_1 \overset{?}{=} R_2$, then $\mathscr{A}$ has successfully determined the values of the attributes possessed by DO. Consider that the universe of discourse has n attributes and that each attribute may possess at maximum m values. Since the first two pairing operations in R_1 and R_2 are the same, they shall be computed only once. Hence, only the last pairing operation of R_1 will have to be computed for each set of attribute values. Therefore, the worst-case complexity to determine the correct set of attribute values of DO is $O(n \times m)$. This allows $\mathscr{A}$, to learn the attribute values with a polynomial-bounded operation time limit.

5 Proposed Solution

As discussed above (Sect. 4), the adversary can learn the attributes of DO by simply checking if $R_1 \overset{?}{=} R_2$. This security vulnerability of the scheme proposed in [13] occurs because there is no uniqueness/randomness assigned to the hash calculated by DO. This enables a polynomial-time adversary to determine the attribute set of DO using the trapdoor components and public parameters. The proposed modified scheme introduces changes to the *KeyGen* and *Trapdoor* algorithms. These changes aim to randomize the hash of attributes computed by the DO, to enhance the security of its attributes.

5.1   Scheme Definition

The scheme is similar to that of Zhang *et al.* scheme [13], and comprises six algorithms, Setup, KeyGen, Encrypt, Trapdoor, Search and Decrypt, each defined as below:

- **Setup** $(1^l, u) \longrightarrow (SGP, PuK, PrK, (ID_u, D_u))$:This algorithm takes a security parameter l as input and outputs the Global Parameters (SGP), Public Key (PuK), Master Private Key (PrK), and identity information (ID_u, D_u). It is run by AC, which acts as a trusted third party.
- **KeyGen** $(SGP, PuK, PrK, AL) \longrightarrow SK_L$: Taking as input the AL, SGP, PuK and MK, the algorithm computes the secret key SK_L, associated with the attribute set AL.
- **Encrypt** $(SGP, PuK, PT_F, W, KS, AP) \longrightarrow (I, CT_F, Sig)$: The encrypt algorithm takes as input the plaintext of file PT_F, Keyword Set KS, access policy AP, along with SGP and PuK, the algorithm outputs encrypted indexes I, ciphertext CT_F, and signature Sig.
- **Trapdoor** $(SGP, PuK, SK_L, KS') \longrightarrow TD_{KS'}$: The trapdoor algorithm takes as input the keyword set KS', along with SGP, PuK and SK_L, the algorithm outputs the search trapdoor $TD_{KS'}$, which functions as a token to perform the search operation.
- **Search** $(SGP, I, TD_{KS'}, UL) \longrightarrow (CT_F, I_2)$ *or* $\perp$: For searching, the algorithm inputs the list of users UL, trapdoor $TD_{KS'}$, SGP, and I. The algorithm outputs the tuple (CT_F, I_2) where I_2 refers to the auxiliary information needed for decryption or $\perp$, which refers to an invalid result.
- **Decrypt** $(SGP, SK_L, (CT_F, I_2)) \longrightarrow PT_F$: The decryption algorithm takes as input SGP, SK_L, and CT_F. The output of this algorithm is the plaintext of file PT_F.

5.2   Threat Model

As the proposed scheme provides a mitigation of the security flaw presented in the scheme of [13], the threat model of the proposed scheme and the scheme of [13] are the same. For the proposed system, the adversary is assumed to be any user inside the system who is not a legitimate receiver of a ciphertext, but because of curiosity, he tries to gain insights of the encrypted index and the access policy. Thereby he tries to reveal the identity of the valid receive of a ciphertext. Here we do not include Attribute Center in the set of Adversary. The Attribute Center is considered to be a fully trusted centralized entity that is responsible for system setup and generating keys for each user in the system.

5.3   Security Model

In the proposed scheme, we provide a mitigation of the security flaw present in the scheme provided by Zhang *et al.* in [13]. As described in Sect. 4, the root of the security flaw lies in the construction of the KeyGen algorithm. To prove that the proposed scheme mitigates that security flaw, we provide the security analysis of the proposed scheme in the IND-CP (Indistinguishability in Ciphertext policy) model.

Init: The adversary $\mathcal{A}$ submits a security parameter l to the challenger $\mathcal{C}$.

Setup: The $\mathcal{C}$ executes the setup and configures the system with secret master and secret public components. The global parameters (SGP) and the public key of system (PuK) are submitted to $\mathcal{A}$.

Phase 1: The $\mathcal{A}$ submits a finite number of the attribute list AL to and for each list, it receives a separate secret key SK_L.

Challenge: The $\mathcal{A}$ submits two attribute lists AL_1 and AL_2 to $\mathcal{C}$ with the condition that none of them has been submitted during Phase 1. The $\mathcal{C}$ randomly selects a list Al_i and runs the trapdoor algorithm to generate a trapdoor $TD_{KS'}^b$, where $b \in \{0, 1\}$.

Phase 2: Again, let $\mathcal{A}$ submit a polynomial number of queries with input of the attribute list AL. In response to each query, he receives a unique secret key SK_L. Here, the condition should be followed that neither AL_1, nor AL_2 should be submitted in any query because they have been submitted during the challenge phase.

Guess: If the $\mathcal{A}$ is able to correctly compute the value of b with more than $1/2 + \epsilon$, where ϵ is a non-negligible quantity, then $\mathcal{A}$ wins the game.

5.4   Detailed Construction

In this modified version, we modified algorithms $KeyGen$ and $Trapdoor$ to overcome security flaws in [13]. The remaining algorithms are similar to those in [13] (as discussed in Sect. 3.2). The elaborative description of the modified algorithms is presented as follows:

- **KeyGen**
  $(SGP, PuK, PrK, AL) \longrightarrow SK_L$: Consider $AL = [AL_1, AL_2, \ldots, AL_n]$ to be the list of attributes possessed by DU. Each AL_i in AL represents the value $v_{i,j}$ that the DU possesses for an attribute i. The AC uses the components SGP, PuK and PrK to generate the secret key of DU in a randomized fashion as follows:$\langle\, D_1 = Y_1^r g_0^r \prod_{i=1}^{n} H_1(i||AL_i)^{\alpha r}, D_2 = g_0^{\frac{r}{\beta}} \,\rangle$. The secret key of the user SK_L is supplied as a tuple (r, D_1, D_2)

– **Trapdoor** $(SGP, PuK, SK_L, KS') \longrightarrow TD_{KS'}$: This algorithm is designed to generate the trapdoor $(TD_{KS'})$ for the set of extracted keywords. The algorithm takes an input of SGP, PuK, SK_L, and the set of keywords KS', in which DU is interested. The DU, then uniformly at random selects $s \in \mathbb{Z}_q$ and generates the components of the trapdoor as follows: $\langle \tilde{D} = D_1^s, \hat{D} = D_2^s,$ $D_0 = Y_1^{rs}, \tilde{D}_{kw'} = H_1(kw')^{\frac{1}{s}}, \hat{D}_{kw'} = H_1(kw')^{\frac{1}{r}} \rangle$. This trapdoor is sent to BC for the search operation.

After receiving the trapdoor, BC proceeds with the matching algorithm. Subsequently, it either performs the search operation or aborts and outputs $\perp$. Finally, DUs perform decryption locally to obtain the plaintext PT_F. These algorithms are in accordance with the scheme proposed in [13]. The detailed constructions of these schemes are provided in Sect. 3.2.

Mathematical Correctness: The correctness of the matching phase in the improved scheme is as follows:

$$\frac{e_0(I_1, \tilde{D})}{e_0(I_0, D_0)e_0(I_2, \hat{D})} = \frac{e_0(g_0^{zv}, D_1^s)}{e_0(g_0^z \prod_{i=1}^n H_1(i||AL_i)^{zv}, Y_1^{rs})e_0(g_0^{zv\beta}, D_2^s)}$$

$$= \frac{e_0(g_0^{zv}, Y_1^{rs} g_0^{rs} \prod_{i=1}^n H_1(i||AL_i)^{\alpha rs})}{e_0(g_0^z \prod_{i=1}^n H_1(i||AL_i)^{zv}, Y_1^{rs})e_0(g_0^{zv\beta}, g_0^{\frac{rs}{\beta}})}$$

$$= \frac{e_0(g_0, g_0)^{zvrs\alpha} e_0(g_0, g_0)^{zvrs} e_0(g_0, \prod_{i=1}^n H_1(i||AL_i))^{zvrs\alpha}}{e_0(g_0, g_0)^{zrs\alpha} e_0(g_0, \prod_{i=1}^n H_1(i||AL_i))^{zvrs\alpha} e_0(g_0, g_0)^{zvrs}}$$

$$= e_0(g_0, g_0)^{zrs\alpha(v-1)}$$

Then,

$$e_0(I_3, D_0) = e_0(g_0^{z(v-1)}, Y_1^{rs}) = e_0(g_0^{z(v-1)}, g_0^{rs\alpha})$$
$$= e_0(g_0, g_0)^{zrs\alpha(v-1)}$$

It is clearly evident from above, that the improved scheme allows blockchain to effectively perform the matching operation.

6   Security Analysis

The proposed scheme provides indistinguishability against ciphertext policy attack. The security lies on the complexity assumption of the Decisional Diffie-Hellman Assumption(DDH).

Theorem 1. *The advantage of adversary to gain the knowledge of User's attributes from the trapdoor function is equivalent to the advantage of solving the Decisional Diffie-Hellman (DDH) assumption.*

Proof. Let the challenger $\mathcal{C}$ prepare two tuples (1)$\{g, X = g^x, Y =g^y$ and $Z =g^z\}$ and (2)$\{g, X = g^x, Y =g^y$ and $Z =g^{xy}\}$. The $\mathcal{C}$ randomly selects a value for ν from the set $\{0,1\}$. If $\nu = 0$, then tuple 1 will be assigned to the simulator $\mathcal{S}$, else tuple 2. The $\mathcal{S}$ uses this tuple in the IND-CP model to identify if it has been given a valid Diffie-Hellman tuple.

Init: The adversary $\mathcal{A}$ submits a security parameter l to the challenger $\mathcal{S}$.

Setup: The $\mathcal{S}$ runs the setup algorithm and sets the following parameters. $g_0 = g$, $Y_1 = g_0^\alpha = g^x = X$ (assuming $\alpha = x$), $H_1(a) = g^{xH'(a)} = X^{H'(a)}$ for any random string a. Here $H' : \{0,1\}^* \to \mathbb{Z}_q^*$. Rest of the parameters are generated as per the regular algorithm.

Phase 1: The $\mathcal{A}$ submits a polynomial number of the attribute list AL to and for each list, it receives a separate secret key SK_L for each AL as follows.

$$D_1 = Y_1^r g_0^r \prod_{i=1}^{n} H_1(i||AL_i)^{\alpha r} = g^r X^{(1+\sum_{i=1}^{n} H'(i||AL_i))r}$$

$$D_2 = g_0^{\frac{r}{\beta}} = g^{\frac{r}{\beta}}$$

Challenge: The $\mathcal{A}$ submits two attribute lists AL^1 and AL^2 to $\mathcal{C}$ with the condition that none of them has been submitted during Phase 1. Here we exclude the submission of keyword kw, because as described in Sect. 4, the components $\widetilde{D}$, $\hat{D}$, and D_0 are responsible to generate the security flaw in original scheme proposed in [13]. In our proposed customized algorithms we have modified these components only, and therefore in security proof we aim to generate these components only.

The $\mathcal{S}$ randomly selects a list AL^b where $b \in \{0, 1\}$, and selects y as the value of s. It generates the following trapdoor values:

$$\widetilde{D} = D_1^s = (Y_1^r \cdot g_0^r \cdot \prod_{i=1}^{n} H_1(i||AL_i^b)^{\alpha r})^s$$

$$= g^{xrs} g^{ry} \cdot g^{(\sum_{i=1}^{n} H'(i||AL_i^b))xry} = Y^r Z^{(1+(\sum_{i=1}^{n} H'(i||AL_i^b)))r}$$

$$\hat{D} = D_2^s = g^{\frac{ry}{\beta}} = Y^{\frac{r}{\beta}}$$

$$D_0 = Y_1^{rs} = g^{xyr} = Z^r$$

Phase 2: Again, the $\mathcal{A}$ is allowed to submit a polynomial number of queries with the input of the attribute list AL. In response to each query, he receives a unique secret key SK_L as defined in Phase - 1. Here, the condition should be followed that neither AL_1, nor AL_2 should be submitted in any query because they have been submitted during the challenge phase.

Guess: Let the $\mathcal{A}$ submits a guess b' for the value of b. If $b' = b$, then $\mathcal{S}$ submits $\nu = 1$ to state that it has been given a true Diffie-Hellman tuple. Else, it prints 0 to represent that the trapdoor is any random element. In this later case, $\mathcal{A}$ has no meaningful clue about b. So, $Pr[b \neq b'|\nu = 0] = \frac{1}{2}$. The $\mathcal{S}$ outputs $\nu' = 0$ when $b \neq b'$. Therefore $Pr[\nu \neq \nu'|\nu = 0] = \frac{1}{2}$. If $b' = 1$, then $\mathcal{A}$ is capable to identify the correct trapdoor components with some advantage ϵ. Therefore, $Pr[b = b'|\nu = 1] = \frac{1}{2} + \epsilon$. The $\mathcal{S}$ prints $\nu' = 1$, when $b = b'$. This implies that $Pr[\nu = \nu'|\nu = 1] = \frac{1}{2} + \epsilon$. Combining the rate of success with all approaches, the advantage of $\mathcal{S}$ for correctly solving the DDH assumption is $\frac{1}{2} \times Pr[\nu \neq \nu'|\nu = 0] + \frac{1}{2} \times Pr[\nu \neq \nu'|\nu = 1] - \frac{1}{2} = \frac{\epsilon}{2}$. This yields that if $\mathcal{A}$ has non-negligible advantage ϵ for winning the security model game of proposed scheme, then we can build a $\mathcal{S}$ that can break the complexity of DDH assumption with non-negligible quantity $\frac{\epsilon}{2}$. Considering the hardness of DDH assumption, the statement cannot be true, therefore the advantage of $\mathcal{A}$ to correctly identify the valid trapdoor content is negligible. Hence Proved.

7   Performance Analysis

The modified algorithm has been executed on a Linux-based system having an 11th Gen, Intel(R) Core(TM)-i5-113G7 processor, running at 2.40 GHz and 8 GB RAM. The algorithm has been implemented using the PBC library framework. To improve the security presented in the scheme proposed by Zhang et al., changes have been made in the algorithms *KeyGen* and *Trapdoor*, respectively. To evaluate the efficiency of our proposed scheme, we compare the execution times of these algorithms with those of Zhang et al.'s. Table 2, provides the execution times for both schemes, when executed for varying number of attributes. It is clearly evident from Table 2, that the execution times for our scheme is similar to that of Zhang et al.'s scheme. This indicates that the enhancements made to our algorithms preserve computational efficiency, with no significant impact on execution time compared to Zhang et al.'s scheme. It is to be noted, that rest of the algorithms(apart from *KeyGen* and *Trapdoor*) perform in accordance with the scheme originally proposed by Zhang et al., hence their computation efficiency remains unaffected.

Table 2. Evaluation of KeyGen and Trapdoor runtime in our scheme versus Zhang *et al.'s* scheme.

Number of attributes	Time for KeyGen (s)		Time for Trapdoor (s)	
	Our Scheme	Zhang *et al.*	Our Scheme	[13]
3	0.0198	0.0228	0.0076	0.0191
5	0.0311	0.0282	0.0133	0.0118
7	0.0350	0.0401	0.0108	0.0094
10	0.0585	0.0659	0.0075	0.0103

8   Conclusion

In this paper, we have discussed the need for Attribute Based Encryption (ABE), Searchable Encryption, Attribute Based Searchable Encryption (ABSE). Subsequently, we also discuss their shortcomings, which has led to the creation of Anonymous ABSE (AABSE) schemes. Upon reviewing the existing AABSE schemes, we found security flaws in the scheme proposed in [13]. The scheme in [13], fails to provide receiver anonymity. An enhanced scheme for the same has been proposed in this paper. The proposed scheme not only maintains the receiver anonymity, but also maintains the integrity of the subsequent matching operations that are to be performed by the blockchain.

Acknowledgments. The proposed work is carried out as part of the research project funded by the Gujarat Council On Science And Technology (GUJCOST) with Project-ID: GUJCOST/STI/2021-22/3867.

Disclosure of Interests. The authors have no competing interests to declare those are relevant to the content of this article.

References

1. Chaudhari, P., Das, M.L.: Keysea: keyword-based search with receiver anonymity in attribute-based searchable encryption. IEEE Trans. Serv. Comput. 15(2), 1036–1044 (2020)
2. Chaudhari, P., Das, M.: Privacy-preserving attribute based searchable encryption. Cryptology ePrint Archive (2015)
3. Goyal, V., Pandey, O., Sahai, A., Waters, B.: Attribute-based encryption for fine-grained access control of encrypted data. In: Proceedings of the 13th ACM Conference on Computer and Communications Security, pp. 89–98 (2006)
4. Khader, D.: Attribute based search in encrypted data: Abse. In: Proceedings of the 2014 ACM Workshop on Information Sharing & Collaborative Security, pp. 31–40 (2014)
5. Li, Y., Zhou, F., Ji, D., Xu, Z.: A hierarchical searchable encryption scheme using blockchain-based indexing. Electronics 11(22), 3832 (2022)

6. Liu, J., et al.: Emk-abse: efficient multikeyword attribute-based searchable encryption scheme through cloud-edge coordination. IEEE Internet Things J. **9**(19), 18650–18662 (2022)

7. Mehla, R., Garg, R.: Anonymous attribute-based searchable encryption for smart health system. SN Comput. Sci. **5**(7), 879 (2024)

8. Meng, L., Chen, L., Tian, Y., Manulis, M.: Fabesa: fast (and anonymous) attribute-based encryption under standard assumption. In: Proceedings of the 2024 on ACM SIGSAC Conference on Computer and Communications Security, pp. 4688–4702 (2024)

9. Mihaljević, M.J., Knežević, M., Urošević, D., Wang, L., Xu, S.: An approach for blockchain and symmetric keys broadcast encryption based access control in iot. Symmetry **15**(2), 299 (2023)

10. Wang, Y., Wang, J., Chen, X.: Secure searchable encryption: a survey. J. Commun. Inf. Netw. **1**(4), 52–65 (2016). https://doi.org/10.1007/BF03391580

11. Yan, L., et al.: Attribute-based searchable encryption: a survey. Electronics **13**(9), 1621 (2024)

12. Yin, H., et al.: CP-ABSE: a ciphertext-policy attribute-based searchable encryption scheme. IEEE Access **7**, 5682–5694 (2019)

13. Zhang, K., Zhang, Y., Li, Y., Liu, X., Lu, L.: A blockchain-based anonymous attribute-based searchable encryption scheme for data sharing. IEEE Internet Things J. **11**(1), 1685–1697 (2023)

14. Zhang, Y., Chen, X., Li, J., Wong, D.S., Li, H.: Anonymous attribute-based encryption supporting efficient decryption test. In: Proceedings of the 8th ACM SIGSAC Symposium on Information, Computer and Communications Security, pp. 511–516 (2013)

Incremental k-Anonymization for Continuously Growing Big Databases

Akifumi Kurumatani, Hiromasa Yoshimoto$^{(\boxtimes)}$ (ID), and Kazuo Goda (ID)

Institute of Industrial Science, The University of Tokyo, Meguro-ku, Tokyo, Japan
yoshimoto@tkl.iis.u-tokyo.ac.jp

Abstract. k-anonymization is a widely used approach to protect the privacy of personal data. In many real-world applications, data is regularly accumulated, and the size of databases continues to grow over time. Anonymizing such dynamic databases poses significant computation time and efficiency challenges. This paper proposes an incremental framework to k-anonymization that processes only newly added records instead of re-anonymizing the whole data, thereby significantly reducing computational time for growing databases. Experiments were conducted on a real-world medical database containing 470,000 individuals' records over 108-month period. Compared to existing methods, our framework achieved a 30 times speedup while keeping the loss information within 9%. This approach provides a scalable solution for privacy-preserving data utilization in continuously updated databases, offering practical benefits for real-world big data applications, such as in healthcare domains.

Keywords: k-anonymization · Incremental framework · Microaggregation · Growing big data

1 Introduction

Continuous data accumulation requires scalable privacy-preserving techniques. Under Japan's universal health coverage system, for example, the government collects all medical billing records to verify the payments. These comprehensive datasets are also used for secondary purposes, such as medical and health services research [9], and have supported numerous academic studies [6,7,19]. Privacy preservation becomes a key issue in enabling such data utilization.

Various privacy-preserving properties have been studied. One of the most prominent properties is k-anonymity. k-anonymity means each record is indistinguishable from at least $k - 1$ others with respect to a set of quasi-identifiers [13,16]. A widely adopted method ensuring k-anonymity is microaggregation [5,8,14], which groups similar records and replaces attribute values within each group with representative values (e.g., replacing individual ages with the group average). The degree of distortion introduced by this replacement is measured as information loss (IL), which reflects how much detail is lost after anonymization; lower IL indicates a higher data utility.

© The Author(s), under exclusive license to Springer Nature Switzerland AG 2026
R. Wrembel et al. (Eds.): DEXA 2025, LNCS 16047, pp. 329–337, 2026.
https://doi.org/10.1007/978-3-032-02088-8_25

Algorithm 1. MDAV$^+(\mathcal{D}, k)$: MDAV$^+$ Algorithm for k-anonymity [14]

Require: Array of records $\mathcal{D}$, anonymity parameter k
Ensure: Array of cluster $\mathcal{X}$, where each cluster is an array of records
1: $c \leftarrow$ centroid of $\mathcal{D}$; $\mathcal{X} \leftarrow \emptyset$
2: **while** $|\mathcal{D}| \geq 2k$ **do**
3:     $r \leftarrow$ find the farthest record from c in $\mathcal{D}$
4:     $\mathcal{N} \leftarrow$ the $(k-1)$ nearest neighbors of r in $\mathcal{D}$
5:     $\mathcal{G} \leftarrow \{r\} \cup \mathcal{N}$
6:     $\mathcal{X} \leftarrow \mathcal{X} \cup \mathcal{G}$; $\mathcal{D} \leftarrow \mathcal{D} \setminus \mathcal{G}$
7: **end while**
8: **for all** $r \in \mathcal{D}$ **do**
9:     $\mathcal{G} \leftarrow$ find the nearest cluster in $\mathcal{X}$
10:     $\mathcal{G} \leftarrow \mathcal{G} \cup \{r\}$
11: **end for**
12: **return** $\mathcal{X}$

Microaggregation is typically formulated as an optimization problem that seeks to minimize IL through suitable clustering. This problem is known to be NP-hard [1,4]. Due to its quadratic computational complexity, $O(N^2)$ with respect to the dataset size N, exact solutions are computationally infeasible for large-scale datasets. Moreover, existing methods generally require complete reprocessing whenever the data is updated, making them increasingly impractical as the database grows.

As data volume and utilization demands increase, efficient anonymization becomes essential. To address this challenge, we focus on an append-only data scenario and propose an incremental framework to k-anonymization tailored for continuously growing datasets. By anonymizing only the newly added records, our framework significantly reduces computational overhead while maintaining low IL. This approach aligns with real-world data accumulation processes and offers a practical solution for scalable privacy protection.

Our key contributions are as follows:

Incremental anonymization design. We design an efficient framework for append-only data scenario.

Dual strategy implementation. We show two practical strategies—Append and Merge—that offer trade-offs in processing speed and anonymization quality.

Comprehensive evaluation. We conduct extensive experiments using real-world and synthetic datasets, demonstrating the scalability and utility preservation of our approach.

The rest of the paper is organized as follows: Sect. 2 reviews standard microaggregation algorithms. Section 3 introduces the proposed incremental framework. Sections 4 and 5 describe the experiments and results. Section 6 discusses implications, and Sect. 7 concludes.

Algorithm 2. ONA($\mathcal{D}, k$): ONA Algorithm for k-anonymity [15]

Require: Array of records $\mathcal{D}$, anonymity parameter k
Ensure: Array of cluster $\mathcal{X}$, where each cluster is an array of records
 1: $\mathcal{X} \leftarrow$ Randomly partitioning $\mathcal{D}$ into clusters, each containing at least k records
 2: **while** the stopping criterion is not met **do**
 3:     **for all** record $r \in \mathcal{D}$ **do**
 4:         $\mathcal{G} \leftarrow$ the cluster in $\mathcal{X}$ to which r currently belongs
 5:         **if** $|\mathcal{G}| > k$ **then**   ▷ Reassigning the record r to decrease cluster size
 6:             $\mathcal{G} \leftarrow \mathcal{G} \setminus \{r\}$
 7:             $\mathcal{T} \leftarrow$ the nearest cluster to r in $\mathcal{X}$
 8:             $\mathcal{T} \leftarrow \mathcal{T} \cup \{r\}$
 9:         **else if** $|\mathcal{G}| = k$ **then**   ▷ Dissolving the cluster $\mathcal{G}$ to reduce IL
10:             $\mathcal{X}_1 \leftarrow$ temporarily reassign each record in $\mathcal{G}$ to its nearest cluster in $\mathcal{X}$
11:             **if** IL of $\mathcal{X}$ ¿ IL of $\mathcal{X}_1$ **then**
12:                 $\mathcal{X} \leftarrow \mathcal{X}_1$ ▷ Dissolve the cluster $\mathcal{G}$ to improve IL
13:             **end if**
14:         **end if**
15:     **end for**
16:     **for all** cluster $\mathcal{G} \in \mathcal{X}$ **do**
17:         **if** $|\mathcal{G}| \geq 2k$ **then**
18:             $\mathcal{N} \leftarrow$ ONA($\mathcal{G}, k$)   ▷ Apply ONA recursively to split $\mathcal{G}$ into $\mathcal{N}$
19:             $\mathcal{X} \leftarrow \{\mathcal{X} \setminus \mathcal{G}\} \cup \mathcal{N}$
20:         **end if**
21:     **end for**
22: **end while**
23: **return** $\mathcal{X}$

---

2   Standard k-Anonymization Algorithms

Various techniques have been proposed to achieve k-anonymity, including generalization, suppression [12,16], noise addition [3], and microaggregation [4,5].

As anonymization modifies original data to protect privacy, it inevitably causes some data distortion. This trade-off is quantified using the IL metric:

$$\text{IL} = \frac{\sum_{i=1}^{N} \sum_{j=1}^{d} (x_{ij} - \bar{x}_{c(i)j})^2}{\sum_{i=1}^{N} \sum_{j=1}^{d} (x_{ij} - \bar{x}_j)^2} \tag{1}$$

where x_{ij} is the j-th attribute of record i, $\bar{x}_{c(i)j}$ is the cluster mean, and $\bar{x}_j$ is the global mean. IL reflects the ratio of within-cluster to total variance; smaller IL implies higher utility.

Among existing methods, microaggregation offers a good balance between privacy and utility and is widely used. It can be viewed as a constrained clustering problem: each cluster must have at least k records, and IL is minimized. As this problem is NP-hard [1,4], heuristic algorithms are commonly employed.

We adopt MDAV$^+$ [14] and ONA [15] as baseline algorithms (Algorithms 1 and 2). Both take a dataset $\mathcal{D}$ and anonymity parameter k and return a set of anonymized clusters $\mathcal{X}$, where each record is replaced with the cluster mean to ensure k-anonymity. MDAV$^+$ is a fast heuristic that repeatedly selects the record farthest from the centroid and forms a cluster with its $k - 1$ nearest

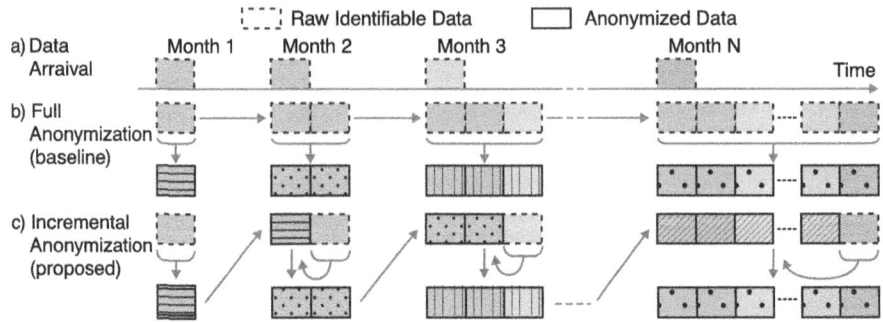

Fig. 1. Overview of the proposed incremental framework. (a) Raw data is appended periodically. (b) Conventional methods re-anonymize the entire dataset upon each update, increasing cost. (c) Our framework incrementally anonymizes new batches using prior results, reducing computation.

neighbors, assigning the rest to their closest clusters. Its complexity is $O(dN^2)$ for d attributes and N records. ONA (Near-Optimal microAggregation) typically achieves lower IL by refining random initial clusters through reassignment, dissolution, and recursive splitting. Its complexity is higher: $O(d(N^2 + Nk^2))$ [18]. We denote these procedures as $\mathrm{MDAV}^+(\mathcal{D}, k)$ and $\mathrm{ONA}(\mathcal{D}, k)$, returning the anonymized cluster set $\mathcal{X}$ for a given input $\mathcal{D}$ and parameter k.

3 Proposed Incremental k-Anonymization Framework

This study targets append-only scenarios and proposes an incremental framework. In practice, data often arrives in periodic batches (e.g., monthly). Conventional methods re-anonymize the entire dataset upon each update, causing computation to grow with data volume (Fig. 1.a–b). In contrast, our framework incrementally processes only the new batch $\mathcal{B}_i$ using the previous anonymized set $\mathcal{X}_{i-1}$, reducing cost significantly (Fig. 1.c).

The framework updates $\mathcal{X}_{i-1}$ incrementally with a new batch $\mathcal{B}_i$. We propose two strategies:

Incremental Append: Assign each record in $\mathcal{B}_i$ to a suitable cluster in $\mathcal{X}_{i-1}$, then split any cluster exceeding size $2k$.

Incremental Merge: First anonymize $\mathcal{B}_i$ using MDAV or ONA to obtain $\mathcal{X}'_i$, then merge it with $\mathcal{X}_{i-1}$ and split oversized clusters if necessary.

Initially, $\mathcal{B}_0$ is anonymized using $\mathrm{MDAV}(\mathcal{B}_0, k)$ or $\mathrm{ONA}(\mathcal{B}_0, k)$ to obtain $\mathcal{X}_0$. For $i \geq 1$, Append assigns each record in $\mathcal{B}_i$ to $\mathcal{X}_{i-1}$ and adjusts oversized clusters, yielding $\mathcal{X}_i$. Merge anonymizes $\mathcal{B}_i$ independently, merges with $\mathcal{X}_{i-1}$, and applies the same adjustment. Repeating this process maintains k-anonymity efficiently over growing data.

4   Experiments

Due to space constraints, we report only the core contributions and key findings. Detailed analysis and theoretical considerations are left for future work.

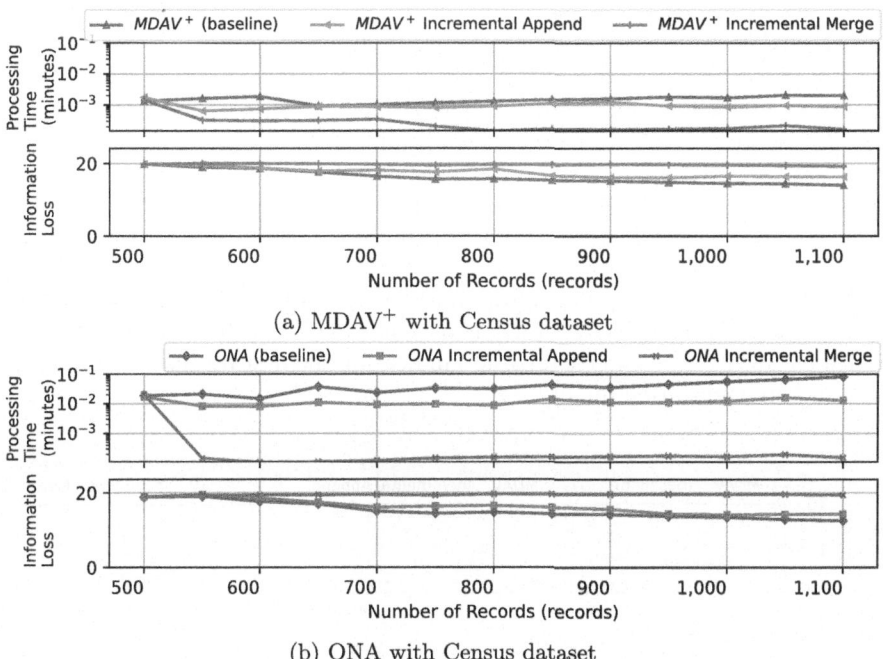

(a) MDAV$^+$ with Census dataset

(b) ONA with Census dataset

Fig. 2. Comparison of processing time and (IL) using the Census dataset. (a) shows results for MDAV$^+$, and (b) for ONA. Baselines are shown with blue in (a) and red in (b); proposed incremental-append and incremental-merge strategies are shown with orange/green in (a), and purple/brown in (b), respectively. The MDAV$^+$ baseline is about 10 times faster than the ONA baseline, with only a slightly higher IL. The proposed MDAV$^+$ incremental-append reduces processing time with only a minimal increase in IL. (Color figure online)

We evaluated processing time and IL using two datasets:

Open dataset (Census) A public dataset widely used in prior evaluations [17].

Real-world Medical dataset (Diabetes) A large-scale dataset of 470,000+ diabetic patients, collected over 108 months (Apr. 2014–Mar. 2023) by Japan's national health systems. It contains no explicit identifiers such as names or contact details, and vectorized into 13 numerical features (e.g., number of medical procedures, dosage of drugs, and healthcare costs).

We simulated a batched data accumulation scenario and measured processing time and IL per batch. Processing time includes only the core anonymization process, excluding data I/O. The anonymity parameter was fixed at $k = 10$. Experiments were run on a Linux with dual Intel Xeon Gold 6132 (2.6 GHz) CPUs and 96 GB RAM. We implemented all components in C++.

5   Results

Figures 2a and 2b show processing time and IL on the Census dataset, while Figs. 3a and 3b present results on real-world medical data (470k+ records). The proposed Incremental Append consistently reduces processing time with minimal IL increase. For example, in Fig. 3a, the baseline method required 44.1 min, whereas Incremental Append completed in just 1.49 min—approximately 30 times faster—with IL increasing by only 8%. Overall, the trends are consistent: Baseline < Append < Merge for IL, and Merge < Append < Baseline for processing time. Figure 4 shows the trade-off curve. MDAV$^+$ Incremental Append appears nearest to the lower-left corner, indicating the best balance of processing time and anonymization quality.

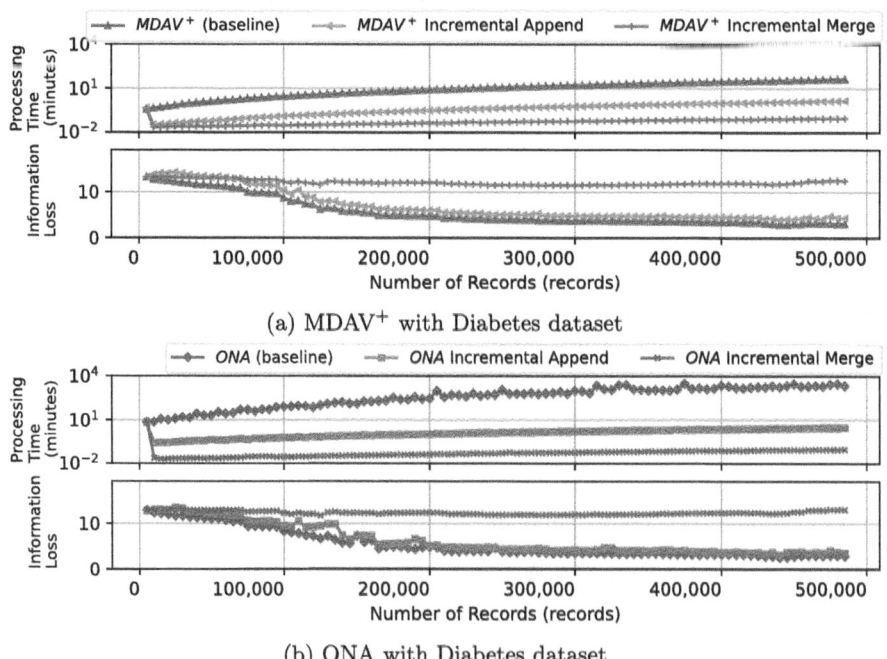

(a) MDAV$^+$ with Diabetes dataset

(b) ONA with Diabetes dataset

Fig. 3. Processing time and IL on diabetes dataset (470k+ records). Baseline (MDAV$^+$, blue) required 44.1 min, whereas the proposed (Incremental Append, green) completed in 1.49 min with comparable IL. (Color figure online)

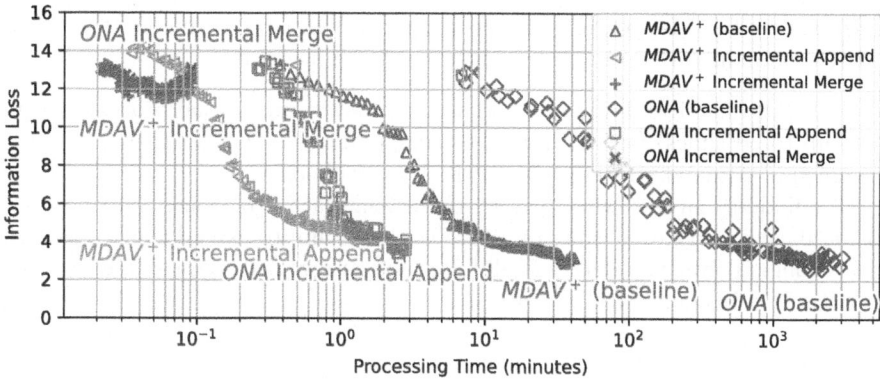

Fig. 4. Trade-off between processing time and IL. Each point represents one batch. MDAV$^+$ Incremental Append achieves the best overall balance.

We acknowledge that a formal analysis of convergence and optimality remains to be established. Preliminary results suggest promising directions, and we plan to elaborate on these findings in a follow-up study. These limitations do not affect the empirical conclusions drawn in this study.

6   Discussion and Limitations

This study proposed an incremental k-anonymization framework that significantly reduces processing time while maintaining data utility. Experiments with large-scale medical data demonstrated that our approach achieves up to 30 times speedup with minimal IL increase, enabling timely updates in real-world settings, such as healthcare.

Prior studies have examined incremental anonymization: [10] focused on privacy leakage in sequential releases but fouced on small datasets, [2] proposed a healthcare method without IL evaluation, and [11] addressed streaming data with latency constraints. In contrast, our batch-wise framework targets practical big data pipelines.

However, two key limitations remain: 1) the lack of analytical processing time bounds under varying data conditions, and 2) limited evaluation of how data distribution (e.g., skewness) affects clustering and IL. Addressing these points is left for the future theoretical and empirical work.

7   Conclusion

We introduced an incremental k-anonymization framework addressing growing datasets, along with two strategies:

Incremental Append Adds new records to existing anonymized clusters, then splits if oversized.

Incremental Merge Anonymizes new records separately, merges them into existing anonymized clusters, and then splits if oversized.

On real-world medical data, our framework achieved substantial processing savings (up to 30 times) with IL increases kept under 8%. Among them, MDAV$^+$ Incremental Append strategy offered the best trade-off.

Our findings suggest that privacy-preserving data processing must account for database growth over time. In future work, we aim to extend this framework with stronger privacy models (e.g., differential privacy) and adaptive mechanisms for data distribution shifts.

Acknowledgments. This work was in part supported by Cross-ministerial Strategic Innovation Promotion Program (SIP) on "Integrated Health Care System". We gratefully acknowledge the support of Dr. Naohiro Mitsutake of the Institute for Healthcare Economics and Research in handling the diabetes dataset.

References

1. Aggarwal, C.C.: On k-anonymity and the curse of dimensionality. In: Proceedings of International Conference on Very Large Data Bases, pp. 901–909 (2005)
2. Aldeen, Y.A.A.S., Salleh, M., Aljeroudi, Y.: An innovative privacy preserving technique for incremental datasets on cloud computing. J. Biomed. Int. **62**, 107–116 (2016)
3. Brand, R.: Microdata Protection through noise addition. In: Domingo-Ferrer, J. (ed.) Inference Control in Statistical Databases. LNCS, vol. 2316, pp. 97–116. Springer, Heidelberg (2002). https://doi.org/10.1007/3-540-47804-3_8
4. Domingo-Ferrer, J., Mateo-Sanz, J.M.: Practical data-oriented microaggregation for statistical disclosure control. Trans. Knowl. Data Eng. **14**(1), 189–201 (2002)
5. Domingo-Ferrer, J., Torra, V.: Ordinal, continuous and heterogeneous k-anonymity through microaggregation. Data Min. Knowl. Disc. **11**(2), 195–212 (2005)
6. Hirayama, K., Kanda, N., et al.: The five-year trends in antibiotic prescription by dentists and antibiotic prophylaxis for tooth extraction: a region-wide claims study in Japan. J. Infect. Chemother. **29**(10), 965–970 (2023)
7. Kanda, N., Hashimoto, H., et al.: Indirect impact of the COVID-19 pandemic on the incidence of non–COVID-19 infectious diseases: a region-wide, patient-based database study in Japan. Public Health **214**, 20–24 (2023)
8. LeFevre, K., DeWitt, D.J., Ramakrishnan, R.: Mondrian multidimensional k-anonymity. In: Proceedings of International Conference on Data Engineering, pp. 25–25 (2006)
9. Ministry of Health, Labour and Welfare (Japan): Data-based health management initiatives roadmap (2021). https://www.mhlw.go.jp/english/policy/health-medical/data-based-health/dl/211124-01.pdf
10. Pei, J., Xu, J., et al.: Maintaining K-anonymity against incremental updates. In: Proceedings of International Conference on Science and Statistics Database Management, p. 5 (2007)
11. Rebollo-Monedero, D., HernáNdez-Baigorri, C., et al.: Incremental k-anonymous microaggregation in large-scale electronic surveys with optimized scheduling. IEEE Access **6**, 60016–60044 (2018)

12. Samarati, P.: Protecting respondents identities in microdata release. Trans. Knowl. Data Eng. **13**(6), 1010–1027 (2001)
13. Samarati, P., Sweeney, L.: Protecting privacy when disclosing information: K-anonymity and its enforcement through generalization and suppression. Technical Report SRI-CSL-98-04 (1998)
14. Solanas, A., Martínez-Ballesté, A.: V-MDAV: a multivariate microaggregation with variable group size. In: Proceedings of Computer Statistics 17th Symposium (2006)
15. Soria-Comas, J., Domingo-Ferrer, J., Mulero, R.: Efficient near-optimal variable-size microaggregation. In: Proceedings of Modeling Decisions for Artificial Intelligence, pp. 333–345 (2019)
16. Sweeney, L.: Achieving K-anonymity privacy protection using generalization and suppression. Int. J. Unc. Fuzz. Knowl. Based Syst. **10**(05), 571–588 (2002)
17. Thaeter, F., Reischuk, R.: Scalable k-anonymous microaggregation: exploiting the tradeoff between computational complexity and information loss:. In: Proceedings of International Conference on Security Cryptography, pp. 87–98 (2021)
18. Thaeter, F., Reischuk, R.: Improving time complexity and utility of k-anonymous microaggregation. E-Bus. Telecommun. 195–223 (2023)
19. Yoshimoto, H., Mitsutake, N., Goda, K.: Predicting medical event occurrence using medical insurance claims big data. In: Proceedings of World Congress Medical Health Information, vol. 310, pp. 654–658 (2024)

Post Quantum Cryptographic Schemes and Libraries Selection

Shubhro Roy⬤, Mangesh Gharote$^{(\boxtimes)}$⬤, Pankaj Sahu⬤, Sutapa Mondal⬤,
M. A. Rajan⬤, and Sachin Lodha⬤

Tata Consultancy Services Limited, TCS Research, Mumbai, India
{shubhro.roy1,mangesh.g,pankajkumar.sahu,
sutapa.mondal,rajan.ma,sachin.lodha}@tcs.com
https://www.tcs.com/what-we-do/research

Abstract. With quantum computers likely to become a reality, applications that rely on classical cryptographic schemes will become vulnerable to attacks. Classical cryptographic schemes such as RSA and ECDSA, which form the backbone of enterprise application security, will be exposed to threats and need to be replaced with Post-Quantum Cryptographic (PQC) schemes, such as Kyber and Dilithium. Several libraries offer implementations of different PQC schemes, but vary in computational overhead, and other supporting features. These factors make selecting the appropriate PQC schemes and libraries a complex decision-making problem. In this work, we propose a comprehensive solution methodology to assist enterprises in selecting the best PQC schemes and libraries by comparing multiple schemes from different categories across several libraries, considering multiple performance criteria.

Keywords: Post Quantum Cryptography · PQC Schemes and Libraries · Combinatorial Optimization · Multi-Objective Optimization

1 Introduction

Enterprise applications rely on cryptographic algorithms to secure sensitive data, protecting them against both insider and external threats. These algorithms ensure integrity, confidentiality, authentication and non-repudiation of enterprise data. These schemes such as RSA, ECC, and AES are designed to withstand attacks from classical computers. However, with the advent of quantum computers, there is a serious threat to enterprise security, as quantum algorithms such as Shor's algorithm and Grover's algorithm can potentially break or weaken traditional cryptographic systems [5].

Despite the current unavailability of functional quantum computers, enterprises must begin planning to migrate the existing cryptographic applications to Post Quantum Cryptography (PQC). This need arises from vulnerabilities associated with 'Harvest Now, Decrypt Later' (HNDL) attacks, as well as the mandatory compliance requirements set by the National Institute of Standards

R. Wrembel et al. (Eds.): DEXA 2025, LNCS 16047, pp. 338–343, 2026.
https://doi.org/10.1007/978-3-032-02088-8_26

and Technology (NIST), which must be met by 2035. NIST has approved four PQC schemes after testing them for performance and vulnerability.

However, more libraries with *pqc* scheme implementations are emerging, including open-source options (e.g., Liboqs, OpenSSL), commercial solutions, and vendor-provided options. These libraries support multiple *pqc* schemes, and their overall performance may vary across libraries [2]. Thus, when selecting *pqc* schemes and libraries, factors such as application-library compatibility, Performance Objective Metrics (POM), and application security requirements, must be considered.

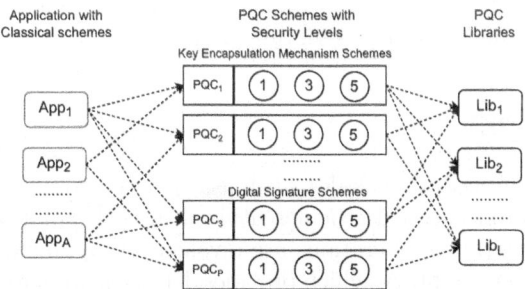

Fig. 1. PQC Library and PQC Scheme Selection Problem

Figure 1 depicts the complexity of the *pqc* selection problem. Consider an enterprise with ($|A|$) applications, consisting of ($|C|$) classical schemes, that need to be replaced with ($|P|$) *pqc* schemes of the same category, supported by ($|L|$) libraries. The total number of combinations to compare is $|P * L|^{|A*C|}$. The evaluation of the best scheme and library selection must be based on multiple criteria and conflicting performance measures, which makes the problem challenging. Hence, in this paper, we propose a novel solution methodology for *pqc* library and scheme selection to generate near-optimal solution in a short time. Our primary contribution includes: (i) A comprehensive methodology for *pqc* scheme and library selection, (ii) Two-phase selection methodology based on multiple performance criteria, (iii) A function - Lowest Bound Maximizer (LBM) to compare two *pqc* schemes with multiple performance criteria.

2   Literature Review

Organizations, such as the European Telecommunications Standards Institute (ETSI), and frameworks like Cryptographic Algorithm Requirements and Framework (CARAF) [7], provide general guidance on transitioning from current cryptographic systems to PQC systems. The steps for migration planning include cryptographic asset identification, security classification, risk assessment, prioritization, remediation, and migration timeline planning [3,6].

Ojetunde et al. [4] proposed a rule-based methodology for selecting *pqc* schemes in 5G applications. Wohlwender et al. [8] developed a tool for context-based *pqc* library selection, using an index that considers 15 library attributes, such as ease of use, security, and documentation. Additionally, the findings by Acar et al. [1] suggest that documentation, code examples, and secure key storage significantly impact the usability of different cryptographic libraries.

Most prior studies, including guidelines provided by organizations (ETSI) and frameworks (CARAF), offer no specific recommendations beyond generic guidelines on *pqc* migration. Hence, there is a need for a comprehensive solution methodology to identify the most suitable *pqc* schemes and libraries, by considering performance metrics, security, and other application requirements. The proposed solution methodology is discussed in the next section.

3   Methodology

The input to the procedure includes, such as: a) A crypto inventory, which provides information on classical schemes utilized in each application, number of times these cryptographic operations are invoked during application execution; b) Application asset inventory, which provides details on application security and other requirements, such as programming language and operating system support; c) Data repository of *pqc* schemes and libraries including various features offered by each library and vendor, with the benchmarked performance data for *pqc* scheme implementations by these library and vendors. Since commercial library performance data is not public, synthesized data can illustrate outcomes. However, actual analysis requires real data from contracted customers. Thus, the library data is synthesized to facilitate the demonstration of experimental outcomes. The proposed comprehensive solution methodology for *pqc* scheme and library selection involves the following four steps.

Step 1 - Feature based Candidate Library Selection: For each application, we identify compatible libraries and eliminate non-compatible ones. The compatibility score between an application (app_i) and a library (lib_l) denoted by ($comp(app_i, lib_l)$) is computed as follows:

$$comp(app_i, lib_l) = \prod_{x=1}^{X} R(app_i, Re_x.r_z) \cdot F(lib_l, Re_x.r_z) \tag{1}$$

A library is considered compatible with an application, if the compatibility score $comp(app_i, lib_l) == \prod_{x=1}^{X} |R(app_i, Re_x.r_z)|$. Otherwise, the library is eliminated from consideration. Here,

- Re_x denotes a stringent requirement attribute (like Language Specification).
- r_z denotes different options available under a given stringent requirement. For example, under the stringent requirement of 'Language Specification', the possible options include {r_1: Java, r_2: Python, r_3: C++}.

- $R(app_i, Re_x.r_z)$ represents the application's requirement for specific options under the stringent requirement. For example, if the application requires Java and Python, it is expressed as $R(app_i, Re_x.r_z) = [1, 1, 0]$
- $F(lib_l, Re_x.r_z)$ denotes the stringent requirement options supported by a library. If a library supports all three programming languages, it is represented as: $F(lib_l, Re_x.r_z) = [1, 1, 1]$
- $|R(app_i, Re_x.r_z)|$ denotes the number of options required by the application under a stringent requirement. For example, if the application requires Java and Python for 'Language Specification', then $|R(app_i, Re_x.r_z)| = 2$.

Step 2 - Shortlisting of PQC Schemes based on Security Level: We first shortlist *pqc* schemes (ps_j) of the same category - KEM, SIG, or SYM - from each candidate library (lib_l) based on application's security requirement (λ_{req}). This is done for each classical scheme category in an application $(app.cat.cs)$. The shortlisted schemes must meet the application's security requirement.

Step 3 - Performance-based PQC Scheme and Library Selection: We design the Performance Objective Metrics (POM) values such as, computation cost, communication cost, execution time and memory usage, for comparing *pqc* schemes to aid in selection. The computation of each POM term is explained as follows.

Computation Cost (C_{comp}^a): It is the consumption of computational resources such as CPU time and memory by an application, post replacement with *pqc* schemes.

$$C_{comp}^a = C_{time}^a * C_{mem}^a \tag{2}$$

Computation Time (C_{time}^a): It is the amount of time required by a *pqc* scheme to perform various cryptographic operations. It includes key generation, encryption, decryption, signing, verification, hashing, and authentication operations. The term N_{ik}^α represents the number of calls for cryptographic operation k of the classical scheme i in application a. The symbol τ_{jkl} denotes the time required to perform operation k using *pqc* scheme j on *pqc* library l. It is computed as follows:

$$C_{time}^a = \sum_{i \in S_a} \sum_{k \in O_i} \sum_{j \in P} \sum_{l \in L} N_{ik}^\alpha * \tau_{jkl} \tag{3}$$

Computation Memory (C_{mem}^a): It is the memory required by a *pqc* scheme during execution of various cryptographic operations. The term μ_{jkl} denotes the memory consumption for operation k of the *pqc* scheme j on *pqc* library l. It is computed as follows:

$$C_{mem}^a = \sum_{i \in S_a} \sum_{k \in O_i} \sum_{j \in P} \sum_{l \in L} N_{ik}^\alpha * \mu_{jkl} \tag{4}$$

Communication Cost (C_{comm}^a): It is the amount of data transmitted between the user and the system during execution of cryptographic operations in applications. Data transmission involves transfer of public key (C_{pk}^a) and ciphertext (C_{ct}^a).

$$C_{comm}^a = C_{pk}^a + C_{ct}^a \tag{5}$$

Public Key Size (C_{pk}^a): The public key is the key generated by *pqc* schemes (KEM and SIG) that is made publicly available to the user for encrypting data and verifying signatures. The term κ_j denotes the public key size of *pqc* scheme j.

$$C_{pk}^a = \sum_{i \in S_a} \sum_{k \in O_i} \sum_{j \in P} \sum_{l \in L} N_{ik}^\alpha * \kappa_j \tag{6}$$

Ciphertext and Signature Size (C_{ct}^a): The ciphertext is the data generated by a key encapsulation mechanism (KEM) scheme. The signature size refers to the amount of data required to represent a cryptographic signature in a digital signature (SIG) scheme. The term γ_j denotes the ciphertext/signature size of *pqc* scheme j.

$$C_{ct}^a = \sum_{i \in S_a} \sum_{k \in O_i} \sum_{j \in P} \sum_{l \in L} N_{ik}^\alpha * \gamma_j \tag{7}$$

Step 4 - Two Phase PQC Scheme and PQC Library Selection: In Phase 1, for each candidate library, the best *pqc* scheme is selected. This selection is done for each classical scheme category (KEM, SIG, SYM) within an application. This selection is done using the Lowest Bound Maximizer (LBM) function (Algorithm 1). LBM compares multiple schemes with multiple objectives in linear time.

At the end of Phase 1, we obtain a list of best *pqc* scheme from each candidate library, for each classical category. But only one *pqc* scheme must be assigned to each classical scheme in an application. Hence, in Phase 2, the best *pqc* scheme is selected by comparing Phase 1 selected *pqc* schemes (of same category) from different candidate libraries. The LBM method is used again to identify the most optimal *pqc* scheme and library in Phase 2.

Lowest Bound Maximizer (LBM). The procedure (Algorithm 1) compares two *pqc* schemes (PS_x, PS_y) based on multiple performance objective values. The objective (o) values for the two schemes are denoted by (obj_o^x, obj_o^y). For example, if compared to the second (y) scheme, the first (x) scheme's performance is better (i.e. lower objective value) for the first objective measure (i.e. $o = 1$), the LBM score is penalized by a factor $(1 - \psi/\phi)$ to discourage selecting the second scheme. It indicates that selecting the second scheme might degrade overall performance. But if, for the same objective, the second scheme performs better, then the LBM score is rewarded by a factor (ψ/ϕ) to encourage selecting the second scheme. It indicates that selecting the second scheme might improve overall performance.

The final LBM score is the sum of all penalties and rewards for all objectives. If the final LBM score is positive, the second scheme becomes the current best. The current best will then be compared with the next scheme by following the same procedure. The process is repeated for all the candidate libraries, ensuring that the best *pqc* schemes across all libraries are identified.

Algorithm 1. Lowest Bound Maximizer

1: **Input:** PQC schemes: (PS_x, PS_y), POM values
2: Initialize LBM score: $\partial = 0$
3: **for** $o = 1$ to $|O|$ **do**
4:     $\phi = obj_o^x$, $\psi = |obj_o^x - obj_o^y|$
5:     **If** $obj_o^x < obj_o^y$: $\partial = \partial + (1 - \psi/\phi)$
6:     **Else** $\partial = \partial + \psi/\phi$
7: **end for**
8: **Return** ∂

4   Summary

In this paper, we propose a solution methodology for selecting *pqc* schemes and libraries based on enterprise application requirements. Our methodology enables the comparison of several available libraries and *pqc* scheme implementations, considering application requirements and performance metrics.

Our proposed methodology could be further enhanced by including additional factors such as the ease of *pqc* implementations, optimized implementations, documentation, community support, and compliance certifications. Furthermore, the subscription costs of commercial libraries may influence the selection of the *pqc* scheme and library. We plan to incorporate these factors into our future research.

References

1. Acar, Y., Backes, M., Fahl, S., Garfinkel, S., Kim, D., Mazurek, M.L., Stransky, C.: Comparing the usability of cryptographic apis. In: 2017 IEEE Symposium on Security and Privacy (SP), pp. 154–171. IEEE (2017)
2. Dam, D.T., Tran, T.H., Hoang, V.P., Pham, C.K., Hoang, T.T.: A survey of post-quantum cryptography: Start of a new race. Cryptography **7**(3), 40 (2023)
3. Joseph, D., Misoczki, R., Manzano, M., Tricot, J., Pinuaga, F.D., Lacombe, O., Leichenauer, S., Hidary, J., Venables, P., Hansen, R.: Transitioning organizations to post-quantum cryptography. Nature **605**(7909), 237–243 (2022)
4. Ojetunde, B., Kurihara, T., Yano, K., Sakano, T., Yokoyama, H.: A selection technique for effective utilization of post-quantum cryptography in 5g application and beyond. In: 2023 15th International Conference on Computer and Automation Engineering (ICCAE), pp. 551–558. IEEE (2023)
5. Sahu, S.K., Mazumdar, K.: State-of-the-art analysis of quantum cryptography: applications and future prospects. Front. Phys. **12**, 1456491 (2024)
6. Thakur, M.S.D., Vidhani, K., Syed, H.B., Rajan, M.: Enterprise post quantum cryptography migration tools. In: 2024 16th International Conference on COMmunication Systems & NETworkS (COMSNETS), pp. 327–329. IEEE (2024)
7. White, B., et al.: Transitioning to quantum-safe cryptography on IBM Z. IBM Redbooks (2023)
8. Wohlwender, J., Huesmann, R., Heinemann, A., Wiesmaier, A.: *crypto_{lib}*: Comparing and selecting cryptography libraries (long version of eicc 2022 publication). arXiv preprint arXiv:2203.16370 (2022)

Benchmarks and Surveys

Workload-Based Clustering of Large Number of Database-as-a-Service Instances

Maciej Zakrzewicz[✉]

Poznan University of Technology, Poznan, Poland
maciej.zakrzewicz@put.poznan.pl

Abstract. Managing tens of thousands of Database-as-a-Service instances presents significant challenges in fine-tuning their configurations to optimize database workload performance. Since individually tuning each instance is impractical, DBAs typically rely on universal, one-for-all configuration templates, which often fail to meet the specific requirements of certain instances. In this paper we explore methods that DBAs could use to automatically divide the Database-as-a-Service instances into a small, manageable number of clusters with similar performance-to-configuration profiles. In this way, each group could share a single configuration template to maximize performance of its instances. We investigate whether such Database-as-a-Service instance clustering can be based solely on workload query frequencies. We validate our approach through an extensive set of experiments.

Keywords: database performance · clustering · workload characterization

1 Introduction

Optimizing database performance is a complex and demanding task for DBAs, requiring a deep understanding of database architecture, query execution, indexing strategies, and system resource management. Performance tuning involves analyzing slow queries, optimizing indexes, adjusting configurations, and balancing workload distribution. Furthermore, DBAs must constantly adapt to evolving application demands, hardware limitations, growing databases, making optimization an continuous process rather than a one-time fix. When a DBA is responsible for a large number of databases, maintaining their optimal performance becomes nearly impossible.

Consider a cloud-based database application deployed in the form of tens of thousands instances/deployments, handling user workloads of various query intensities (Fig. 1). The database schemas and query sets are identical across all the instances, because users run a set of predefined template queries. Differences between the instances include: data sizes and data distributions, query rates, query distributions, numbers of sessions (concurrency level), read/write ratios, numbers of records/blocks read/written, etc. These metrics may gradually change over time. Performance of the database query workloads heavily depends on the configuration settings applied to the database

© The Author(s), under exclusive license to Springer Nature Switzerland AG 2026
R. Wrembel et al. (Eds.): DEXA 2025, LNCS 16047, pp. 347–359, 2026.
https://doi.org/10.1007/978-3-032-02088-8_27

servers, including cache sizes, parallelism degrees, query optimizer hints, checkpointing frequency, and more. Fine tuning of the configuration parameters is one of DBA's responsibilities.

Due to a large number of deployments, it is not feasible for DBAs to individually optimize (fine tune) database server configurations for each of those deployments. Therefore, a common approach is to use the same generalized database server configuration settings across all deployments. This is far from optimal, but seems to be a reasonable solution due to the scale of the problem.

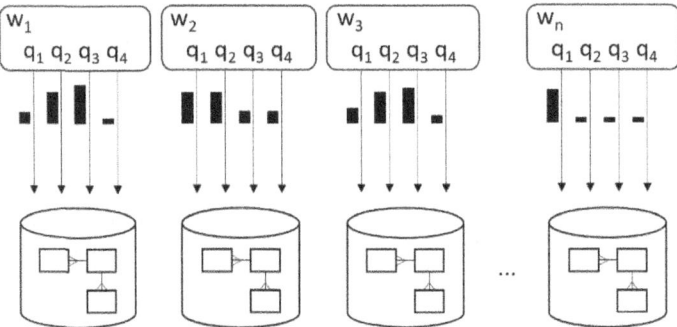

Fig. 1. Cloud-based database application deployment

The main drawback of the one-for-all database configuration template approach is that when the database configuration requires fine tuning to improve performance of one specific kind of workload, other kinds of workload may be hurt with performance degradation. For example, adjusting RAM memory allocations between the Buffer Cache and Sort Memory may optimize either DSS queries or OLTP commands while negatively impacting the other. The final result of such fine tuning will depend on the actual shares of DSS and OLTP operations in the database workload. See [16] for an example of a complex behavior of a TCP-H query performance in the context of just two PostgreSQL configuration parameters. Figure 2 shows an example of how identical database configuration changes can differently impact the performance of various query workloads executed using the same database size, same database server, same machine: w_i are different workloads based on the same query set, and the bars show total accumulated response times observed for different database configurations, represented by colors. For example, changing from the *blue* configuration to *orange* improves performance of workloads w_2, w_3, w_4, w_6, w_7, w_{10} while degrading performance of workloads w_1, w_8, w_9. If we were able to guess which workloads would follow a consistent performance-to-configuration profile, we might use separate, dedicated database configuration templates for groups of deployments that execute those workloads. Unfortunately, verification of performance-to-configuration profiles of the workloads would require extensive calibration by replaying the workloads against a number of database configurations. It could be done for a small sample of workloads/deployments but definitely not for tens of thousands of them.

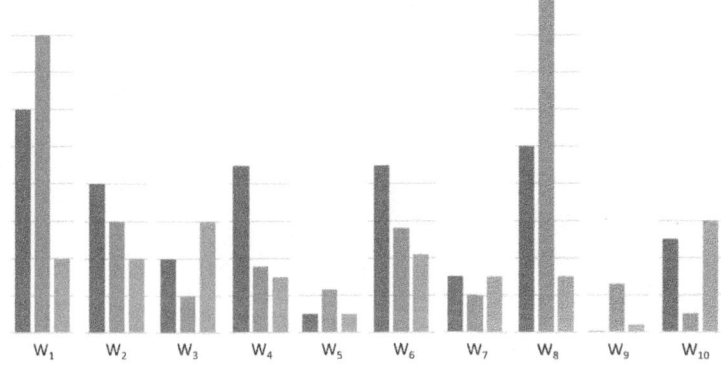

Fig. 2. Database workload performance responses to identical configuration changes

In this paper we introduce the problem of partitioning a set of database server instances into a manageable number of *management groups* such that each group is characterized by a consistent performance-to-configuration profile and therefore it can use a single database configuration template. Due to this partitioning, future changes to database configuration templates will uniformly impact all workloads in a group, eliminating the need for DBAs to balance performance trade-offs. We propose unsupervised and supervised learning approaches to create ML models that assign instances/deployments to management groups based on their workload properties. We experimentally verify the effectiveness of the analyzed approach.

Our Approach. We describe database server's performance by means of total accumulated time spent by database server actively working on queries within a time unit (inspired by [17]). Two database workloads have similar performance-to-configuration profiles if their performances are similarly affected by identical database configuration changes, i.e. switching to another database configuration either causes both workloads to run similarly faster or causes both workloads to run similarly slower. Based on such workload similarity definition, we cluster database workloads into management groups. See Fig. 3 for an example of such clustering - the "shape" similarities visually demonstrate performance-to-configuration profile similarities.

In a real-life scenario, however, we do not collect performance-to-configuration profiles for all deployments and all possible database configuration changes. The straightforward idea of performing shape-based clustering of the workloads/deployments is hence not feasible. What we can easily collect are the numbers of database workload queries executed and their respective execution times. We argue that performance-to-configuration profiles are statistically connected to distribution of query execution frequencies. Therefore, we use the workloads' distribution of query execution frequencies to split the set of their database servers into a small number of *management groups*.

Our choice to base clustering solely on workload query frequencies is intended to eliminate any configuration bias. Database queries can be characterized by various performance metrics that, unfortunately, depend on the current database configuration. When the configuration changes, the metrics - such as query execution times, the number

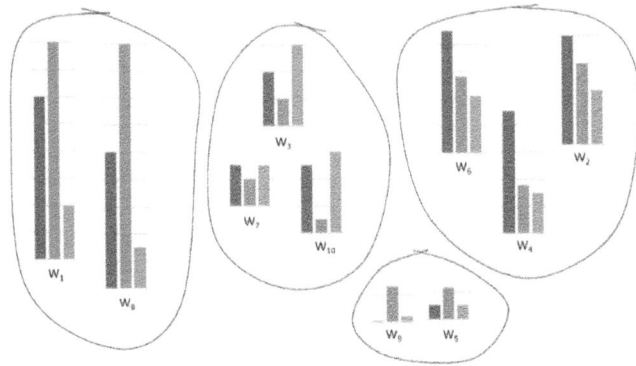

Fig. 3. Sample clustering of database workloads based on performance-to-configuration profiles

of I/O operations, in-memory sort percentages, query execution plan costs, parallelization levels - may also change. In this context, query execution frequencies appear to be
the only objective metric.

2   Basic Definitions and Problem Formulation

We define a *Database Configuration* c as a mapping of tunable database server parameters to their assigned values: $c = \{(p_1, v_1), (p_2, v_2),..., (p_k, v_k)\}$, where p_i represents a
configuration parameter (e.g. Buffer cache size, sort memory size) and v_i represents the
parameter's value.

Example. The the following database configuration contains three tunable parameters,
buffer_cache_size = 128MB, sort_memory_size = 4MB, max_parallel_workers = 16:

$$c_1 = \{(buffer\_cache\_size, 128MB), \ (sort\_memory\_size, 4MB), \ (max\_parallel\_workers, 16))\}$$

A *Database Workload Template* T is defined as $T = \{q_1, q_2,..., q_n\}$, where q_i is a
database query. A *Database Workload* w is defined as $w = \{f_1, f_2,..., f_n\}$, where f_i is a
frequency of appearance of q_i from T in workload w (percentage).

$$\sum_{i=1}^{n} f_i = 1$$

A *Performance Response* $P(w,c)$ for the database workload w executed using the
database configuration c is defined as the total weighted accumulated time spent by
database server actively working on queries:

$$P(w, c) = \sum_{i=1}^{n} f_i \times t(q_i, c)$$

where $t(q_i,c)$ is the average execution time of the query q_i when using the database
configuration c.

Example. The following example presents calculation of a performance response for the given database workload template T, database workload w, and sample execution times for current database configuration.

$T = \{$*"insert into customers..."*, *"select sum(quantity) from orders..."*$\}$.
$w = \{75\%, 25\%)\}$.
$t($*"insert into customers..."*$, c_1) = 0.17$ s.
$t($*"select sum(quantity) from orders..."*$, c_1) = 5.3$ s.
$P(w,c_1) = 75\% \times 0.17\ s + 25\% \times 5.3\ s = 1.4525\ s$.

We define a measure $d(w_A, w_B)$ to represent similarity of normalized performance responses of two database workloads executed using various database configurations c_i.

$$d(w_A, w_B) = \sqrt{\sum_i \left(\frac{P(w_A, c_i)}{\sum_k P(w_A, c_k)} - \frac{P(w_B, c_i)}{\sum_k P(w_B, c_k)}\right)^2}$$

A low value of $d(w_A, w_B)$ indicates that the workloads w_A and w_B respond to database configuraton changes in a very similar way. Configuration tuning techniques which have been successful on the database server that handles w_A can be directly transferred to the database server that handles w_B. Therefore, both database servers should belong to the same management group.

A high value of $d(w_A, w_B)$ means that the workloads w_A and w_B respond very differently to database configuraton changes. Configuration tuning techniques are not compatible and both database servers require separate DBA attention. The database servers should not belong to the same management group.

We will use the measure $d()$ to asses the quality of clustering based on distribution of query execution frequencies.

Example. The following example illustrates calculation of $d(w_A, w_B)$ using database configurations c_1 and c_2.

$c_1 \quad = \quad \{($*buffer_cache_size*, $\quad$ *128MB*$)$, $\quad$ $($*sort_memory_size*, $\quad$ *4MB*$)$, $($*max_parallel_workers*, *16*$))\}$.
$c_2 \quad = \quad \{($*buffer_cache_size*, $\quad$ *64MB*$)$, $\quad$ $($*sort_memory_size*, $\quad$ *8MB*$)$, $($*max_parallel_workers*, *16*$))\}$.

$T = \{$*"insert into customers..."*, *"select sum(quantity) from orders..."*$\}$.
$P(w_A,c_1) = 75\% \times 0.17\ s + 25\% \times 5.3\ s = 1.4525\ s$.
$P(w_A,c_2) = 75\% \times 0.24\ s + 25\% \times 8.45\ s = 2.2925\ s$.
$P(w_B,c_1) = 60\% \times 0.21\ s + 40\% \times 4.92\ s = 2.094\ s$.
$P(w_B,c_2) = 60\% \times 1.98\ s + 40\% \times 6.6\ s = 3.828\ s$.
$d(w_A,w_B) = \sqrt{(1.4525\ s - 2.094\ s)^2 + (2.2925\ s - 3.828\ s)^2} = 1.66$.

To describe cohesion of database workload clusters W_A, W_B,..., we evaluate total of variances of normalized performance responses across all database configurations. The lowest the total variance, the more similar shapes of the workloads' performance responses.

$$\sum var = \sum_i var\left(\left\{\frac{P(w_A, c_i)}{\sum_k P(w_A, c_k)}, \frac{P(w_B, c_i)}{\sum_k P(w_B, c_k)}, \ldots\right\}\right)$$

Example. The following example illustrates calculation of total variance of normalized performance responses $P(w_A,c_1) = 5$ s, $P(w_B,c_1) = 10$ s, $P(w_A,c_2) = 10$ s, $P(w_B,c_2) = 20$ s, $P(w_A,c_3) = 20$ s, $P(w_B,c_3) = 30$ s:

$$\sum var = var\left(\left\{\frac{5}{5+10+20}, \frac{10}{10+20+30}\right\}\right)$$
$$+ var\left(\left\{\frac{10}{5+10+20}, \frac{20}{10+20+30}\right\}\right)$$
$$+ var\left(\left\{\frac{20}{5+10+20}, \frac{30}{10+20+30}\right\}\right)$$
$$= 0.000141723 + 0.000566893 + 0.00127551 = 0.001984127$$

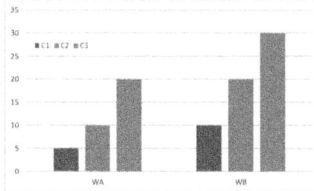

Our problem of *Database Workload Clustering* is defined as follows. Given are: a database workload template T, a set of database workloads $W = \{w_A, w_B,..., w_N\}$, a set of database configurations $C = \{c_1, c_2,..., c_m\}$, and as set of performance responses $P(w_i,c_j)$ for all combinations of database workloads and database configurations. The goal is to use information about T and W only in order to divide W into k subsets (clusters) of similar database workloads, so that total of variances of normalized performance responses across all database configurations would be minimized. We emphasize that the database workload clustering should rely on SQL queries and their frequences (T and W), because: (1) they can be easily gathered by DBAs, (2) they are objective, not influenced by current database configuration or hardware specifics. Notice that we do use the total of variances of normalized performance responses across all database configurations to describe the optimal clustering, which becomes our experimental reference.

3   The Methods

We represent database workloads as normalized vectors, describing frequencies of all database workload pattern queries. Using a $d()$-distance based clustering algorithm, we divide the set of workloads into non-overlapping groups (Method 1). To assess the effectiveness (quality) of the clustering, we evaluate total of variances of normalized performance responses in the groups across all database configurations ($\sum var$). Additionally, we generate a reference clustering scheme by using known performance responses, experimentally generated for real database configurations. We then use the Adjusted Rand Index (ARI) [7, 4] to compare our grouping results to the reference clustering scheme.

Workload query frequencies do not, however, provide us with any semantic information about amount of system resources needed to execute the queries, which may affect the performance-to-configuration profiles as well. Therefore, an alternative method of dividing the set of workloads into non-overlapping groups can be considered (Method 2). Using a relatively small sample of experimentally generated performance responses from real database configurations, we train classifiers on the normalized query frequency vectors to predict one of the reference clusters.

Both methods allow us to assign a database workload to a management group of similarly behaving workloads/deployments. When database workloads change over time, we can easily reassign them to new management groups.

4 Experimental Analysis

In order to assess the effectiveness of workload clustering based on distribution of query execution frequencies, we performed an extensive series of experiments using synthetic data sets. The experiments were executed using 8GB RAM Docker containers running PostgreSQL database server on Intel Core i5–12500. We used four databases of sizes 2.8 GB - 7.7 GB that contained both TPC-H [19] and TPC-C [18] tables, generated using scale factors of 1,3 and 10, 30 respectively. The database workloads were simulated using a mix of 22 standard TPC-H query types and three write-heavy TPC-C transactions: New Order, Payment and Delivery. The frequencies of the queries were randomized to generate 50 test workloads. We manipulated 10 database configuration parameters to build 32 test database configurations as described in Table 1. Thus, we ran 4 x 50 x 32 = 6400 experiments.

Table 1. Database configuration generation details.

Configuration parameter	Value
shared_buffers	random value between 10MB and 1GB
work_mem	random value between 1MB and 100MB
checkpoint_timeout	random value between 30s and 10 min
random_page_cost	random integer between 1 and 10
vacuum_cost_limit	random integer between 200 and 2000
autovacuum_max_workers	random integer between 1 and 10
min_parallel_table_scan_size	random value between 1M and 10M
max_parallel_workers	16
max_parallel_workers_per_gather	16
max_worker_processes	32

4.1 Clustering Based on Distribution of Query Execution Frequencies

Database workloads were represented by normalized vectors, describing frequencies of all database workload pattern queries. We clustered the workloads into 4, 6, 8 and 10 management groups using popular clustering methods, implemented in Weka 3.8.6 [5]: K-means Clustering, Expectation Maximization Clustering, Hierarchical Clustering, and Canopy Clustering. Next, using combinations of 5 randomly chosen database configurations and 4 database sizes, we validated the generated workload clusters by evaluating total variance of normalized performance responses within each group and by evaluating Adjusted Rand Index (ARI) against reference clusters generated using normalized performance responses. As an illustration of sample clustering results obtained during the experiments, see Table 2 containing generated management groups (*Grp ID*), database workload IDs (*w*), normalized performance responses for five configurations (c_1-c_5), total variances of normalized performance responses (Σ var) and performance response charts. Notice specific shape similarities of the performance response charts in each group, which confirm similar performance-to-configuration profiles.

Table. 2. Sample clusters based on frequencies of workload queries

Grp ID	Performance Responses for Database Configurations						Σ var	Performance Response Charts
	w	c_1	c_2	c_3	c_4	c_5		
1	w_{42}	0.186	0.223	0.228	0.176	0.184	0.0200	
	w_{16}	0.190	0.217	0.225	0.178	0.186		
	w_{33}	0.195	0.257	0.197	0.158	0.191		
	w_{32}	0.115	0.219	0.189	0.119	0.355		
	...							
2	w_6	0.192	0.204	0.213	0.194	0.194	0.0032	
	w_{87}	0.210	0.200	0.189	0.198	0.200		
	w_{37}	0.198	0.204	0.196	0.200	0.199		
	w_{12}	0.193	0.201	0.202	0.199	0.201		
	...							
3	w_{19}	0.183	0.211	0.274	0.155	0.176	0.0151	
	w_4	0.184	0.201	0.268	0.166	0.178		
	w_{55}	0.174	0.231	0.259	0.161	0.173		
	w_1	0.174	0.187	0.316	0.156	0.165		

Table 3 shows our experimental results on effectiveness of clustering based on distribution of query execution frequencies. As a reference, we also provide results for random clustering of the database workloads. The table shows the clustering method, target number of clusters, total variance of normalized performance responses and Adjusted

Rand Index (ARI) against reference clusters generated using normalized performance responses. The achieved ARI levels of 0.25–0.37 show that our clustering has some alignment with the reference groups but is obviously far from a strong match. On the other hand, the comparison with random grouping of database workloads shows significant improvements in terms of total variance of normalized performance responses, and of course, ARI. We also performed auxiliary experiments (not enclosed here) based on weighted query frequency vectors: by using query execution plan costs, query complexity metrics, numbers of blocks/rows read, etc., but no further improvement was observed.

Table 3. Effectiveness of clustering based on distribution of query execution frequencies

Workload clustering method	Number of clusters	Σ variance	ARI
Random clusters	4	0.0328	−0.0324
	6	0.0488	−0.0518
Expectation Maximisation	4	0.0221	0.2428
	6	0.0222	0.1951
K-means	4	0.0221	0.2428
	6	0.0227	0.3664
Hierarchical Clustering	4	0.0219	0.1949
	6	0.0217	0.1876
Canopy	4	0.0222	0.2922
	6	0.0220	0.2731

For better understanding of the total variance of normalized performance responses, we compared it to K-means reference clusters generated using workload performance responses gathered for various database configurations - see Table 4. These reference values can be considered the optimal solution to the problem.

Table 4. Total variance in clusters generated using performance responses

Clustering method	Number of clusters	Σ variance
K-means	4	0.0156
	6	0.0205

4.2 Learning to Group Deployments Using Distribution of Query Execution Frequencies

We found that clustering based on distribution of query execution frequencies could provide DBAs with hints on how to split a large number of deployments into a small set of management groups that can use identical fine-tuned database configurations. Yet, it should not be expected that results of this approach will always lead to the optimal way of grouping the deployments, because distance-based similarities between query execution frequencies do not have to directly map to performance-to-configuration profiles of the workloads.

Although we assumed that performance-to-configuration profiles for all deployments and all database configuration changes are not readily available, we can consider gathering a small sample of performance-to-configuration profiles for *some* of the deployments. If DBAs recorded workload performance responses for some tested database configuration changes, we could use those recorded performance responses to calibrate our workload grouping method by using machine learning algorithms. In order to assess the effectiveness of this extended method, we generated and validated several classifiers on a training dataset containing both query execution frequencies and reference clusters generated using five normalized performance responses each. The training dataset covered 30% (60) of all the database workloads. We used popular Weka classification algorithms: Random Forest, C4.5 (J48), Naive Bayes, Logistic Regression, and KNN. Table 5 shows the classifier quality, determined using 10-fold cross validation. To facilitate the results comparison of the classification-based approach to the former clustering-based approach, Table 6 shows total variance of normalized performance responses and Adjusted Rand Index (ARI) against reference clusters generated using normalized performance responses, as achieved by the classifier trained with J48.

Table 5. Database workload classification for deployment group prediction

Classification method	Number of clusters/classes	Correctly classified instances
Random Forest	4	59,3%
J48	4	70,4%
Naive Bayes	4	63,0%
Logistic Regression	4	55,6%
K-NN	4	62,9%
Random Forest	6	59,3%
J48	6	62,9%
Naive Bayes	6	51,8%
Logistic Regression	6	51,9%
K-NN	6	62,9%

5   Related Work

To the best of our knowledge, the particular problem of clustering of large number of Database-as-a-Service instances, all running the same query templates but with varying query frequencies, has not been addressed before. There are two related topics that

Table 6. Effectiveness of workload classification for C4.5 (J48)

Workload classification method	Number of clusters	Σ variance	ARI
J48	4	0.01812	0.6714
	6	0.02047	0.6957

seem to be partially connected to our research: workload characterization and general workload clustering in cloud environments.

Broad aspects of workload characterization have been thoroughly studied. In [1], query performance prediction was discussed. Query performance prediction predictive modeling techniques were proposed to learn query execution behavior at different granularities, ranging from coarse-grained plan-level models to fine-grained operator level models. [2] showed that training a classifier to compare the cost of two query execution plans can be more accurate than learning to predict the cost for a plan and then comparing the cost. The authors presented a technique to featurize query plans into vectors and studied model alternatives. [3] describes a system that uses machine learning to accurately predict the performance metrics of database queries. [9] focuses on estimating resource consumption of SQL queries by combining knowledge of database query processing with statistical models. In [10], the authors apply deep learning to the query performance prediction problem, and introduce a neural network architecture for the task, called plan-structured neural network. SageDB database system that learns the structure of the data and optimal access methods and query plans was presented in [8]. [20] presents an automated approach that leverages past experience and collects new information to tune DBMS configurations, using a combination of supervised and unsupervised machine learning methods to select the most impactful knobs, map unseen database workloads to previous workloads from which we can transfer experience, and recommend knob settings. [15] considers MapReduce workloads that are produced by analytic applications and proposes a technique that predicts the runtime performance for a fixed set of queries running over varying input data sets.

Workload clustering of ad-hoc queries and requests in cloud environments has also been researched in the past. In [12], a K-means-based clustering method of IaaS cloud workloads types was proposed. [21] compares and analyzes multiple clustering methods for workload categorization based on usage of CPU, memory and disk I/O operations. [13] compares K-Means and Gaussian Mixture Model to evaluate cluster representativeness of two methods for heterogeneity in resource usage of cloud workloads. The authors of [14] proposed a workload characterization method based on query plan encoders that learn essential features and their correlations from query plans. The pretrained encoders captured structural and computational performance of queries independently. In [6] the authors attempt to discover underlying relationships that characterize OLTP performance over a wide range of configurations. They derive the "iron law" of database performance - that both the average instructions executed per transaction and the average cycles per instruction are critical to the transaction-throughput performance. One of the earliest publications, [22] describes a relational database workload analyzer (REDWAR), developed to characterize workload in a DB2 environment, analyzing structure and complexity of SQL statements, makeup and run-time behavior of transactions/queries, and composition of relations and views. [11] suggests an interesting idea of using idleness detection to characterize database workload.

6   Conclusions

In this work, we studied methods of dividing Database-as-a-Service instances into a small, manageable number of clusters with similar performance-to-configuration profiles that could share single configuration templates to maximize performance of their instances. We investigated whether such clustering can be based solely on workload query frequencies. We followed and experimentally validated two approaches: (1) unsupervised clustering based on workload query frequencies, and (2) using a pretrained classifier to assign workloads, described with their query frequencies, to predefined clusters. We showed that workload's query frequencies alone can help determine a management group for a deployment that runs the workload. Even when using unsupervised clustering methods to provide a form of preliminary recommendations, we were able to assist DBAs in reducing the complexity of their database optimization tasks. More promising results were observed when classifiers were trained on a small sample of performance responses gathered for different database configurations.

References

1. Akdere, M., Cetintemel, U., Riondato, M., Upfal, E., Zdonik, S.B.: Learning-based query performance modeling and prediction. In 2012 IEEE 28th International Conference on Data Engineering, pp. 390–401. IEEE (2012)
2. Ding, B., Das, S., Marcus, R., Wu, W., Chaudhuri, S., Narasayya, V.R.: Ai meets ai: Leveraging query executions to improve index recommendations. In: Proceedings of the 2019 International Conference on Management of Data, pp. 1241–1258 (2019)
3. Ganapathi, A., Kuno, H., Dayal, U., Wiener, J. L., Fox, A., Jordan, M., Patterson, D.: Predicting multiple metrics for queries: Better decisions enabled by machine learning. In: 2009 IEEE 25th International Conference on Data Engineering, pp. 592–603. IEEE (2009)
4. Halkidi, M., Batistakis, Y., Vazirgiannis, M.: Cluster validity methods: Part I. SIGMOD Record 31(2), 40–45 (2002)
5. Hall, M., Eibe, F.E., Holmes, G., Pfahringer, B., Reutemann, P., Witten, I.H.: The WEKA data mining software: an update. SIGKDD Explorations 11(1), 10–18 (2009)
6. Hankins, R., et al.: Scaling and characterizing database workloads: bridging the gap between research and practice. In: Proceedings of 36th Annual IEEE/ACM International Symposium on Microarchitecture, MICRO-36., San Diego, CA, USA, pp. 151–162 (2003)
7. Hubert, L., Arabie, P.: Comparing partitions. J. Classif. 2, 193–218 (1985)
8. Kraska, T., et al.: Sagedb: a learned database system. In: Conference on Innovative Data Systems Research (2019)
9. Li, J., Konig, A.C., Narasayya, V., Chaudhuri, S.: Robust estimation of resource consumption for sql queries using statistical techniques. Proc. VLDB Endowment 5(11), 1555–1566 (2012)
10. Marcus, R., Papaemmanouil, O.: Plan-structured deep neural network models for query performance prediction. In: Proceedings of the VLDB Endowment, ACM, pp. 1733–1746 (2019)

11. Mateen, A., Mahmood, K.T., Nam, S.Y.: DB workload management through characterization and idleness detection. In: 26th International Conference on Advanced Communications Technology (ICACT), Pyeong Chang, Korea, pp. 226–231 (2024)
12. Orzechowski, P., Proficz, J., Krawczyk, H., Szymanski, J.: Categorization of cloud workload types with clustering. In: Proceedings of the International Conference on Signal, Networks, Computing, and Systems, vol. 1, 395, pp. 303–313 (2017)
13. Patel, E., Kushwaha, D.S.: Clustering cloud workloads: K-means vs gaussian mixture model. Procedia Comput. Sci. **171**, 158–167 (2020)
14. Paul, D., Cao, J., Li, F., and Srikumar, V.: Database workload characterization with query plan encoders. Proc. VLDB Endowment **15**(4), 923–935 (2021)
15. Popescu, A.D., Ercegovac, V., Balmin, A., Branco, M., Ailamaki, A.: Same queries, different data: Can we predict runtime performance? In: 2012 IEEE 28th International Conference on Data Engineering Workshops, pp. 275–280. IEEE (2012)
16. Thummala, V., Babu, S.: iTuned: a tool for configuring and visualizing database parameters. In: Proceedings of the 2010 ACM SIGMOD International Conference on Management of data, pp. 1231–1234. ACM (2010)
17. Time Model Statistics, Oracle Database Documentation. https://docs.oracle.com/en/database/oracle/oracle-database/23/tdppt/time-model-statistics.html
18. TPC Benchmark C: Standard specification 5.11. https://www.tpc.org/tpc_documents_current_versions/pdf/tpc-c_v5.11.0.pdf
19. TPC Benchmark H: Standard Specification 2.18.0. https://www.tpc.org/tpc_documents_current_versions/pdf/tpc-h_v2.18.0.pdf
20. Van Aken, D., Pavlo, A., Gordon, G.J., Zhang, B.: Automatic database management system tuning through large-scale machine learning. In: Proceedings of the 2017 ACM International Conference on Management of Data, pp. 1009–1024. ACM (2017)
21. Visweshwaran, P.M.S., Sathiya, R.R.: Cloud workload clustering. In: 8th International Conference on Smart Structures and Systems (ICSSS), Chennai, India, pp. 1–4 (2022)
22. Yu, P., Chen, M.-S., Heiß, H.U., Lee, S.: Workload characterization of relation database environments. IEEE Trans. Softw. Eng. **18**, pp. 347–355 (1992)

Accelerating Python Code
with Parallel I/O

Robin Varghese[1], Hashirul Quadir[1], Ladjel Bellatreche[2],
and Carlos Ordonez[1(✉)]

[1] Department of Computer Science, University of Houston, Houston, USA
{hquadir,carlos}@central.uh.edu,co@cs.uh.edu
[2] LIAS/ISAE-ENSMA, Poitiers, France

Abstract. Python is the dominating data science language, leaving behind other languages like C++, Java and R. Python libraries wrap highly tuned, efficient, accurate C++ and C code for linear algebra, numerical methods and data manipulation. Moreover, the Python runtime works flawlessly across diverse operating systems (Linux, Windows) and CPU architectures, including x86 and ARM. From an accelerator perspective, Python code is processed on multi-core CPUs and GPUs, whose power is not fully exploited. In this paper, we study how to improve Python I/O bottlenecks. We focus on data set summarization to compute a model on a large data set, stored on a CSV file (the most common format used in practice). Heeding these challenges, we introduce two simple, but fundamental, I/O optimizations: parallel multi-threaded read and chunk-based scan (similar to reading file blocks in a DBMS). An experimental validation on different cloud servers, provides a realistic scenario. We show our optimized Python code can work faster than existing Python functions, it exhibits almost linear speed up I/O as more threads are used (up to a limit), but it can still leverage parallel processing for the CPU-intensive floating point computations. To round up our study, we justify chunk size is a critical performance parameter that depends on data set size as well as cloud server configuration.

1 Introduction

Our study aims to highlight I/O on text files as a major bottleneck in AI and ML computations processed by Python [6] in a typical server in the cloud. Most analytic projects target is a predictive machine learning (ML) model. Logistic regression stems from linear regression [4] and logistic regression is a foundation model for deep neural networks [1]. Therefore, we study linear regression (LR) as a representative and fundamental AI model representing a demanding computation (I/O + floating point operations) on a large input matrix, perhaps exceeding main memory limits. Our paper advances [3], where summarization is proposed to accelerate ML computations in the R language, with serial I/O. A major step forward is the parallel computation of summarization in Python,

R. Wrembel et al. (Eds.): DEXA 2025, LNCS 16047, pp. 360–366, 2026.
https://doi.org/10.1007/978-3-032-02088-8_28

with parallel I/O, with chunked files, in a typical multi-core cloud server. This paper borrows query processing ideas to compute summarization in parallel in a DBMS with SQL queries combining joins and aggregations [5].

2 Accelerating and Scaling Data Summarization

2.1 Definitions: Input Matrix and Machine Learning Model

The input data set is defined as X, a $d \times n$ matrix consisting of a set of n column vectors each with d dimensions. We refer to the machine learning model as Θ, in this case vector $\hat{\beta}$.

Our data summarization works well for multiple ML models Θ such as PCA, NB (Naive Bayes), K-Means clustering, and LR (Linear Regression). Given the mathematical importance of LR, and being used as a theoretical foundation for Neural Networks we focus on its solution.

2.2 Data Set Summarization

In order to compute data summarization matrix Γ for LR, first we augment X with a row of ones to produce a $(d + 1) \times n$ matrix $\mathbf{X}$. $\mathbf{Y}$ is an n-row vector corresponding to the output for each n observation. Given a $(d + 1) \times n$ input matrix $\mathbf{X}$, and a $(d+1) \times 1$ column vector $\hat{\beta}$ of coefficients, the predicted outputs $\hat{\mathbf{Y}}$, an n-row vector corresponding to the predicted outputs of each n observation produced by the LR model, are computed as $\hat{\mathbf{Y}} = \hat{\beta}^T \mathbf{X} + \epsilon$, where ϵ represents error. For computing Γ, augment $\mathbf{X}$ with $\mathbf{Y}$. In general this $(d + 2) \times n$ matrix is defined as $\mathbf{Z}$, but to optimize I/O we create sub-matrix of size $(d + 2) \times c$ where c is the chunk size (a block of vectors). Therefore, c can be considered a hyper-parameter and it must be tuned to achieve optimal performance. The serial computation of a model Θ becomes a two-phase algorithm as follows:

- Phase 1: Summarize X: compute $\Gamma = \sum_{i=1}^{n} z_i \otimes z_i^T$;
- Phase 2: Compute model Θ: Solve $\hat{\beta}$ exploiting Γ.

Phase 1: Begin by reading a chunk of the input data set of size $d \times c$. Augment this chunk in main memory with a row vector of c ones and $\mathbf{Y}$ (dependent variable) also a c row vector. This will produce $\mathbf{Z}$, a $(d + 2) \times c$, a dense matrix whose size is $\Theta(d^2)$. Compute $\mathbf{Z}\mathbf{Z}^T$ to produce the partial gamma Γ_c.

The sufficient statistics (SS) L, n and Q are defined as follows: $n = |X|$, $L = \sum_{i=1}^{n} x_i$, and $Q = XX^T = \sum_{i=1}^{n} x_i \cdot x_i^T$, n is total number of points in the data set, L is the linear sum of x_i and Q is the sum of vector outer products of x_i. It should be noted that Phase 1 takes most computation time. Phase 2: Exploit the sufficient statistics integrated into the single matrix Γ_c to further compute an ML model. In our target case, the goal is to compute the regression coefficients $\hat{\beta}$ whose solution by the least square method is $\hat{\mathbf{Y}} = \hat{\beta}^T \mathbf{X} + \epsilon$. The "quadratic" sufficient statistic matrix Q and the matrix-vector product XY^T are exploited by substituting them into $\hat{\beta} = (\mathbf{X}\mathbf{X}^T)^{-1}\mathbf{X}\mathbf{Y}^T = Q^{-1}(\mathbf{X}\mathbf{Y}^T)$, where the second expression is much faster to compute because it is based only on Γ.

The serial algorithm has time complexity $\Theta(d^2 n)$ for Phase 1, but $\Theta(d^3)$ for Phase 2 (much lower as $n \to \infty$). That is, Phase 1 dominates time growth.

Data summarization is quadratic in data set dimensionality (i.e. demanding) and therefore, it has a significantly higher time complexity than stochastic gradient descent, the workhorse of AI models.

2.3   Hardware and Software Evolution

Modern computing has evolved to feature multi-core CPUs, with high-end models like the Xeon Granite Rapids offering up to 128 cores and support for vectorized instructions, significantly enhancing computational efficiency. Server memory configurations range from 16 to 32 GB, scaling up to terabytes to meet the demands of Big Data and Deep Learning. Advances in storage, with SSDs providing 3X, NVMs providing 10X, faster read speed than HDDs, have significantly narrowed the performance gap between RAM and secondary storage, though I/O from secondary storage remains a bottleneck, albeit less so with modern PCI-connected NVMs.

Processing of Python code is done by multi-core CPUs supporting vectorized instructions, which accelerate matrix computations, but only in memory. In systems without distributed memory (i.e. a cluster of computers), multi-threaded processing is the primary parallelization strategy, involving "coarse" threads (e.g., Python) mapped to hardware CPU threads (embedded in the chip). This setup emphasizes the importance of multi-threading processing for parallel execution in shared RAM scenarios and the alignment of software with hardware threads to optimize efficiency. However, practical limitations such as Python's Global Interpreter Lock (GIL) create a need for alternative parallelism, highlighting the complexity of optimizing parallel processing, especially for I/O operations.

2.4   Parallel I/O Efficient Algorithm

Let p be the number of processors (cores, machines), under a partitioned memory architecture, where each processor has its independent main memory and persistent storage. We assume $d \ll n$ and $p \ll n$, but d is independent from p. Our parallel algorithm follows:

- Data set X is uniformly partitioned among the p processors with $\approx n/p$ points x_i per partition. Initialize: $\Gamma = [0]$
- Phase 1 in parallel with p processors: Compute Γ in parallel, dynamically building z_i in main memory, updating Γ_j in main memory reading X_j from persistent storage in blocks. Send the partial matrices to the master processor and aggregate the p partial summary matrices into the global matrix Γ.
- Phase 2 at master processor: Compute Θ using Γ in intermediate computations. Iterate method until convergence exploiting Γ in intermediate matrix computations.

Notice we opt for a partitioned memory model with parallel I/O [2] instead of PRAM, which would require locking mechanisms in Phase 1, introducing significant overhead. Parallelization is accomplished by computing separate summaries on each partition. That is, threads do not share variables in main memory. The main reason this parallelization is feasible is because matrix multiplication is distributive and additive, meaning we can compute separate matrix multiplications and add them at the end (with negligible overhead to lock the global summarization). Phase 2 is very fast since it does not depend on n.

3   Experimental Evaluation

As explained in Sect. 2 Phase 1 takes around 99% of time and Phase 2 takes only 1%-2%. Therefore, we focus on studying Phase 1, which is I/O bound.

3.1   Evaluation Setup

Cloud Server Configuration. We ran experiments on three cloud servers, provided by Amazon AWS. Our results should be similar on other cloud providers, which also use VMware for virtualization. Our server specifications were as follows: Server 1 (2 vCPUs) is a Xeon-based HVM domU instance with an Intel(R) Xeon(R) E5-2686 v4 CPU, 2 cores per socket - 1 socket - 1 thread/core, 8GB RAM. Server 2 (4vCPUs), an Amazon EC2 g4dn.xlarge instance, with an Intel Xeon CPU with 24 cores and 48 physical threads, 16 GB RAM. 16 GB RAM, 2 cores per socket - 1 socket - 2 threads/core. Server 3 (8 vCPUs) is an Amazon EC2 g4dn.xlarge instance, with an Intel Xeon CPU with 18 cores and 36 physical threads, 32 GB RAM. All servers had detached SSD storage, connected via a fast network, with limited space around 200 GBs and they were running Linux Ubuntu under VMware.

Python Data Science Libraries. Our prototype setup leverages Python libraries: `Pandas` for I/O operations, `NumPy` for numerical computations, and `threading` for multi-threaded parallelism to efficiently read large data sets, eliminating RAM limitations. Data sets were chunked (divided into blocks similar to an SQL table) and partitioned across p files for scalable processing: parallel multi-threaded I/O, lock-free, without RAM limitations. Each thread worked in parallel computing Γ_c matrices. As explained above, to mitigate the I/O bottleneck, we utilize Python's built-in `threading` library to spawn p threads equal to the number of data set partitions, stored on p files. Notice we also exploit CPU parallel processing for floating point operations.

3.2   Profiling Computation Steps to Identify Bottlenecks

To verify I/O is the main bottleneck in the computation, we profiled each computation step on three widely different cloud servers, as shown in Table 1. We analyzed how the number of threads affects speed. Table 1 shows that 90% of

Table 1. Profiling each step, highlighting significant time is spent on I/O.

vCPUs	#threads p	size n	chunk size c	I/O time	summarization time	total time	total speedup
2	1	16M	100k	491	13	504	1.0X
2	2	16M	100k	242	10	252	2.0X
2	4	16M	100k	173	6	179	2.8X
2	8	16M	100k	173	fail	fail	-
4	1	16M	100k	421	10	431	1.0X
4	2	16M	100k	219	7	227	1.9X
4	4	16M	100k	165	9	174	2.5X
4	8	16M	100k	165	fail	fail	-
8	1	1M	1k	30	8	38	1.0X
8	2	1M	1k	22	4	26	1.5X
8	4	1M	1k	275	3	278	0.1X
8	8	1M	1k	885	2	887	0.1X

the whole computation time is due to I/O and this fraction is bigger for large n, when the data set size exceeds RAM size. Increasing the number of threads accelerates the summarization computation, but to a limit: for the smaller cloud servers (2 vCPUs, 4 vCPUs) 8 threads fail, whereas for the larger server (8 vCPUs) the speedup stops at 2 threads. These time results indicate there is a complex interaction among number of vCPUs, number of threads, data set size and chunk size. Unfortunately, we could not determine the specific storage device information (model, block size, PCI vs SATA connection).

3.3   Accelerating I/O Speed with Parallelization and Chunks

Our next experiments aim to find out how many threads "saturate" the CPU cores. Figure 1 compares CPU core saturation with 1 thread versus the optimal number of threads, obtained from Table 1. That is, we want to use cores as close as possible to 100% to avoid cores, and vCPUs in consequence, being idle. As can be seen, 4 threads are optimal for the 4 vCPU server, but only 2 threads are optimal for the 8 vCPU server. Notice the virtualization software and the operating system consume CPU cycles anyway on some CPU cores.

Our last experiments explore chunk size on two cloud servers, shown in Fig. 2. For the 4 vCPUs server increasing c decreases time, but the impact is not look significant. A large chunk size $c = 100k$ gives optimal time, but it does not improve when chunk size c approaches $n = 1M$. On the other hand, a smaller chunk size around $c = 100$ is best for the 8 vCPUs server and then performance decreases as n grows. These results indicate there exists a chunk size that minimizes I/O time and they highlight that d, n alone are not sufficient to determine

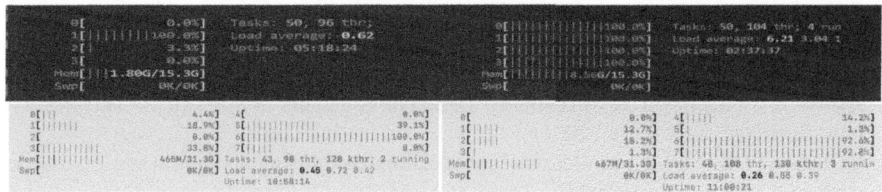

Fig. 1. Achieving full vCPU utilization with multi-threaded I/O (black = 4 vCPUs, gray = 8 vCPUs): left/up = 1 thread, 4 vCPUs; right/up = 4 threads, 4 vCPUs; left/down = 1 thread, 8 vCPUs; right/down = 2 threads, 8 vCPUs. (Color figure online)

chunk size. Why? because the number of vCPUs and the storage device also have an impact.

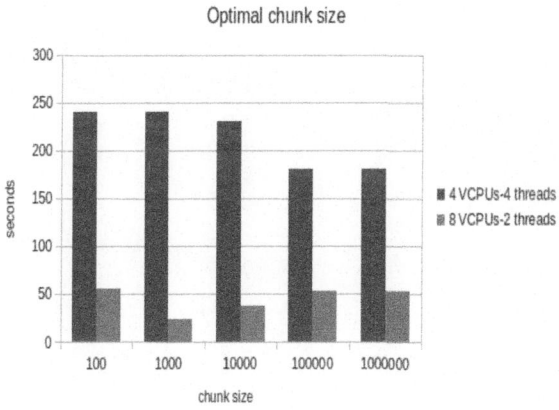

Fig. 2. Finding optimal chunk size on 2 cloud servers with their best number of threads: 4 vCPUs and 4 threads; 8 vCPUs and 2 threads.

References

1. Charu, C.: Aggarwal. Springer, Neural Networks and Deep Learning - A Textbook (2023)
2. Bender, M.A., Brodal, G.S., Fagerberg, R., Jacob, R., Vicari, E.: Optimal sparse matrix dense vector multiplication in the i/o-model. In: SPAA 2007: Proceedings of the 19th Annual ACM Symposium on Parallelism in Algorithms and Architectures, pp. 61–70. ACM (2007)

3. Chebolu, S.U.S., Ordonez, C., Al-Amin, S.T.: Scalable machine learning in the R language using a summarization matrix. In: Database and Expert Systems Applications DEXA, pp. 247–262 (2019)
4. Hastie, T., Tibshirani, R., Friedman, J.H.: The Elements of Statistical Learning, 1st edn. Springer, New York (2001)
5. Ordonez, C.: Scalable parallel machine learning computing a summarization matrix with SQL queries. In:IEEE Big Data, pp. 151–160 (2017)
6. Sarkar, D., Bali, R., Sharma, T.: Practical machine learning with Python. Apress, A Problem-Solvers Guide To Building Real-World Intelligent Systems. Berkely (2018)

Benchmarking Embedding Techniques for Modeling User Navigation Behavior on Task-Oriented Software

Ikram Boukharouba[1,2]($\boxtimes$) (ID), Florence Sèdes[1] (ID), Benoit Verhaeghe[2] (ID),
and Christophe Bortolaso[2] (ID)

[1] IRIT, Université de Toulouse, Toulouse, France
florence.sedes@irit.fr
[2] Berger-Levrault, Labège, France
{ikram.boukharouba,benoit.verhaeghe,
christophe.bortolaso}@berger-levrault.com

Abstract. Understanding user navigation patterns from clickstream data is crucial for improving business software, yet remains challenging due to the complexity and variability of real-world environments. Unlike controlled settings, real-world clickstreams are noisy, fragmented, and often incomplete, due to session timeouts, network issues, caching, or third-party interactions—making it difficult to reconstruct coherent user journeys. Additionally, the absence of labeled data hinders the use of supervised learning, pushing researchers toward unsupervised or heuristic-based approaches that struggle to fully capture user behavior. In this paper, we present a benchmark of embedding techniques for modeling user navigation behavior on task-oriented software. We identify distinct user behaviors across three real-world case studies. Results show that Pattern2Vec outperforms Word2Vec in capturing meaningful task-based navigation patterns, confirming its suitability for clickstream analysis.

Keywords: User Navigation Pattern · Clickstream analysis · Clickstream embeddings · Task-oriented software

1 Introduction

The rise of digital platforms has transformed user interaction, making it essential for businesses to understand navigation patterns. Clickstream data, which logs sequences of user actions, offers valuable insights into preferences and behaviors, supporting interface design and personalization [9]. However, real-world clickstreams pose major challenges: they are often noisy, fragmented, and lack labels [15]. Factors like session interruptions, multi-device usage, and interactions with third-party systems complicate data interpretation. Traditional supervised methods rely on labeled datasets, which are rarely available in practice [3]. In response,

© The Author(s), under exclusive license to Springer Nature Switzerland AG 2026
R. Wrembel et al. (Eds.): DEXA 2025, LNCS 16047, pp. 367–375, 2026.
https://doi.org/10.1007/978-3-032-02088-8_29

unsupervised techniques—particularly clustering—have become popular for pattern discovery. Yet, these too face limitations: high-dimensional and sparse clickstreams hinder cluster clarity, and most algorithms ignore temporal dependencies. Additionally, privacy regulations like GDPR must be respected when modeling user paths. Embedding techniques help address these issues by transforming clickstreams into low-dimensional representations, reducing noise and preserving behavioral structure [11]. This paper benchmarks several embedding methods combined with clustering to uncover navigation patterns in task-oriented software across three case studies. Section 2 reviews related work, Section 3 details our methodology, Section 4 presents results, and Section 5 concludes with future directions.

2   Related work

Uncovering user navigation patterns has evolved significantly, from early statistical analyses to modern machine learning and reinforcement learning approaches. Initial work in the late 1990s and early 2000s [4] relied on frequency analysis, correlation, and data mining techniques such as clustering and association rule mining. For example, Mobasher et al. [10] used association rules on clickstream data to improve recommendation precision and coverage efficiently. With the rise of machine learning in the mid-2000s, more sophisticated models emerged. Eirinaki et al. [6] leveraged ML for personalized recommendation systems, while Borges et al. [2] improved Markov models through clustering and state cloning for better navigation behavior modeling. By the 2010s, reinforcement learning (RL) gained traction as a way to model user navigation as sequential decision-making. Derhami et al. [5] proposed RL-based ranking algorithms like RL_Rank, simulating web surfers. Later, Todi et al. [14] applied model-based RL for adaptive user interfaces using predictive HCI models. More recently, graph-based learning has become prominent for capturing complex relationships in user activity [1]. In summary, research on user navigation has progressed from simple statistical tools to sophisticated ML, RL, and graph-based techniques, each deepening our understanding of user interaction in digital environments.

3   Benchmark Approach

Our benchmark approach, illustrated in Fig. 1, aims to evaluate the effectiveness of various embedding methods in uncovering user navigation patterns within task-oriented software. Not all embedding techniques are equally suited for modeling clickstream data, especially in structured, task-driven environments where capturing sequential dependencies and contextual nuances is essential. This study assesses how well different methods preserve meaningful behavioral signals when transformed into low-dimensional representations suitable for clustering. **The distinguishing feature of our work is the use of a multi-method framework**, which leverages three distinct embedding methods: Pattern2Vec [12], Sequence Graph Transform (SGT) [13], and Word2Vec [8], to transform

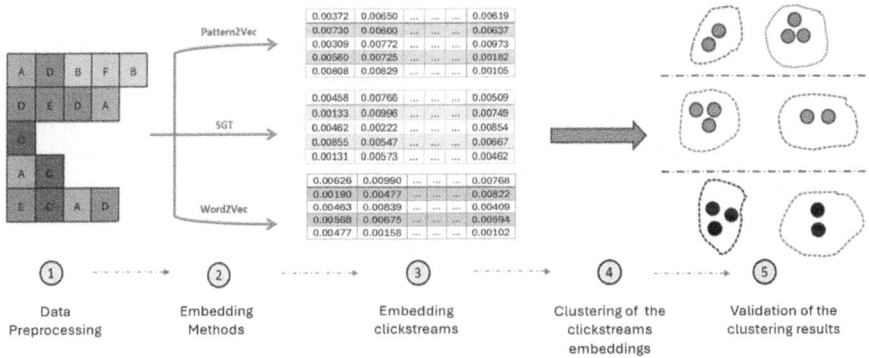

Fig. 1. Overview of our benchmark approach for uncovering user navigation pattern through clustering clickstream embeddings.

clickstream data into meaningful low-dimensional embeddings. These embeddings are then clustered to reveal patterns in user navigation behavior. In the following, we outline the 5 steps of our benchmark study.

① **Data Preprocessing** The first step involves preprocessing raw clickstream data to prepare it for embedding generation. This includes: noise Reduction to ensure that the resulting clickstreams reflect genuine user behavior. Then standardizing page names and other contextual information to ensure consistency across clickstreams.

②.③ **Embedding Methods and clickstream transformation** To transform clickstreams into low-dimensional embeddings, we use three complementary methods: Pattern2Vec, SGT, and Word2Vec. Each method brings specific strengths, and their combined application enables a comprehensive evaluation on real-world software clickstream data.

– **Pattern2Vec**: Mad for clickstream embedding, it captures temporal and sequential dependencies by learning contextual relationships and transition probabilities between pages, preserving meaningful navigation sequences.
– **Sequence Graph Transform (SGT)**: A graph-based method that models clickstreams as graphs, capturing both local and global structure. It focuses on connectivity and is effective for both common and rare navigation.
– **Word2Vec**: Originally designed for natural language, it models clickstreams as sequences of pages to capture co-occurrence-based semantic similarities. While efficient, it lacks the ability to represent temporal order and sequence dynamics critical for behavioral analysis.

④ **Clustering of the Clickstream Embeddings** Once the clickstream embeddings are generated using the three methods, we apply clustering algorithms to group similar clickstreams. Depending on the nature of the embeddings and the required granularity, we use K-Means, Agglomerative Hierarchical Clustering, and Spectral Clustering. A key aspect of our approach is the use of varied distance metrics and linkage types to capture diverse relationships within the data. For K-Means, we employ both Euclidean and Cosine distances. In Agglomerative Clustering, we explore different linkage criteria (e.g., single, complete, Ward) combined with multiple distance metrics to reflect the hierarchical structure of user behaviors. This methodological flexibility ensures a robust clustering process, enabling the discovery of meaningful task-oriented patterns and improving the interpretability of clickstream structures.

⑤ **Validation of Clustering Results** Evaluating the quality of the clusters ensures that the identified patterns reflect genuine user navigation behavior. In the absence of ground truth labels, we rely on unsupervised clustering evaluation metrics that assess intra-cluster cohesion and inter-cluster separation. In this study, we use the Silhouette Score to determine the optimal number of clusters. This score ranges from -1 to 1, with higher values indicating better defined clusters. To complement this, we employ dimensionality reduction techniques, such as t-distributed Stochastic Neighbor Embedding (t-SNE), to project high-dimensional embeddings into a lower-dimensional space, enabling visual assessment of cluster separability.

4   Experiments and results

Our objectives are 1) to benchmark clickstream embedding methods to determine which embedding technique captures better user navigation patterns in task-oriented software, 2) Group users with similar navigation patterns. Our experiments are conducted on 3 datasets SEDIT, EGRC and MSNBC, using 3 embedding techniques: Pattern2Vec, SGT and Word2Vec.

4.1   Dataset Description

The datasets used in this study come from distinct sources, capturing diverse user interactions and navigation complexities on real-world software. Two datasets originate from Berger-Levrault's enterprise applications, and one is a publicly available dataset.

- **SEDIT RH Dataset:** comes from SEDIT, a human resources management software of the company Berger-Levrault, that is used by administration in the public sector. It consists of 239 sequences, of 1 to 1,362 action per sequence, representing the diversity of daily administrative tasks.

- **EGRC Dataset:** Originating from an electoral data management software issued by Berger-Levrault, gathering interactions made by agents who manage and synchronize electoral data. It includes 119 sequences of 1 to 18,509 actions, reflecting the complexity of user interactions.
- **MSNBC Dataset** [7]: Contains anonymized user clickstream sequences from the MSNBC media website. We used 500 sequences ranging from 6 to 14,795 actions. These clickstreams reflects the varied user behaviors.

4.2 Embedding and Clustering of Clickstreams

We follow the same procedure on the 3 different datasets. We created 3 embedding models corresponding to Pattern2Vec, SGT and Word2Vec. We run 100 iterations on each model, with the following parameters:

Pattern2Vec: embedding_dim =1000, num_filters=32, filter_size=3.

SGT: embedding_dim =16, kappa=1, lengthsensitive=True.

Word2Vec: embedding_dim =1000, window_size=5, min_count=1, workers=4.

Once embeddings generated for each clickstream, we applied clustering algorithms to group similar user sessions. Using three different clustering algorithms: Kmeans, Agglomerative and Spectral clustering. Through clustering, we aim to uncover distinct navigation patterns across user sessions. We iteratively tested different cluster numbers, analyzing their impact on clustering quality. The final choice of clusters was based on the configuration that maximized the Silhouette Score ensuring well-defined and interpretable clusters (See Fig.2).

Table 1. Results of silhouette scores on the different datasets.

Dataset	Embedding method	Nb_clusters	Clustering method	Distance
EGRC	Pattern2Vec	6	Agglomerative	N/A
EGRC	SGT	8	Spectral	N/A
EGRC	Word2Vec	14	Spectral	N/A
MSNBC	Pattern2Vec	6	Agglomerative	N/A
MSNBC	SGT	6	Agglomerative	N/A
MSNBC	Word2Vec	17	KMeans	Cosine
SEDIT	Pattern2Vec	6	Agglomerative	N/A
SEDIT	SGT	6	Agglomerative	N/A
SEDIT	Word2Vec	6	Agglomerative	N/A

4.3 Results and Discussion

The results visualized in Fig. 2 present a comparative analysis of clustering performance and quality across the three datasets (EGRC, SEDIT, and MSNBC).

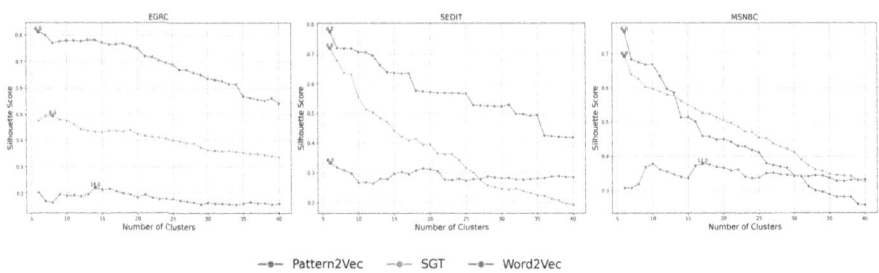

Fig. 2. Choosing the optimal number of clusters for each dataset based on the variation of the Silhouete Score.

The x-axis represents the number of clusters, while the y-axis shows the corresponding Silhouette Score. Across all datasets, Pattern2Vec consistently achieves the highest Silhouette Scores, indicating superior clustering quality and its effectiveness in capturing underlying navigation structures. SGT follows as the second-best performer, demonstrating moderate clustering quality but still significantly outperforming Word2Vec, which struggles to form well-defined clusters. This aligns with expectations, as Word2Vec was not originally designed for sequential clickstream data, making it less effective in preserving navigation patterns. In contrast, Pattern2Vec, specifically designed for sequential user behaviors, achieves superior clustering performance, successfully distinguishing distinct navigation trends. Fig. 2, further highlights that the optimal number of clusters varies across datasets, with the selected values marked on the plots. Despite this variation, Pattern2Vec consistently produces the best results. These findings highlight the importance of using embedding methods tailored for sequential navigation patterns to achieve high quality clustering outcomes.

Clustering Visualization and Interpretation. Figure 3 shows t-SNE visualizations of clustering results across the three datasets (EGRC, SEDIT, MSNBC) using Pattern2Vec, SGT, and Word2Vec. Each subplot corresponds to a specific dataset-embedding combination. Consistent with Silhouette Scores, Pattern2Vec produces the most well-separated clusters, indicating strong capability in capturing navigational structures. SGT shows mixed results—performing well on MSNBC but yielding less distinct clusters on EGRC—suggesting sensitivity to dataset characteristics. Word2Vec exhibits scattered, less cohesive clusters, highlighting its limitations for modeling task-oriented navigation.

Implications for User Navigation Modeling. These findings underscore the critical role of embedding selection. Pattern2Vec's strong performance makes it well-suited for downstream applications such as recommendations, anomaly detection, and behavior analytics. SGT remains promising but requires further validation across varying datasets. Word2Vec's weaker results suggest it is less effective for structured, sequential navigation data.

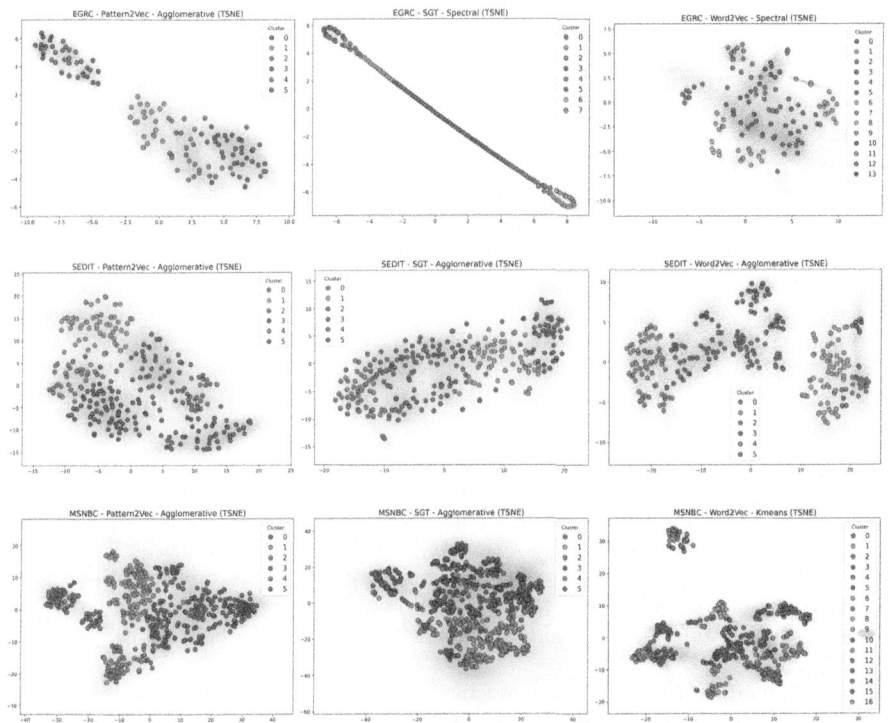

Fig. 3. Visualization of the clustering results.

Summary of the Discussion

- Pattern2Vec consistently outperforms SGT and Word2Vec in both clustering quality and visual clarity.
- SGT performs well in specific contexts but lacks consistent generalization.
- Word2Vec fails to separate clusters effectively, limiting its utility for structured navigation modeling.

These results establish a strong baseline for future research on clickstream embeddings, highlighting the importance of task-specific representation learning.

5   Conclusion

We proposed a benchmark for evaluating clickstream embeddings to uncover user navigation patterns in task-oriented software. Our study shows that Pattern2Vec is the most effective method for this context, followed by SGT. Word2Vec underperforms due to its lack of temporal awareness. This work addresses real-world complexities noisy, unstructured, unlabeled clickstreams and validates unsupervised methods for behavior modeling. Future work includes extending this to graph-based navigation representations.

References

1. Beutel, A., Akoglu, L., Faloutsos, C.: Graph-based user behavior modeling: from prediction to fraud detection. In: Proceedings of ACM SIGKDD International Conference on Knowledge Discovery and Data Mining, pp. 2309–2310 (2015). https://doi.org/10.1145/2783258.2789985

2. Borges, J., Levene, M.: A dynamic clustering-based markov model for web usage mining (2004). https://doi.org/10.48550/ARXIV.CS/0406032. arXiv:cs/0406032

3. Boukharouba, I., Sèdes, F., Bortolaso, C., Mouysset, F.: From user activity traces to navigation graph for software enhancement: an application of gnn on a real-world non-attributed graph. In: Proceedings ACM International Conference on Information and Knowledge Management, pp. 5236–5237 (2023). https://doi.org/10.1145/3583780.3615998

4. Cooley, R., Mobasher, B., Srivastava, J.: Data preparation for mining world wide web browsing patterns. Knowl. Inf. Syst. **1**(1), 5–32 (1999). https://doi.org/10.1007/BF03325089

5. Derhami, V., Khodadadian, E., Ghasemzadeh, M., Zareh Bidoki, A.M.: Applying reinforcement learning for web pages ranking algorithms. Appl. Soft Comput. **13**(4), 1686–1692 (2013). https://doi.org/10.1016/j.asoc.2012.12.023

6. Eirinaki, M., Vazirgiannis, M., Kapogiannis, D.: Web path recommendations based on page ranking and markov models. In: Proceedings of ACM International Workshop on Web Information and Data Management, pp. 2–9 (2005). https://doi.org/10.1145/1097047.1097050

7. Heckerman, D.: Msnbc.com anonymous web data. uCI Machine Learning Repository (1999). https://doi.org/10.24432/C5390X

8. Mikolov, T., Chen, K., Corrado, G., Dean, J.: Efficient estimation of word representations in vector space (2013). https://doi.org/10.48550/ARXIV.1301.3781. arXiv:1301.3781

9. Misra, G., Migliavacca, M., Otero, F.E.B.: Behavioural user identification from clickstream data for business improvement. In: Bramer, M., Ellis, R. (eds.) SGAI-AI 2021. LNCS (LNAI), vol. 13101, pp. 341–354. Springer, Cham (2021). https://doi.org/10.1007/978-3-030-91100-3_27

10. Mobasher, B., Dai, H., Luo, T., Nakagawa, M.: Effective personalization based on association rule discovery from web usage data. In: Proceedings of International Workshop on Web Information and Data Management, pp. 9–15 (2001). https://doi.org/10.1145/502932.502935

11. Olmezogullari, E., Aktas, M.S.: Representation of click-stream data sequences for learning user navigational behavior by using embeddings. In: Proceedings of IEEE International Conference on Big Data, pp. 3173–3179 (2020). https://doi.org/10.1109/BigData50022.2020.9378437

12. Olmezogullari, E., Aktas, M.S.: Pattern2vec: representation of clickstream data sequences for learning user navigational behavior. Concurr. Comput. Pract. Exp. **34**(9), e6546 (2022). https://doi.org/10.1002/cpe.6546

13. Ranjan, C., Ebrahimi, S., Paynabar, K.: Sequence graph transform (sgt): a feature embedding function for sequence data mining (2016). https://doi.org/10.48550/ARXIV.1608.03533. arXiv:1608.03533

14. Todi, K., Bailly, G., Leiva, L., Oulasvirta, A.: Adapting user interfaces with model-based reinforcement learning. In: Proceedings of CHI Conference on Human Factors in Computing Systems, pp. 1–13 (2021). https://doi.org/10.1145/3411764.3445497
15. Wittenbach, J., d'Alessandro, B., Bruss, C.B.: Machine learning for temporal data in finance: Challenges and opportunities (2020). https://doi.org/10.48550/ARXIV.2009.05636. arXiv:2009.05636

The Wrecking SQL Incremental Validation Methodology

Ruanitto Docini[✉], Eduardo C. de Almeida[✉], and Luiz S. Oliveira[✉]

Departamento de Informática (DINF), Universidade Federal do Paraná (UFPR),
Curitiba, Brazil
{ruanito.docini,eduardo.almeida,luiz.oliveira}@ufpr.br

Abstract. Large language models (LLMs) have made significant strides
in text-to-SQL performance, transforming natural language queries into
SQL queries. The BIRD benchmark, a cross-domain dataset with 12,751
question-SQL pairs across 95 databases, is currently the most challenging
benchmark in the field. However, its leaderboard is dominated by solutions that rely on closed-source LLMs, creating a financial barrier for
researchers. This paper focuses on leveraging open-source models and
presents an incremental validation methodology, called Wrecking SQL,
that in six steps, incrementally modifies the schema of datasets by replacing meaningful column and table names with meaningless ones—a real-
world problem found in legacy SQL systems. We explore how meaningless
names affect LLM accuracy and demonstrate that inferring meaningful
names improves translation accuracy.

Keywords: Text-to-SQL · Large Language Models · Schema
Nomenclature

1 Introduction

LLMs have significantly advanced text-to-SQL performance. The BIRD [7]
benchmark represents the most challenging benchmark, with top solutions
achieving 77% execution accuracy. However, most solutions rely on closed-source
LLMs, creating financial barriers for its application and development. BIRD
lacks real-world database variability, particularly found in legacy systems with
poor naming conventions—common in banking and old mission-critical systems.
We hypothesize that meaningless column and table names significantly hinder
LLM performance. We present Wrecking SQL, an incremental methodology with
six steps that modifies BIRD by replacing meaningful names with meaningless
ones iteratively. Our methodology addresses: **1)** How does schema nomenclature
affect text-to-SQL translation? **2)** Can schema inference improve translation
in legacy databases? Our contributions: **1)** We analyze nomenclature impact

This work is supported by the CNPQ grants 302909/2022-2, 303298/2022-7,
444192/2024-7, 441610/2023-4, and 446030/2023-5.

R. Wrembel et al. (Eds.): DEXA 2025, LNCS 16047, pp. 376–382, 2026.
https://doi.org/10.1007/978-3-032-02088-8_30

on text-to-SQL translation using open-source models. **2)** We demonstrate that poor naming degrades performance across all difficulties, with column naming having the most significant impact. **3)** We show that inferring meaningful names improves performance compared to irrelevant names.

2   State of the Art

Early text-to-SQL systems relied on template-based or rule-based solutions. Today, state-of-the-art leverage LLMs with techniques such as fine-tuning and few-shot prompting. More recently the BIRD benchmark became a key benchmark for measuring advancements in the field, with leading solutions like Ask-Data + GPT-4o achieving 77.14% accuracy. However, these models are evaluated on optimized datasets that don't fully reflect real-world database complexities.

The community has made several attempts to infer schemas using SQL functions [2], applying techniques to NoSQL databases [4], and conducting schema extraction attacks [6]. Three publications laid the foundation for our work: SNAIL [8] investigates schema labeling based on naming convention "naturalness"; STCExtract [1] proposes a two-phase algorithm to reconstruct table structures; and SeLaB [9] leverages BERT [3] for semantic labeling. We build upon these findings, proposing in-context learning techniques with LLMs to capitalize on modern language models' expanded capabilities.

3   The Wrecking SQL Incremental Validation Methodology

In the Wrecking SQL pipeline we create six dataset versions, that are also the stages of our incremental methodology, each version is incrementally modified by removing specific parts of the schema description or inferring the columns using an LLM. Our goal is analyzing the degradation in execution accuracy. The incremental modifications are: original, columns with meaningless names (MC), tables with meaningless names (MT), both being meaningless (MCT), and columns inferred with simple (INFS) or complex prompts (INFC). By "meaningless," we refer to names lacking semantics or connection to content commonly found in legacy systems (e.g., Filler0001, Filler0002) [5]. We apply similar naming with schema mapping: original dataset schema points to modified schema (e.g., playername becomes column1, restaurant_menu becomes table1). As a basis for our methodology and modifications we used the BIRD dataset, which consists of 12,751 SQL questions categorized into simple (30%), moderate (60%), and challenging (10%) levels across 95 databases with tables varying from 3 to 64 and columns from 4 to 115. BIRD is more challenging than SPIDER [10], with challenging queries involving multiple joins and nested queries like:

```
SELECT AVG(single_bond_count) FROM (SELECT T3.molecule_id, COUNT(T1.bond_tp) AS single_bond_ctd
FROM bond AS T1 INNER JOIN atom AS T2 ON T1.molecule_id = T2.molecule_id INNER JOIN
molecule AS T3 ON T3.molecule_id = T2.molecule_id WHERE T1.bond_tp = '-' AND T3.label = '+'
GROUP BY T3.molecule_id) AS subquery
```

Listing 1.1. Example of our renaming on BIRD databases

```
{"debit_card_specializing": {
    "columns": {"customers": {"CustomerID": "column_0",
        "Segment": "column_1", "Currency": "column_2"}
        ,
    "gasstations": {"GasStationID": "column_0","Country
        ": "column_2",}
    },
    "tables": {"customers": "table_0","gasstations": "
        table_1",}
}}
```

For the text-to-SQL translation employed zero-shot prompt engineering with triplet structure: instruction, schema, and question. Each prompt guides the model toward correct solutions by providing relevant information directly in input without fine-tuning. For each question in the benchmark, we use this structured prompt to generate the corresponding SQL query automatically.

```
1   Based on the database schema below and the question, create a SQL query that will return
2   the desired result:
3   ---------------------
4   DATABASE SCHEMA
5   CREATE TABLE table_0 (column_0 text, ... column_n column_type PRIMARY KEY column_0));
6   ...
7   CREATE TABLE table_4 (column_0 integer, ... column_n column_type FOREIGN KEY (column_0)
8   REFERENCES table_0(column_0));
9   QUESTION: Which country had the gas station that  sold the most expensive product id No.2
10  for one unit?
11  ---------------------
```

To evaluate our performance in each dataset we employed execution accuracy, measuring the proportion of queries where predicted SQL output matches ground truth query results:

$$\frac{\sum_{n=1}^{N} f(Q_n, \hat{Q}_n)}{N}$$

where $f()$ is an indicator function, which can be represented as:

$$f = \begin{cases} & \text{if } Q = \hat{Q} \\ & \text{if } Q \neq \hat{Q} \end{cases}$$

Q^n stands for results obtained by the nth ground truth SQL, $\hat{Q}^n$ stands for results by the nth SQL predicted by the model, and N is the overall number of text-to-SQL pairs.

4   Results

We followed the incremental steps of the Wrecking SQL methodology and present the result with the abbreviations introduced in Sect. 3 (Table 1).

Table 1. Performance comparison of models across dataset variants

Model	Default	MT	MC	MCT
Command-R7b [7B]	23.01%	16.29%	6.45%	2.34%
Gemma2 [2B]	10.03%	7.17%	2.73%	1.49%
Gemma2 [9B]	22.09%	21.05%	6.38%	2.54%
Granite3.1-dense [2B]	5.01%	4.23%	1.69%	1.10%
Granite3.1-dense [8B]	11.92%	10.69%	2.15%	1.89%
Llama3.1 [8B]	20.92%	17.99%	6.06%	2.47%
Llama3.2 [1B]	7.36%	3.58%	2.99%	2.15%
Phi4 [14B]	23.46%	22.29%	8.08%	1.89%
Qwen2.5-coder [1.5B]	12.38%	7.88%	2.60%	1.89%
Qwen2.5-coder [14B]	29.92%	29.59%	11.79%	3.25%
Qwen2.5 [14B]	22.03%	23.14%	7.43%	2.28%

4.1 How Does Schema Nomenclature Affect Text-to-SQL Translation with LLMs?

Results enabled by the incremental methodology show consistent patterns: best performance comes from default database, followed by databases with non-significant table names. Performance declines for altered column names, with lowest performance when both column and table names are modified.

Key findings: **1)** Table naming: Modest yet significant performance drop after modifying table names, indicating LLMs are sufficiently robust for legacy databases with suboptimal table naming **2)** Column naming: Significant performance drop when models tested on altered column names, with all models struggling to generalize **3)** Combined impact: Combined misnaming exacerbates negative impact on model performance

All difficulty tiers suffer when column names are altered, indicating that proper column naming is crucial for effective text-to-SQL systems (Table 2).

4.2 Can Schema Naming Inference Improve Text-to-SQL Translation in Legacy Databases?

To address column naming importance, we developed a method to automatically generate meaningful names based on table content using two prompts: simple (providing sample column content) and complete (providing both column content and full rows) (Table 3).

```
You are supplied with the content of a specific column from a database. You are tasked with
rephrasing the  name of the column to better reflect its content. Remember that this name
should be simple and also descriptive. The content of the column is as follows: "{content}"
Your response should be a simple new string that represents the content of the column. Just
that, nothing more is accepted. Respond with a single string, without explanations. String
must be a single phrase linked by underscores, like "new_column_name". Give just the name,
```

Table 2. Performance comparison of models across different query difficulty levels (Simple, Moderate, Challenging)

Model	Simple		Moderate		Challenging	
	Default	MC	Default	MC	Default	MC
Command-R7b [7B]	30.91%	8.21%	11.39%	3.87%	9.72%	3.47%
Gemma2 [2B]	14.81%	3.13%	2.79%	2.58%	2.7%	0.69%
Gemma2 [9B]	28.54%	8.10%	12.47%	4.08%	11.80%	2.77%
Granite3.1-dense [2B]	6.91%	2.37%	1.72%	0.86%	3.47%	0.0%
Granite3.1-dense [8B]	16.97%	2.70%	4.08%	1.72%	4.86%	0.0%
Llama3.1 [8B]	27.02%	7.45%	10.75%	4.51%	14.58%	2.08%
Llama3.2 [1B]	10.48%	3.24%	2.58%	3.44%	2.77%	0.0%
Phi4 [14B]	30.59%	10.27%	13.54%	4.30%	9.72%	6.25%
Qwen2.5-coder [1.5B]	17.18%	2.81%	4.94%	2.58%	5.55%	1.38%
Qwen2.5-coder [14B]	38.48%	15.67%	16.55%	6.66%	18.05%	3.47%
Qwen2.5 [14B]	29.29%	9.40%	10.96%	4.08%	11.11%	5.55%

```
NOTHING MORE IS ACCEPTED
follow the examples: content: "1, 2, 3, 4, 5" new column name: "numbers"
content: "apple, banana, orange, pear" new column name: "fruits"
now based on the previous examples, do the inference for following information:
content: "{content}" new column name:
```

And a more **complete** one including both column content and full rows for reasoning:

```
You are supplied with the content of a specific column from a database and also the content
of the table at which this column is located. Based on this information, you are tasked with
rephrasing the name of the column to better reflect its content. I want you to consider the
content of the table and the column to come up with a new name. The content is of extreme
importance, so make sure to consider it when coming up with the new name. Remember that this
name should be simple and descriptive. Give just the name, NOTHING MORE IS ACCEPTED.
table content: "{table_content}" | column content: "{column_content}"
follow the examples bellow:
1) table content: [["hot", "sunny", "south", "beach"],
["cold", "rainy", "north", "mountain"], ["mild", "cloudy", "west", "forest"]]
column content: ["hot", "cold", "mild"] -> new column name: "temperature"
2) table content: [["hot", "sunny", "south", "beach"], ["cold", "rainy", "north", "mountain"],
["mild", "cloudy", "west", "forest"]] column content: ["sunny", "rainy", "cloudy"]
-> new column name: "weather"

Think of this new name as a way to summurize what is in the column, the ideia is that,
by reading in you will have a good idea of what is in the column. Because of that it should
be semantically related to the content of the column and the table, very content specific.
Your response should be a simple new string that represents the content of the column.
Just that, nothing more is accepted. Respond with a single string, without explanations
Avoid names that are devoid of semantic meaning and very general. The name should be specific
to the content of the column and the table. String must be a single phrase linked by
underscores, like "new_column_name".
```

Both methods were applied to each column in the dataset to create new column name mappings for accuracy testing. Most LLMs showed a 24% accuracy

Table 3. Performance of models across datasets with column modifications

	MC	INFS	INFC
Command-R7b [7B]	6.45%	8.53%	10.62%
Gemma2 [2B]	2.73%	3.06%	3.58%
Gemma2 [9B]	6.38%	8.86%	10.43%
Granite3.1-dense [2B]	1.69%	2.02%	1.95%
Granite3.1-dense [8B]	2.15%	4.04%	5.41%
Llama3.1 [8B]	6.06%	7.62%	7.88%
Llama3.2 [1B]	2.99%	1.56%	2.02%
Phi4 [14B]	8.08%	9.64%	9.77%
Qwen2.5-coder [1.5B]	2.60%	2.67%	4.49%
Qwen2.5-coder [14B]	11.79%	13.88%	14.40%
Qwen2.5 [14B]	7.43%	9.12%	9.51%

improvement with inferred names over meaningless ones. The comprehensive method using full row data generally performed better, though it still falls short of the default dataset's accuracy.

5   Conclusion

Our research shows that open-source text-to-SQL methods struggle with suboptimal database naming and structure. Key findings: (1) Column naming is vital for execution accuracy, while (2) table naming has no significant effect. In-context learning can improve column naming, enhancing compatibility with text-to-SQL. The Wrecking SQL approach confirms proper column naming is essential, though inference can partly offset poor naming in legacy databases. Future work should focus on domain-specific knowledge to improve inference.

References

1. Cvijetić, B., Radivojević, Z.: Restoration of data structures using machine learning techniques. IEEE Access **PP**, 1, January 2023
2. Dani, C., Jahangiri, S., Hütter, T.: Introducing schema inference as a scalable SQL function [Extended Version]. arXiv (2024)
3. Devlin, J., Chang, M.W., Lee, K., Toutanova, K.: BERT: pre-training of deep bidirectional transformers for language understanding. In: Proceedings of the 2019 Conference of the North American Chapter of the Association for Computational Linguistics: Human Language Technologies, Volume 1 (Long and Short Papers), pp. 4171–4186 (2019)
4. Frozza, A., Defreyn, E., Mello, R.: A process for inference of columnar NoSQL database schemas. In: Anais do XXXV Simpósio Brasileiro de Bancos de Dados, pp. 175–180. Evento Online (2020)

5. IBM: Filler definition. IBM Rational Programming Patterns Documentation, version 9.7.2. https://www.ibm.com/docs/en/rpp/9.7.2?topic=tab-filler-definition. Accessed 09 June 2025

6. Klisura, D., Rios, A.: Unmasking database vulnerabilities: Zero-knowledge schema inference attacks in text-to-SQL systems. arXiv (2024)

7. Li, J., et al.: Can LLM already serve as a database interface? A big bench for large-scale database grounded text-to-SQLs. Advances in Neural Information Processing Systems **36** (2024)

8. Luoma, K., Kumar, A.: SNAILS: schema naming assessments for improved LLM-based SQL inference. Proceedings of the ACM on Management of Data **3**(1) (Feb 2025)

9. Trabelsi, M., Cao, J., Heflin, J.: SeLaB: semantic labeling with BERT. In: 2021 International Joint Conference on Neural Networks (IJCNN), pp. 1–8 (2021)

10. Yu, T., et al.: Spider: a large-scale human-labeled dataset for complex and cross-domain semantic parsing and text-to-sql task (2019). https://arxiv.org/abs/1809.08887

A Survey of Control Technologies for Autonomous Underwater Vehicles

Janette Christin Kaspar$^{(\boxtimes)}$ ⓘ

Salzburg University of Applied Sciences, Urstein Süd 1, 5412 Puch, Austria
`janettechristin.kaspar@fh-salzburg.ac.at`

Abstract. Autonomous Underwater Vehicles (AUVs) require robust control strategies to ensure stable trajectory tracking in dynamic and uncertain environments. This paper presents a systematic literature review of recent advancements in AUV control methodologies. A total of 52 peer-reviewed papers from 2023 to 2024 were analyzed, categorized by control techniques, including Backstepping, Sliding Mode Control, Proportional-Integral-Derivative (PID) controllers, Reinforcement Learning (RL), Neural Networks (NN), and Model Predictive Control (MPC). Results indicate that MPC is the most commonly applied method, followed by Backstepping and hybrid approaches integrating PID and RL. The Lyapunov function was widely used for stability analysis. However, most studies relied on simulations rather than real-world implementations. Future research should emphasize experimental validation on physical AUVs to bridge the gap between theoretical advancements and practical applications.

Keywords: AUV · trajectory tracking control · control theory

1 Introduction

Autonomous Underwater Vehicles (AUVs) play a crucial role in ocean exploration, environmental monitoring and military applications. These vehicles operate in dynamic and uncertain environments, their trajectory tracking has to be stable and reliable. To ensure this during operation, various control strategies are used. However, AUV control presents several challenges, such as hydrodynamic model uncertainties, energy limitations and communication constraints [1].

This paper presents a comprehensive literature survey that categorizes and analyzes recent advancements in AUV control strategies. The primary objective is to provide a structured overview of the current research landscape, highlight emerging trends, and identify persistent challenges in the field. The reviewed control strategies are grouped into classical methods (e.g., PID), advanced data-driven techniques (machine learning, e.g., NN and RL) and model-based techniques. While AUVs comprise multiple interdependent subsystems, including navigation, localization and energy management, this survey focuses specifically on software-based control methods for trajectory tracking. Other key research areas, such as SLAM, mission planning, or underwater communication protocols, are acknowledged but outside the scope of this paper.

© The Author(s), under exclusive license to Springer Nature Switzerland AG 2026
R. Wrembel et al. (Eds.): DEXA 2025, LNCS 16047, pp. 383–388, 2026.
https://doi.org/10.1007/978-3-032-02088-8_31

2   Methodology

This survey is based on a systematic literature review approach that was proposed by Snyder [21]. It includes only peer-reviewed papers published from 2023 to 2024. The IEEE database was consulted, with two separate queries focusing on titles, abstracts and keywords.

The first query, "underwater robotics AND PID control", was formulated to identify studies focusing on the classical control method of PID loops, resulting in 70 papers. The second query, "AUV control strategies", took a broader approach resulting in 165 papers. Thus, the queries resulted in a total of 235 papers.

Subsequently, the titles and abstracts were manually reviewed to ensure relevance to AUVs and their trajectory tracking control. Duplicates were also eliminated during this screening process. In total, 52 papers were further investigated. Due to space constraints, the full list of the 52 reviewed papers (along with their titles, bibliographic details, and statistical classification) is available in a dedicated GitLab repository. This dataset also includes scripts for citation and keyword network analysis. Interested readers can access it at:
https://gitlab.com/janettechristin.kaspar/dexa-2025-auv-literature-survey.

As an initial step, the AUVs in the identified articles were classified as either underactuated or fully actuated. Underactuated AUVs have more Degrees of Freedom (DOF) than the number of independent control inputs [9]. This showed that 20 papers explicitly focused on underactuated AUVs.

Further, the surveyed papers applied various control techniques, some employing multiple methods. Thus, the articles were clustered by the methods that were used. The following techniques were identified: Backstepping, Sliding Mode Control, PIDs, RL, NN and MPC. These approaches were also combined with each other and, in addition, the following methods were integrated with these control strategies: Lyapunov Stability function [19], Fuzzy Logic and Gaussian Process.

To assess research influence and identify leading contributors, a bibliometric analysis was also performed using citation metrics and keywords co-occurrence networks. This analysis enables the identification of frequently explored themes, the evolution of research focus over time, and key publications that have shaped the field.

3   Results and Discussion

The majority of studies modeled and controlled AUVs with all six DOFs active, aiming for full spatial maneuverability. While this offers comprehensive control, it also increases system complexity and computational demand. Few studies addressed the potential benefits of simplifying control by reducing the number of controlled DOFs, which could offer practical advantages in constrained or resource-limited applications. Among studies employing a single control method, the majority of those focusing on either MPC, Backstepping or NN used underactuated AUVs. MPC is most commonly used with seven studies exclusively using constrained MPC approaches [2, 7, 8, 12, 20, 22, 25]. Five papers implemented Backstepping control [13, 14, 16–18], leveraging a Lyapunov function as the foundation. Four papers incorporated NNs as function approximators within broader control frameworks [4, 6, 24, 25]. In most cases, NNs were used to estimate system dynamics or adapt controller parameters and were often part of RL or adaptive control schemes. Three papers relied solely on the classical PID controller, while another three papers explored a hybrid approach that combined PID and RL [15, 23, 26].

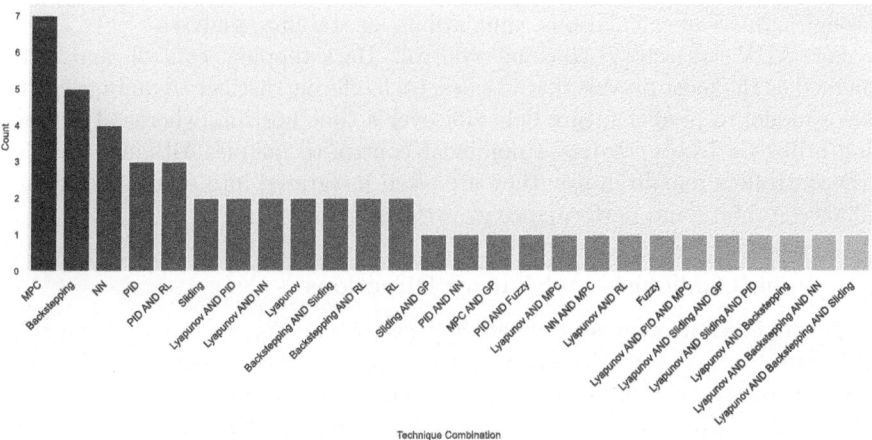

Fig. 1. is based on the classification of 52 reviewed papers. Each paper was assigned to one or more categories based on its primary control method.

The Lyapunov function was used in 14 papers in order to prove the stability of the control strategies (e.g., in source [25]). Figure 1 illustrates the distribution of control strategies among the reviewed papers. Two papers employed distinct approaches: one proposed a novel approach to linear active disturbance rejection control (LADRC) [10]. This adaptive LADRC is designed to handle system uncertainties and external disturbances without requiring an accurate mathematical model, dynamically adjusting observer parameters online to improve disturbance rejection. The second study [3] applied a two-player zero-sum game, in which the AUV aims to minimize the error and the disturbances aim to maximize it. This approach, designed for fully-actuated AUVs, utilized a critic NN and solved the Hamilton-Jacobi-Isaacs (HJI) equation using online policy iteration, an RL-based technique. Stability was once again verified by the Lyapunov function.

3.1 Trends and Insights

The variety of control techniques reflects the multifaceted challenges associated with AUV operations in dynamic, nonlinear, and partially known ocean environments. Ocean conditions such as currents, turbulence, and pressure variability introduce disturbances that necessitate adaptive and resilient control mechanisms [1]. Notably, the fusion of classical methods like PID with intelligent techniques (e.g., NNs and RL) marks a transition toward hybrid control strategies that aim to balance interpretability, computational efficiency, and adaptability [5]. This synergy between traditional control techniques and advanced machine learning frameworks [11] offers novel approaches for more efficient and resilient AUV control systems, especially in the face of complex and unpredictable ocean dynamics. It is noteworthy that the majority of studies did not validate their strategies on physical AUVs but instead demonstrated efficacy through numerical calculations, simulations, or stability analyses.

For AUV trajectory tracking control, Backstepping control and MPC emerged as the most prevalent strategies, each offering distinct advantages. MPC uses a model to predict future behavior over a time horizon, whereas Backstepping utilizes a Lyapunov-based nonlinear control technique. Although classical PID controllers remain in use, they are often integrated into other strategies to enhance stability and performance.

3.2   Open Challenges and Future Directions

Despite significant progress, several challenges remain open. A major limitation in the current literature is the lack of experimental validation on real AUV platforms. Most studies rely solely on simulations of idealized ocean currents. This gap restricts their applicability to real-life scenarios, particularly in more complex or constrained environments such as lakes, rivers or waste water systems. Additionally, energy efficiency, real-time processing capabilities, and robustness against model uncertainties remain key areas for future development. The integration of learning-based methods introduces challenges related to safety, and training stability, particularly in safety-critical underwater missions. Furthermore, little attention has been paid to standardizing evaluation benchmarks across studies, making it difficult to compare performance metrics directly.

Future research should prioritize the application of these control strategies in real-world scenarios with physical AUVs in varied environments. Moreover, the development of hybrid control strategies that combine the simplicity of classical controllers with the adaptability of learning-based methods represents a promising direction. Establishing standardized testing frameworks and shared simulation environments would further benefit the research community, facilitating more transparent and comparable evaluations of control performance. Future work will involve applying these control strategies to real-world AUVs in a practical research setting, bridging the gap between simulation-based results and field validation.

Acknowledgments. The author gratefully acknowledges the *innovatHER* project at the Salzburg University of Applied Sciences and the Information Technologies Department for supporting this work. The contributions by these funding sources have been indispensable to the realization of this paper.

References

1. Aldhaheri, S., Masi, G.D., Èric Pairet, Ardón, P.: Underwater robot manipulation: advances, challenges and prospective ventures (2022). https://arxiv.org/abs/2201.02954

2. Bhat, S., Stenius, I.: Controlling an underactuated auv as an inverted pendulum using nonlinear model predictive control and behavior trees. In: 2023 IEEE International Conference on Robotics and Automation (ICRA), pp. 12261–12267 (2023). https://doi.org/10.1109/ICRA48891.2023.10160926

3. Che, G.: Zero-sum game based tracking control for fully-actuated auv with unknown disturbances using critic network. In: 2024 43rd Chinese Control Conference (CCC), pp. 2462–2467 (2024). https://doi.org/10.23919/CCC63176.2024.10662079

4. Chen, G., Dong, J.: Approximate optimal adaptive prescribed performance fault-tolerant control for autonomous underwater vehicle based on self-organizing neural networks. IEEE Trans. Veh. Technol. **73**(7), 9776–9785 (2024). https://doi.org/10.1109/TVT.2024.3381338

5. Chen, H., Bo, T., Wang, S., Tang, G.: Deep reinforcement learning based backstepping control for underactuated auv. In: 2023 International Conference on Advanced Robotics and Mechatronics (ICARM), pp. 239–243 (2023). https://doi.org/10.1109/ICARM58088.2023.10218400

6. Chen, Y., Sun, Y., Wang, D., Ji, M., Sun, K.: Observer based fixed-time backstepping tracking control with the prescribed performance for auvs. In: 2024 43rd Chinese Control Conference (CCC), pp. 640–645 (2024). https://doi.org/10.23919/CCC63176.2024.10662797

7. Fontaine, A.F., Zhu, D., Chen, N., Pan, Y.J.: Nonlinear model predictive control for autonomous underwater vehicle trajectory tracking. In: 2023 IEEE 2nd Industrial Electronics Society Annual On-Line Conference (ONCON), pp. 1–6 (2023). https://doi.org/10.1109/ONCON60463.2023.10431297

8. Hao, L.Y., Wang, R.Z., Shen, C., Shi, Y.: Trajectory tracking control of autonomous underwater vehicles using improved tube-based model predictive control approach. IEEE Trans. Industr. Inf. **20**(4), 5647–5657 (2024). https://doi.org/10.1109/TII.2023.3331772

9. He, B., Wang, S., Liu, Y.: Underactuated robotics: a review. Int. J. Adv. Robot. Syst. **16**, 172988141986216 (2019). https://doi.org/10.1177/1729881419862164

10. Ji, D., Wang, X., Xu, M., Zhang, Q., Sharma, S., Sutton, R.: Trajectory tracking of auv under current disturbance based on adaptive ladrc. In: 2023 2nd International Conference on Automation, Robotics and Computer Engineering (ICARCE), pp. 1–5 (2023). https://doi.org/10.1109/ICARCE59252.2024.10492523

11. Ji, D., Xu, M., Wang, X., Ye, Z.: Adaptive path following control based on nonlinear disturbance observer for a six thrusters auv. In: 2023 WRC Symposium on Advanced Robotics and Automation (WRC SARA), pp. 156–161 (2023). https://doi.org/10.1109/WRCSARA60131.2023.10261821

12. Li, S., Wang, S., Luo, X.: Depth control of autonomous underwater vehicles based on constrained model predictive control. In: 2023 42nd Chinese Control Conference (CCC), pp. 2707–2712 (2023).https://doi.org/10.23919/CCC58697.2023.10239875

13. Liu, J., Du, J.: Dynamic event-triggered robust adaptive practical fixed-time trajectory tracking control for underactuated auvs. In: 2023 6th International Conference on Intelligent Autonomous Systems (ICoIAS), pp. 192–197 (2023). https://doi.org/10.1109/ICoIAS61634.2023.00039

14. Liu, J., Lin, D., Du, J.: Predefined-time 3-d trajectory tracking control of underactuated auvs with predefinedtime disturbance observer. In: 2024 14th Asian Control Conference (ASCC), pp. 897–902 (2024)

15. Liu, R., Cui, Z., Lian, Y., Li, K., Liao, C., Su, X.: Auv adaptive pid control method based on deep reinforcement learning. In: 2023 China Automation Congress (CAC), pp. 2098–2103 (2023). https://doi.org/10.1109/CAC59555.2023.10451043

16. Liu, Y., Wang, Q., Wang, L., Ding, Q., Fan, Z.: Predefined time trajectory tracking control of underactuated autonomous underwater vehicle*. In: 2024 China Automation Congress (CAC), pp. 768–773 (2024). https://doi.org/10.1109/CAC63892.2024.10864675

17. Liu, Y., Liu, J., Wang, Q.G., Yu, J.: Adaptive command filtered backstepping tracking control for auvs considering model uncertainties and input saturation. IEEE Trans. Circuits Syst. II Express Briefs **70**(4), 1475–1479 (2023). https://doi.org/10.1109/TCSII.2022.3221082

18. Ma, T., Chen, Z.: Trajectory tracking controller design for underactuated auvs based on adaptive robust control. In: 2024 9th International Conference on Automation, Control and Robotics Engineering (CACRE), pp. 305–309 (2024). https://doi.org/10.1109/CACRE62362.2024.10635061

19. Mawhin, J.: Alexandr Mikhailovich Liapunov, The general problem of the stability of motion (1892), pp. 664–676. Elsevier (01 2005)

20. Rahmadiansyah, M., et al.: Hybrid control strategy for underactuated auvs to overcoming environmental uncertainties. In: 2023 International Conference on Radar, Antenna, Microwave, Electronics, and Telecommunications (ICRAMET), pp. 325–330 (2023). https://doi.org/10.1109/ICRAMET60171.2023.10366642

21. Snyder, H.: Literature review as a research methodology: An overview and guidelines. J. Bus. Res. **104**, 333–339 (2019). https://doi.org/10.1016/j.jbusres.2019.07.039. https://linkinghub.elsevier.com/retrieve/pii/S0148296319304564

22. Wang, S., Er, M.J., Liu, T., Gong, H.: Path following control of underactuated auv based on improved model predictive control. In: 2023 6th International Conference on Intelligent Autonomous Systems (ICoIAS), pp. 222–227 (2023). https://doi.org/10.1109/ICoIAS61634.2023.00044

23. Wozniak, G., Bhat, S., Stenius, I.: Using reinforcement learning for hydrobatic maneuvering with autonomous underwater vehicles. In: OCEANS 2024 - Singapore, pp. 1–8 (2024). https://doi.org/10.1109/OCEANS51537.2024.10682215

24. Yan, H., Xiao, Y., Zhang, H.: Practical fixed-time adaptive nn fault-tolerant control for underactuated auvs with input quantization and unknown dead zone. IEEE Access **11**, 118973–118982 (2023). https://doi.org/10.1109/ACCESS.2023.3326442

25. Zhang, K., Shi, Y.: Tube mpc-based tracking control of auvs using contraction metric. In: 2024 American Control Conference (ACC), pp. 905–910 (2024). https://doi.org/10.23919/ACC60939.2024.10645038

26. Zhang, Y., et al.: Depth control of cable patrol autonomous underwater vehicle based on reinforcement learning. In: 2023 WRC Symposium on Advanced Robotics and Automation (WRC SARA), pp. 534–540 (2023). https://doi.org/10.1109/WRCSARA60131.2023.10261868

Author Index

© The Editor(s) (if applicable) and The Author(s), under exclusive license
to Springer Nature Switzerland AG 2026
R. Wrembel et al. (Eds.): DEXA 2025, LNCS 16047, pp. 389–391, 2026.
https://doi.org/10.1007/978-3-032-02088-8

The manufacturer's authorised representative in the EU is Springer
Nature Customer Service Centre GmbH, Europaplatz 3, 69115 Heidelberg,
Germany. If you have any concerns regarding our products, please
contact ProductSafety@springernature.com

Printed and bound by CPI Group (UK) Ltd, Croydon, CR0 4YY
28/04/2026
02098515-0007